Cell and Molecular Biology in Action

A series published by Pearson Education

Edited by: Dr Ed Wood, Department of Biochemistry and Molecular Biology, University of Leeds, UK

The series aims to provide introductions to key, exciting areas of cell and molecular biology, stimulating students imaginations and initiative to bridge the gap between memorising concepts and the active approach needed for research and literature review projects. This active learning series also introduces students to experimental design and information retrieval and analysis, including exploration of the World Wide Web.

Each text in the series will cover key theory concisely and use boxes to highlight skills, techniques and applications of the theory covered. Each text will also have its own web page providing updates and useful links to relevant sites.

For details of forthcoming titles in the series please visit the Pearson Education website at http://www.pearsoneduc.com/biology

CELL AND MOLECULAR BIOLOGY IN ACTION SERIES

Human Molecular Genetics

Second edition

Peter Sudbery

Prentice
Hall

An imprint of **Pearson Education**

Harlow, England · London · New York · Reading, Massachusetts · San Francisco
Toronto · Don Mills, Ontario · Sydney · Tokyo · Singapore · Hong Kong · Seoul
Taipei · Cape Town · Madrid · Mexico City · Amsterdam · Munich · Paris · Milan

Pearson Education Limited
Edinburgh Gate
Harlow
Essex CM20 2JE
England

and Associated Companies throughout the world

Visit us on the World·Wide Web at:
www.pearsoneduc.com

First published under the Addison Wesley Longman imprint 1998
Second edition 2002

ISBN 0 130 42811 6

1 0 0 3 4 1 5 1 4 7

British Library Cataloguing-in-Publication Data
A catalogue record for this book is
available from the British Library

Library of Congress Cataloging-in-Publication Data
Sudbery, Peter.
  Human molecular genetics / Peter Sudbery.-- 2nd ed.
    p. cm. -- (Cell and molecular biology in action series)
  Includes bibliographical references and index.
  ISBN 0-13-042811-6 (pbk.)
    1. Human genetics. 2. Molecular genetics. I. Title. II. Series.

  QH431 .S932 2002
  611'.01816--dc21

                                                    2 001056017

10 9 8 7 6 5 4 3 2
07 06 05 04 03

Typeset in Concorde BE by 30
Printed in Great Britain by Henry Ling Ltd., at the Dorset Press, Dorchester, Dorset

Contents

Acknowledgements   viii
Preface   x
Abbreviations   xiii

1 Human genetic disease   1
1.1 Introduction   1
1.2 Single-gene defects   2
1.3 Multifactorial or complex disorders and traits   13
1.4 Chromosomal mutations   14
1.5 Mitochondrial mutations   17
1.6 The Human Genome Project   22
1.7 Summary   23
     Further reading   25

2 An introduction to the structure of the human genome   27
2.1 Introduction   27
2.2 Sequence architecture of the human genome   28
2.3 Structure of chromosomes   42
2.4 Summary   47
     Further reading   49

3 Mapping the human genome   52
3.1 Introduction   52
3.2 Sequence tagged sites   55
3.3 Genetic maps   58
3.4 Physical maps   67
3.5 An integrated physical and genetic map of the human genome   81
3.6 Genome mapping is big science   83
3.7 Summary   84
     Further reading   85

4 The sequence of the human genome 87
4.1 Introduction 87
4.2 Basic technology for sequencing the human genome 89
4.3 The IHGSC clone-by-clone sequencing strategy 94
4.4 The Celera shotgun sequencing 98
4.5 Single nucleotide polymorphisms 100
4.6 Accessing the data 101
4.7 Model organisms 103
4.8 Analysis of the human genome sequence 104
4.9 Exploiting the human genome sequence 122
4.10 Summary 135
Further reading 137

5 Single-gene disorders 140
5.1 Introduction 140
5.2 Cloning disease genes 141
5.3 Cystic fibrosis 147
5.4 Duchenne muscular dystrophy 152
5.5 Trinucleotide repeat expansion mutations 154
5.6 Haemoglobinopathies 159
5.7 Inherited predisposition to cancer 164
5.8 Summary 175
Further reading 177

6 The genetic components of complex diseases 180
6.1 Introduction 180
6.2 Evidence for a genetic component in complex disorders and phenotypic traits 182
6.3 The genetic architecture of complex diseases 185
6.4 Identifying genes involved in complex disease 193
6.5 Conclusions: complex diseases are complex 220
6.6 Summary 220
Further reading 222

7 Gene therapy 225
7.1 Introduction 225
7.2 Types of gene therapy 226
7.3 Methods of transferring transgenes into target cells 230
7.4 Gene therapy for cystic fibrosis 237
7.5 Gene therapy for Duchenne muscular dystrophy 239
7.6 Gene therapy for bleeding disorders 240
7.7 Gene therapy for Severe Combined Immunodeficiency Syndrome 241
7.8 Gene therapy for non-heritable disorders 241
7.9 Summary 245
Further reading 247

8 Genetic testing 249
8.1 Introduction 249
8.2 Testing for known mutations 255
8.3 Scanning genes for unknown mutations 261
8.4 Summary 268
Further reading 270

9 Human population genetics and evolution 271
9.1 Introduction 271
9.2 Basic principles in human population genetics 273
9.3 Polymorphism in protein encoding genes 281
9.4 Mitochondrial DNA polymorphisms 285
9.5 Y chromosome variation 290
9.6 Single nucleotide polymorphisms 293
9.7 Disease frequencies in different populations 295
9.8 Summary 298
Further reading 301

10 DNA fingerprinting 304
10.1 Introduction 304
10.2 Use of minisatellites for DNA fingerprinting 305
10.3 Single-locus probes 307
10.4 DNA profiling based on STRs 308
10.5 Summary 310
Further reading 311

11 Human genetics and society 312
11.1 Introduction 312
11.2 Genetic testing 315
11.3 Human rights 322
11.4 Patents 325
11.5 Gene therapy 328
11.6 Genetic determinism 329
11.7 Summary 330
Further reading 332

Glossary 334
Index 356

Further information for use with this book can be accessed via a link in the catalogue entry for *Human Molecular Genetics* on the publisher's website at **http://www.pearsoneduc.com/biology**. The aim of this facility is to provide updates and news of major advances in human genetics since publication. It also includes study guides for chapters, and cites additional website addresses that provide information of interest to human genetics. We would encourage readers to use the website regularly.

Acknowledgements

We are grateful for permission to reproduce copyright material in the following illustrations:

Plates 1–3 and Figure 2.8 Maltby, E., Langhill Centre for Human Genetics, Sheffield, UK; **Table 1.1** Weatherall, D.J. (1990) The New Genetics and Clinical Practice, 3rd edition, copyright of Oxford University Press; **Figure 2.9** Caskey, C.A. and Rossiter, B.J.F, (1992) *Journal of Pharmacy and Pharmacology* **44** (suppl. 1) 198–203, copyright of Charles Fry Publishers; **Figure 3.1** Schuler, G.D. *et al.* (1996) A gene map of the human genome, *Science* **274** 540–1, copyright of American Association for the Advancement of Science; **Figure 3.12** Chumakov *et al.* (1992) Continuum of overlapping clones spanning the entire chromosome 21q, *Nature* **359** 380–7, reprinted with permission, copyright 1992 Macmillan Magazines Ltd; **Figure 3.17** from http://www-genome.wi.mit.edu/ with permission from Dr T.H. Hudson; **Figures 4.4 and 4.10** The International Human Genome Sequencing Consortium (2001) Initial sequencing and analysis of the human genome, *Nature* **409** 860–921, reprinted with permission from *Nature* and The International Human Genome Sequencing Consortium, copyright 2001 Macmillan Magazines Limited; **Figure 4.5** The BAC Consortium (2001) Integration of cytogenetic landmarks into the draft sequence of the human genome, *Nature* **409** 953–8, reprinted with permission, copyright 2001 Macmillan Magazines Ltd; **Figures 4.9 and 4.11** Venter, J.C. *et al.* (2001) The sequence of the human genome, *Science* **291** 1304–51, copyright 2001 of American Association for the Advancement of Science; **Figures 4.15 and 4.17** Shoemaker, D.D. *et al.* (2001) Experimental annotation of the human genome using microarray technology, *Nature* **409** 922–7, reprinted with permission, copyright 2001 Macmillan Magazines Ltd; **Figure 4.20** Evans, W.E. and Relling, M.V. (1999) Pharmacogenetics: translating functional genomics into rational therapeutics, *Science* **286**

487–91, copyright 1999 of American Association for the Advancement of Science; **Figure 5.3** Collins, F.S. (1992) Cystic Fibrosis: Molecular biology and therapeutic implications, *Science* **256** 774–9, copyright of American Association for the Advancement of Science; **Figure 5.4** Nawrotski *et al.* (1996) The genetic basis of neuromuscular disorders, *Trends in Genetics* **12** 294–8, copyright of Elsevier Science Ltd; **Figure 5.5** The Huntington's disease collaborative research group (1993) A novel gene containing a trinucleotide repeat that is expanded and unstable on Huntington's disease chromosomes, *Cell* **72** 971–83, copyright of Cell Press; **Figure 5.9** Roses, A.D. (2000) Pharmacogenetics and the practice of medicine, *Nature* **405** 857–65, reprinted with permission from *Nature* and GlaxoSmithKline, copyright 2000 Macmillan Magazines Ltd; **Figure 6.4** Risch, N.J. (2000) Searching for genetic determinants in the new millennium, *Nature* **405** 847–56, reprinted with permission, copyright 2000 Macmillan Magazines Ltd; **Figure 6.6** Cordell, H.J. and Todd, J.A. (1995) Multifactorial inheritance of type I diabetes, *Trends in Genetics* **11** 499–503, copyright of Elsevier Science Ltd and P.E. Applied Biosystems; **Figure 6.7** Bennet, S.T. and Todd, J.A. (1996) Human Diabetes and the Insulin Gene: Principle of mapping polygenes, *Annual Review of Genetics* **30** 343–70, copyright of Annual Reviews Inc.; **Figures 8.1, 8.3, 8.4, 8.5, 8.6, 8.9** North Trent Molecular Genetics Laboratory, Sheffield, UK; **Figure 9.1** Efstratiadis, A. *et al.* (1980) The structure and evolution of the human β globin family, *Cell* **21** 653–68, copyright of Cell Press; **Figures 9.4 and 9.5 (top and middle)** Cavalli Sforza, L.L. *et al.* (1996) The history and geography of human genes, abridged paperback edition, copyright of Princeton University Press; **Figure 9.5 (bottom)** Devoto *et al.* (1990) Gradient of distribution in Europe of the world CF mutation and its associated haplo-types, *Human Genetics* **85** 436–41, copyright of SpringerVerlag; **Figure 10.3** from http://www2.perkin-elmer.com:80/fo/773201/773201.html copyright of PE Applied Biosystems.

We would especially like to thank the following for their valuable reviews on draft forms of the manuscript:

John Armour (University of Nottingham)
Kerry Bloom (University of North Carolina)
Keith Brown (University of Bristol)
Dave Curtis (Royal College of London)
Ed Wood (University of Leeds)

Preface

This book is intended to provide an introduction to human molecular genetics that is more focused and up-to-date than that found in large, general textbooks on genetics, and yet remains accessible to second- and third-year science undergraduates, medical students and other health-care specialists. It assumes the background knowledge of genetics and molecular biology that would normally be taught in the first year of a university course. Not only is this a necessity to keep both the size and cost down, but it also seems futile to duplicate the coverage in the generous selection of excellent general textbooks to which most readers will have access. Each chapter ends with a list of further reading for those who would like to investigate a topic in more detail. For the most part the references are also the sources for most of the material presented in the chapter. For a more detailed and comprehensive treatment I also recommend the excellent textbook by Professors Strachan and Read, now also in its second edition (Strachan, T. and Read, A.P. (eds) (1999) *Human Molecular Genetics*, 2nd edn, Bios Scientific Publishers, Oxford).

An introductory text to such a large and complex field must restrict itself to illustrating general principles with selected examples. Inevitably work on many important genes must be omitted, and I can only apologise to those who feel their favourite gene has not received the attention it deserves. Equally, a certain amount of simplification is essential to make it accessible to an undergraduate audience. This is particularly true of those areas such as population genetics that are mathematically based, which experience shows can be a cause of difficulty to biology students. I can only hope that I have succeeded in spreading enlightenment without oversimplifying.

Since the first edition of this book was written in 1997 there have been several major advances. In 1997 the mapping phase of the human genome project was largely complete but sequencing had barely begun. Now not one, but two draft sequences have been produced and analysed. There still remains much to do in this area, including filling the gaps and identifying the genes, but it is clear that a major scientific milestone has been passed. In terms of this book the genome sequence has moved to centre stage and I have written a completely new chapter detailing the methodology used, what was revealed about genome structure and evolution and how the genome sequence will be exploited in diagnosing and treating common diseases. The analysis of genome structure from the sequence itself presented the dilemma of whether to retain a chapter introducing this topic. In principle, a description of genome sequencing could have been followed by a definitive account of genome structure. The problem was that knowledge of genome structure is necessary to understand the sequencing methodology. Because of this I elected to retain the introductory chapter at the cost of a certain amount of duplication between Chapters 2 and 4. Where genome statistics are used in Chapter 2, they have been taken from the analysis of the draft genome.

The strategies used in searching for genes involved in complex disease has also undergone a paradigm shift since the first edition was written. Previously, attention was focused on non-parametric genome scans of affected sib pairs using polymorphic microsatellites. Current strategies are based on population surveys using SNPs. Because of this I have also completely rewritten the chapter on complex disease. Of course, there have been many false dawns in this field. I have endeavoured to provide a balance between an appreciation of the difficulties and complexities of this area and the exciting prospects afforded by the new technologies.

Lastly, the human genome project opens up new prospects in population genetics and evolution. This area remains crucial in informing the design of population surveys to map complex disease alleles. It also focuses on the fascinating fundamental issue of the evolution of modern humans and the history of past population movements. Central to this field is the fierce debate between the proponents of the Out-of-Africa versus Multiregional theories of human origin. I have also rewritten this chapter in the light of new developments.

The rest of the book has been revised and updated throughout, while keeping the basic structure of the chapters in the first edition.

I am grateful for the tolerance and support of those around me while preparing this second edition, particularly my wife Carol. I much appreciate the rapid and detailed reading of draft versions of Chapters 4 and 6 by John Armour (University of Nottingham) and Keith Brown (University of Bristol) respectively; of course all deficiencies that remain are entirely my own responsibility. Anne Dalton, Steve Evans and Richard Kirk (North Trent Molecular Genetics Laboratory) generously provided examples of their work. Finally I am grateful to the staff at Pearson Education, particularly Alex Seabrook and Pauline Gillett, for their help and support.

Abbreviations

AAV	adeno-associated virus	**CMD**	congenital muscular dystrophy
ABC	ATP-binding cassette		
ACGT	Advisory Committee for Gene Testing	**CPEO**	chronic progressive external opthalomoplegia
AD	Alzheimer's disease	**CRE**	cyclic AMP response element
ADA	adenine deaminase		
AGE	agarose gel electrophoresis	**DASC**	dystrophin-associated sarcoglycan complex
APM	affected pedigree member		
APP	amyloid precursor protein	**DGGE**	denaturing gradient gel electrophoresis
ARMS	amplification refractory mutation system		
		DHHS	Department of Health and Human Services (USA)
ASO	allele-specific oligonucleotide		
		DMAHP	dystrophia myotonica associated homeobox protein
ASP	affected sib pair		
AT	ataxia telangiectasia		
BAC	bacterial artificial chromosome	**DMD**	Duchenne muscular dystrophy
BMD	Becker muscular dystrophy	**DMPK**	dystrophia myotonica protein kinase
bp	base pairs		
CBAVD	congenital bilateral absence of the vas deferens	**DOE**	Department of the Environment (USA)
CDK	cyclin-dependent kinase	**EBI**	European Bioinformatics Centre
CEPH	Centre d'Étude Polymorphism Humain		
		ELSI	ethical, legal and social issues
CF	cystic fibrosis		
CFTR	cystic fibrosis transmembrane conductance regulator	**EMBL**	European Molecular Biology Laboratory
		EMC	enzyme mismatch cleavage
cM	centimorgans	**EPC**	European Patent Convention
CMC	chemical mismatch cleavage	**ePCR**	electronic PCR

EPO	European Patent Office	**LOAD**	late onset Alzheimer's disease
ERV	endogenous retroviruses		
EST	expressed sequence tag	**LOD**	log ratio of odds
ETS	external transcribed spacer unit	**LOH**	loss of heterozygosity
		LTR	long terminal repeat
FAP	familial adenomatous polyposis	**Mb**	mega base pairs
		MD	myotonic dystrophy
FBC	familial breast cancer	**MDR**	multiple drug resistance
FISH	flourescence *in situ* hybridisation	**MELAS**	mitochondrial encephalomyopathy, lactic acidosis and stroke-like episodes
GAIC	Genetics and Insurance Committee		
GAP	GTP-activating protein	**MHC**	major histocompatibility complex
GRR	genotype relative risk		
GTAC	Gene Therapy Advisory Committee	**MLS**	maximum likelihood score
		MODY	maturity onset diabetes of the young
HBC	hereditary breast cancer		
HD	Huntington's disease	**MRCA**	most recent common ancestor
HGAC	Human Genetics Advisory Committee		
		mtDNA	mitochondrial DNA
HGC	Human Genetics Commission	**MVR-PCR**	multivariant repeat-polymerase chain reaction
HLA	human leucocyte antigens	**mya**	million years ago
HNPCC	hereditary non-polyposis colorectal cancer	**NARP**	neurogenic muscle weakness, ataxia and retinitis pigmentosa
HnRNA	heterogeneous nuclear RNA		
		NBF	nucleotide binding fold
HPFH	hereditary persistence of foetal haemoglobin	**NCBI**	National Centre for Biotechnology Information
HRE	heat-shock response element		
		NCHGR	National Council for Human Genetic Research
HUGO	Human Genome Organisation		
		NF	neurofibromatosis
IBD	identical by descent	**NIDDM**	non-insulin dependent diabetes mellitus
IBS	identical by state		
IDDM	insulin-dependent diabetes mellitus	**NIH**	National Institutes of Health (USA)
IDE	insulin degrading enzyme	**NMR**	nuclear magnetic resonance
IGI	Initial Gene Index		
IHGSC	International Human Genome Sequencing Consortium	**NOD**	non-obese diabetic (mouse)
		NRY	non-recombining region of the Y chromosome
ITS1	international transcribed spacer 1	**ORF**	open reading frame
		PAC	P1-derived artificial chromosome
kb	kilobase		
kya	thousand years ago	**PAR**	pseudoautosomal region (of the Y chromosome)
LCR	locus control region		
LD	linkage disequilibrium	**PC**	principal component
LHON	Leber's hereditary optic neuropathy	**PCR**	polymerase chain reaction
		PFGE	pulse field gel electrophoresis
LINE	long interspersed element		

PIC	polymorphism information content	**SSCP**	single-stranded conformational polymorphism
rDNA	DNA coding for ribosomal RNA molecules	**SSTR**	simple sequence tandem repeat
RFLP	restriction fragment length polymorphism	**STC**	sequence tagged connector
		STR	short tandem repeat
RH	radiation hybrid	**STS**	sequence-tagged site
RT-PCR	reverse transcribed PCR	**TBF**	TATA-box binding factor
SACGT	The Secretary's Advisory Committee on Genetic Testing (USA)	**TBP**	TATA-binding protein
		TDT	transmission disequilibrium test
SCID	severe combined immune deficiency syndrome	**TNF**	tumour necrosis factor
		TRE	trinucleotide repeat expansion
SD	standard deviation		
SINE	short interspersed element	**TSC**	the SNP consortium
SNP	single nucleotide polymorphism	**USPTO**	The United States Patent Office
		UTR	untranslated region
snRNP	small nuclear ribonucleoprotein	**VNTR**	variable number tandem repeat
SRE	serum response element	**YAC**	yeast artificial chromosome

---

A Companion Web Site accompanies *Human Molecular Genetics*, 2nd edition, by Dr Peter Sudbery

Visit the *Human Molecular Genetics* Companion Web Site at www.booksites.net/sudbery to find valuable teaching and learning material including:

For Students:
- Study material designed to help you improve your results
- Study guides for each chapter
- Self-assessment questions to aid understanding of difficult concepts, with worked solutions available to course teachers

For Students and Lecturers:
- Updates and news of major advances in human genetics since publication. For example it is expected that shortly after publication it will be announced that the human genome has been sequenced to "finished standard". An account of this development will be posted when it is published in the scientific press
- URLs for websites listed in the book and updates where necessary

For Lecturers:
- Password protected worked solutions to the self-assessment questions
- A syllabus manager that will build and host a course web page

Human genetic disease

Key topics

- Frequency and types of genetic disease
- Single-gene disorders
  - Complexity in single-gene disorders
  - Autosomal recessive
  - Autosomal dominant
  - Sex-linked
- Multifactorial or complex disorders
  - Evidence for genetic factors in common diseases
  - Genetic influences on personality disorders and phenotypic traits
- Chromosomal imbalances
- Mitochondrial disorders
- The Human Genome Project

1.1 Introduction

About 5% of liveborn babies suffer from a significant medical disorder, which may be life-threatening or, at the very least, require hospital treatment. A disorder present at birth is said to be **congenital**. Most congenital disorders will have a genetic component in their **aetiology**. This may take the form of a **single-gene** or **monogenic defect**, a mitochondrial disorder, a chromosomal imbalance or a multifactorial condition, which is partly genetic and partly environmental. Table 1.1 shows estimates of the genetic burden affecting the health of the newborn imposed by these various genetic factors.

A survey of admissions to a paediatric hospital in Montreal showed that one-third of all admissions were for diseases with a genetic component. Moreover, 70% of patients admitted more than once had a disorder with a genetic component. Such disorders are responsible for an immense amount of suffering, reduced quality of life, shortened life expectancy and distress to

Table 1.1 The genetic load in the newborn. It is difficult to give precise figures because of variation between different populations, so the figures are in the form of ranges. The genetic load of congenital defects is estimated as half the overall incidence of such disorders. No overall figures are available for mitochondrial disorders.

Category	Frequency (per 1000 live births)
Single gene	
Autosomal dominant	1.8–9.5
Autosomal recessive	2.2–2.5
X-linked	0.5–2.0
Chromosomal	6.8
Congenital malformations	19–22

family members and carers. Moreover, it is becoming increasingly clear that many of the common diseases of later life, perhaps up to two-thirds, have a genetic component in their aetiology. Clearly our genetic constitution plays a major part in determining our lifetime health. This chapter considers the ways in which this comes about and introduces some of the major genetic disorders that are considered in more detail in later chapters.

1.2 Single-gene defects

Single-gene or monogenic disorders, as their name implies, are traceable to a defect in a single gene. They follow simple patterns of inheritance, predictable from the Mendelian laws of genetics. These patterns are classified according to whether the affected gene is located on the X chromosome or one of the 22 autosomes and whether the trait is recessive or dominant. That is, they are said to be **X-linked**, **autosomal recessive** or **autosomal dominant**, respectively. The Online Mendelian Inheritance in Man database (see Further reading at the end of the chapter) currently lists over 5000 disorders that have been definitely traced to single-gene defects. This represents a significant fraction of the total number of human genes, estimated to be between 30 000 and 35 000 (see Chapter 2). Single-gene disorders that are both severe and relatively common are known as **major disorders**.

The frequency of mutant alleles responsible for a monogenic disorder in a population is specified by the **Hardy–Weinberg distribution**, which states that:

$$p^2 + 2pq + q^2 = 1$$

where p is the allele frequency of the more common allele and q the frequency of the less common allele. This equation holds only if certain conditions are met, such as that mating is random and there is no migration into or from the population. This is discussed in more detail in Chapter 9. From this equation the carrier (heterozygote) frequency for recessive disorders can be calculated ($2pq$) from the observed frequency with which the disease occurs in a population. This can give surprising results. For example, the frequency of cystic fibrosis in the UK is 1 in 2000, from which the carrier frequency can be calculated as 1 in 22.

1.2.1 Autosomal recessive disorders

Autosomal recessive disorders require the inheritance of two defective alleles. This means that there are no functioning copies of the gene and implies that the disorder results from a loss of function. An example of an autosomal recessive pattern of inheritance is shown in Figure 1.1 and the frequencies of some major autosomal recessive disorders are shown in Table 1.2.

Typically more than one sibling of unaffected parents may be affected; children of cousin marriages are also at risk (Figure 1.1). Both parents must be carriers or **heterozygous** at the locus concerned (Figure 1.2); as a result, for each child there is a 25% chance that it will be affected, a 50% chance that it will be a carrier and a 25% chance that it will be completely normal. If the normal allele is dominant, the ratio of unaffected to affected children will be 3:1. To put it another way, for each child born there is a 25% chance that it will be affected.

As shown in Figure 1.1, the children of a cousin marriage may be **homozygous** for a deleterious allele that is heterozygous in a common grandparent. In fact, the chance that this will happen is 1 in 64 for each such allele. Because everyone is likely to be heterozygous for about five different deleterious alleles, cousin marriages are more likely to result in children affected by an autosomal recessive disorder than marriages

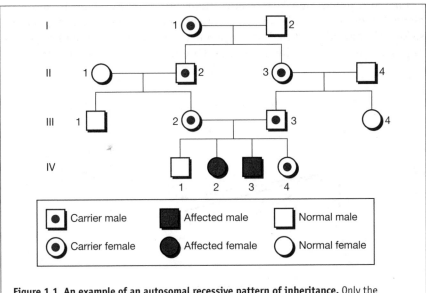

Figure 1.1 An example of an autosomal recessive pattern of inheritance. Only the homozygous recessive individuals (IV-2 and IV-3) are affected. Note how the heterozygous state in I-1 becomes homozygous in IV-2 and IV-3 as a result of the **consanguineous** marriage of III-2 and III-3.

Table 1.2 Some major monogenic recessive disorders.

Disease	Frequency	Symptoms
Cystic fibrosis	1 in 2000 (N. Europeans)	Recurrent lung infections, pancreatic exocrine deficiency, male sterility
α_1- Antitrypsin deficiency	1 in 5000 to 1 in 10 000 (N. Europeans)	Liver failure, emphysema
Phenylketonuria	1 in 2000 to 1 in 5000 (Europeans)	Mental retardation
Tay–Sachs disease	1 in 3000 (Ashkenazi Jews)	Neurological degeneration, blindness and paralysis
Sickle cell anaemia	1–2 in 100 (Africa where malaria is endemic)	Anaemia
Haemochromatosis	1 in 500	Excessive iron accumulation in adults, resulting in diabetes, liver cirrhosis and heart failure
Thalassaemias	1–2 in 100 (Mediterranean and Asia where malaria is endemic)	Anaemia

between unrelated parents. Indeed, many rare monogenic disorders are only observed in the children of cousin marriages.

Some diseases caused by single-gene defects are present in certain populations at a much higher frequency than would be expected by mutation alone. Important examples of this are cystic fibrosis in European populations, sickle cell anaemia and thalassaemias in Asian and African populations, and Tay–Sachs disease in **Ashkenazi Jews**. One reason for an elevated frequency

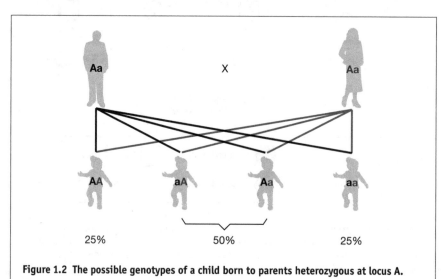

Figure 1.2 The possible genotypes of a child born to parents heterozygous at locus A.

is that the heterozygote may enjoy some advantage, resulting in the allele frequency being elevated by selection. In the case of sickle cell anaemia or thalassaemias, it has been demonstrated that heterozygotes are more resistant to malaria, which is endemic in those countries where such disorders are prevalent. This variation means that the ethnic origin of a patient may be relevant in arriving at a correct diagnosis. Other possible reasons for elevated disease frequencies are discussed in Chapter 9.

1.2.1.1 Cystic fibrosis

Cystic fibrosis (CF) is a common single-gene disorder among people of European extraction. It affects approximately 1 in 2000 babies, and about 1 in 22 are carriers. The affected gene encodes a protein known as the **cystic fibrosis transmembrane conductance regulator** (CFTR; see Chapter 5). The CFTR protein is responsible for the export of chloride ions across the plasma membrane of epithelial cells that line the lung airways. CF alleles impair or obliterate this function and, as water molecules follow the ions because of osmosis, insufficient water is secreted onto the cell surface. This results in a thick and sticky mucus that causes congestion in the lung airways. In a healthy person, the cilia of the epithelial cells continually move the mucus to the top of the airways and into the digestive system. In this way foreign bodies including bacteria are removed from the lungs. In a CF patient, the mucus is too thick to be moved by the cilia. Consequently this cleansing mechanism fails, resulting in repeated bacterial infections. The cumulative effect of these causes long-term damage to the lungs, and even with the best treatment the maximum life expectancy of someone with CF is 30 years. CF also affects other bodily systems. The pancreatic duct may become blocked, resulting in digestive difficulties. This is known as pancreatic exocrine deficiency. In some male patients, the vas deferens does not form properly, resulting in sterility. This is known as **congenital bilateral absence of the vas deferens** (CBAVD); sometimes this can occur without any other obvious symptoms of CF. Finally, CF results in an excess of chloride ion secretion in sweat glands, resulting in abnormally salty sweat. This can be simply recognised by measuring sweat electrolyte levels.

1.2.1.2 The haemoglobinopathies

Sickle cell anaemia, α and β **thalassaemias** and **glucose-6-phosphate dehydrogenase deficiency** (in some environments) affect the formation of haemoglobin, causing a class of disease known as the **haemoglobinopathies**. Haemoglobin is a tetramer of two β-globin molecules and two α-globin molecules, each complexed with a molecule of haem. Sickle cell anaemia results from a single-base mutation affecting the gene encoding β-globin. Study of this disease played an important part in the development of modern molecular genetics, because in 1956 Ingram demonstrated that the β-globin of sickle cell patients differed from the normal protein by one amino acid, providing experimental evidence for the first time that the sequence of amino acids in proteins was determined by genes. It is also important because it was the first, and is still the clearest, example in humans of heterozygous advantage. Carriers of the sickle cell gene are more resistant to malaria. This results in selection for the mutation and consequently an increased occurrence of the recessive homozygotes who suffer from the disease. Such a situation is known as a **balanced polymorphism**. The α and β thalassaemias result in a decrease or absence of α-globin and β-globin respectively. Like sickle cell anaemia, heterozygotes are more resistant to malaria so a balanced polymorphism results in thalassaemias being more common where malaria is, or was, endemic.

1.2.1.3 Tay–Sachs disease

Tay–Sachs disease is a progressive neurological degeneration that starts in the first year of life, characterised by developmental and mental retardation, progressive muscle weakness and paralysis, and blindness. Death usually results by the age of 5 years. It is caused by a mutation in the *HEXA* gene, which encodes the α-subunit of hexosaminidase A, a lysosomal enzyme required for the breakdown of a complex **glycolipid** called **ganglioside** GM_2 to a simpler molecule called ganglioside GM_3. The build-up of ganglioside GM_2 impairs neurone function and results in the disease. It is unusually common in the Ashkenazi Jew population, where it affects 1 in 3600 of the population.

1.2.1.4 Phenylketonuria

Phenylketonuria is characterised by mental subnormality. It results from a lack of phenylalanine hydroxylase, which converts phenylalanine to tyrosine, the first step in the phenylalanine degradation pathway. Phenylalanine consequently accumulates in the bloodstream. The damage to the developing nervous system is actually caused by phenylpyruvic acid, to which phenylalanine spontaneously converts. Most babies in the UK and the USA are now screened at birth for this deficiency because it can be almost completely controlled by a low-phenylalanine diet. It is thus an interesting example of a disorder that may be either completely genetically determined or completely environmentally determined.

1.2.1.5 α_1-Antitrypsin deficiency

Antitrypsin is an inhibitor of elastase. Deficiency results in unregulated breakdown of connective tissue that particularly affects the elasticity of the lungs, resulting in emphysema. It is another disorder that is particularly common among people of European descent. The protein can be readily made by recombinant methods, for example transgenic sheep have been produced that secrete large amounts of the protein in their milk. This provides the prospect of a cheap and effective treatment of the disease.

1.2.1.6 Haemochromatosis

This is probably the most common recessive disorder in the UK. It is caused by excessive iron accumulation and can be treated by blood letting. The symptoms are very variable and it is often not correctly diagnosed. It commonly affects women after the menopause, because the blood loss during menstruation is protective.

1.2.2 Autosomal dominant disorders

Autosomal dominant disorders result from the inheritance of only one mutant allele. This usually results in a protein that has gained a novel function or is expressing its normal function in an unregulated fashion. If affected individuals are able to reproduce, the disorder is likely to be manifested in every generation of a pedigree in which it is segregating. Thus 50% of the children of an affected person are likely to suffer from the disease (Figure 1.3).

If the disease is sufficiently serious to prevent reproduction, it follows that most cases will arise from *de novo* mutation during **gametogenesis** and the occurrence of the disease will be apparently sporadic. Some autosomal dominant disorders are only manifested later in life, after an individual is likely to have finished reproducing. Such mutations will escape the effect of selection. Examples of this are Huntington's disease, familial breast cancer (see Chapter 5), and hereditary Alzheimer's disease (see Chapter 6). Some examples of major autosomal dominant disorders are shown in Table 1.3.

1.2.2.1 Huntington's disease

Huntington's disease (HD) is a progressive neurological degeneration that affects patients in middle and later life, from their fifth decade onwards. Typical symptoms include dementia, severe depression and a characteristic involuntary, dance-like movement known as chorea. It is a particularly distressing condition, because by the time the symptoms are evident an affected person is likely to have already had children, which each have a 50% chance of suffering from the disease. The HD gene proved difficult to clone and this was only achieved in 1993. The mutation that causes the dis-

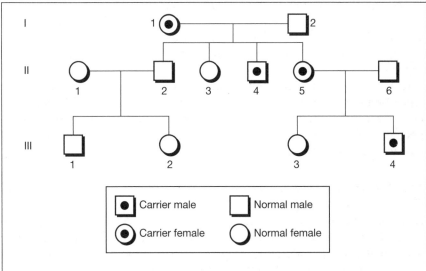

Figure 1.3 Example of an autosomal dominant pattern of inheritance. All heterozygous individuals are affected (I-1, II-4, II-5 and III-4). Because 50% of children will be affected, the disease is likely to be manifested in each generation.

Table 1.3 Examples of some major autosomal dominant disorders.

Disorder	Frequency	Symptoms
Familial hypercholesterolaemia	1 in 500	Premature heart disease
Familial breast cancer		
BRCA1	1 in 800 (USA)	High lifetime risk of breast and
BRCA2	1 in 100 (Ashkenazi Jews[1])	ovarian cancer. Earlier onset than sporadic cases
Familial Alzheimer's disease	10% lifetime risk at age 80[2]	Dementia. Earlier onset than sporadic cases
Hereditary non-polyposis colorectal cancer	1 in 400[3]	Colon cancer not associated with polyps
Familial adenomatous polyposis	1 in 8000	Bowel polyps that may become malignant
Neurofibromatosis type 1	1 in 4000	Tumours of peripheral nerves. *Café-au-lait* pigment spots on skin
Huntington's disease	1–2 in 10 000	Involuntary choreiform movements. Dementia. Late onset
Myotonic dystrophy	1 in 8500	Myotonia, heart defects and cataracts
Familial retinoblastoma	1 in 14 000	Multiple unilateral and bilateral tumours of the retina

[1] *BRCA1* and *BRCA2* each have a frequency of 1 in 100 in Ashkenazi Jews.
[2] Alzheimer's disease is normally a complex disease with genetic and environmental components. A subset of cases show early onset and simple autosomal dominant inheritance. What proportion of total cases is represented in this category is not known.
[3] Wide range of estimates from 1 in 200 to 1 in 10 000 (see Section 5.8.4).

ease is known as a trinucleotide repeat expansion (see Chapter 5). In normal individuals a sequence is found in which the triplet CAG is repeated about 15 times (the actual number can vary slightly from this value). In HD patients the number of repeats increases to 36 or more. This type of mutation has been found in a number of other genes that have been cloned recently. The existence of this type of mutation was entirely unexpected and could only have been discovered by the cloning and characterisation of the genes involved.

1.2.2.2 Familial breast cancer

Genetic influences are not thought to be an important factor in the occurrence of most breast cancer cases (see below). However, a subset, perhaps about 5% of all cases, follow a pattern of inheritance characteristic of an autosomal dominant mutation. Two genes, *BRCA1* and *BRCA2*, have been identified and cloned (see Chapter 5). They were recognised by studying cases of breast cancer that occurred before the patient was 40 years old, an unusually early age for the onset of the disease. Inheritance of *BRCA1* results in

80% lifetime risk of breast cancer. As well as leading to predisposition to breast cancer, both *BRCA1* and *BRCA2* (to a lesser extent) also cause a predisposition to other cancers, particularly ovarian cancer. Inheritance of the *BRCA2* allele in males carries a 1 in 100 risk of male breast cancer. Although responsible for only a small fraction of breast cancer cases, *BRCA1* is a relatively common disease allele, estimated in US populations to occur at a frequency of 1 in 800. In Ashkenazi Jewish populations the frequency for both *BRCA1* and *BRCA2* is much higher (1 in 100).

1.2.2.3 Myotonic dystrophy

Myotonic dystrophy (MD) is characterised by myotonia (delayed muscle relaxation) and degeneration of various organs including heart and eyes. It is variable in its expression, ranging from minimally affected late onset through classic adult onset to congenitally affected children of affected mothers. Although characterised as a Mendelian disorder of the autosomal dominant type, its pattern of inheritance shows some deviation from the classic pattern. In particular it may show **anticipation**, where a mother may be only slightly affected but have a severely affected child. Often the mother's symptoms are so mild that the disease is only diagnosed when her child is found to be affected. MD is another example of a trinucleotide repeat expansion. It is now known that anticipation is a result of the instability of the expansion (see Chapter 5). MD is a relatively common disorder, the allele frequency being 1 in 8500.

1.2.3 X-linked disorders

X-linked disorders affect genes located on the X chromosome. Because males only have one copy of this chromosome and therefore one copy of each of the genes located on it, they are much more likely to suffer from such disorders. Thus X-linked disorders affect male children of unaffected parents. Most X-linked disorders are recessive, so for the most part females are unaffected but act as carriers. Thus in a pedigree in which a sex-linked mutation is segregating, the mutation is inherited through the female line and the resultant disorder can apparently skip generations (Figure 1.4). Another characteristic of X-linked disorders is that uncles and nephews are often affected. One rare documented example of an X-linked dominant condition is hypophosphataemic (vitamin D-independent) rickets. In this case an affected father always passes the disorder to his daughters, but never to his sons.

Of course, it is possible for a female to be homozygous for an X-linked mutation and therefore suffer from the disease. Sometimes, a heterozygous female may show some symptoms of the disease. This may happen because of **X-chromosome inactivation** or **Lyonisation**. Named after its discoverer, Mary Lyon, this is a process that occurs early in development to randomly inactivate one or other of the X chromosomes. Cells descended from this progenitor cell abide by the decision made. Thus the body of a heterozygous female is a mosaic, consisting of patches of different clones, some of which will lack the product of the defective gene. If the function is cell autonomous, i.e. the function cannot be supplied by another cell expressing

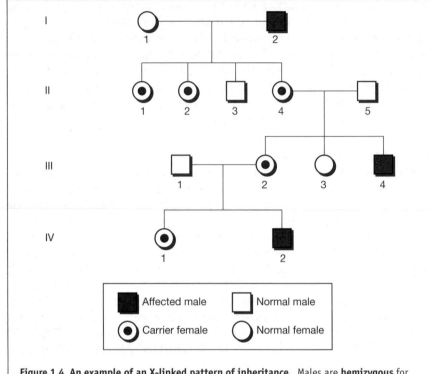

Figure 1.4 An example of an X-linked pattern of inheritance. Males are **hemizygous** for the X chromosome and have only one copy of genes on the X chromosome; males that inherit the affected gene are shown with a filled square. Note that the great-grandfather of IV-2 was affected, but not his mother or grandmother. The disease therefore skipped generations II and III with respect to IV-2 and the disorder was passed through the female line. However, IV-2's uncle was also affected, a typical pattern in X-linked disorders. Finally, note that daughters of affected males are obligate carriers.

the gene, it is possible that the lack of function will affect the overall phenotype of the carrier. This may be the reason why, for example, one-third of female carriers of fragile-X syndrome are mildly affected (see below).

The major X-linked disorders are shown in Table 1.4.

1.2.3.1 Haemophilia

Haemophilia is an inability to form blood clots, resulting in prolonged bleeding upon injury and spontaneous internal bleeding. At a molecular

Table 1.4 Some examples of X-linked disorders.

Disorder	Frequency	Symptoms
Haemophilia A and B	1 in 10 000	Abnormally prolonged bleeding after trauma
Duchenne muscular dystrophy	1 in 3000 to 1 in 4000	Muscle wastage in teenage years
Fragile-X syndrome	1 in 1000	Mental retardation

level it results from a lack of either blood clotting factor VIII (haemophilia A, 80% of cases) or blood clotting factor IX (haemophilia B or Christmas disease). It is famous for affecting the Victorian royal family.

1.2.3.2 Duchenne muscular dystrophy

Duchenne muscular dystrophy (DMD) results in progressive muscle wasting or dystrophy, starting in the teenage years. It soon results in confinement to a wheelchair and affected boys normally die in their twenties. There is a milder form called Becker muscular dystrophy which maps to the same gene. DMD is notable for the size of the gene affected, 2.5 Mb, equivalent to 60% of the entire genome of *Escherichia coli* (see Chapter 5). However, after splicing the mRNA is 14 kb in size; thus 99% of the gene is in the form of introns and only 1% actually encodes amino acid sequence.

1.2.3.3 Fragile-X syndrome

Fragile-X syndrome results in mental subnormality. After Down's syndrome, it is the most common cause of mental subnormality in boys and is said to be responsible for the excess of boys in institutions for the mentally handicapped. It derives its name from a cytogenetic observation: when cells are cultured in medium that is deficient in precursors of DNA metabolism the X chromosome appears to have a break in metaphase spreads. At least five X-linked loci have been identified, FRXA–FRXE. FRXA has been shown to be caused by a trinucleotide repeat expansion of the type discussed above for MD and HD (see Chapter 5). Fragile-X syndrome mildly affects one-third of female carriers, so is partially dominant.

1.2.4 Complexity in single-gene disorders

There could be a temptation to regard analysis of monogenic disorders as straightforward, i.e. that the sole factor in their occurrence is the disease gene, whose inheritance follows simple and predictable patterns. In fact, this is often very far from the case. There are many factors that may modify the pattern of inheritance or symptoms of the disorder and these are discussed below. Such complexity provides a warning of the intricacies that may be encountered in the analysis of multifactorial disorders, which involve a combination of **polygenic** and environmental factors.

1.2.4.1 Genetic heterogeneity

This describes a situation where apparently clinically similar disorders are caused by mutations in different genes (non-allelic) or where mutations in the same gene result in clinically diverse conditions (allelic). Some examples of each are given in Table 1.5.

1.2.4.2 Penetrance

This refers to the frequency with which the disorder or phenotype is manifested in an individual who has inherited the disease allele. A mutation that does not inevitably cause the disorder is said to show incomplete penetrance. An example is split hand syndrome, a claw-like deformity of the hand caused by an autosomal dominant allele. Some individuals inherit the allele, but have normally formed hands.

1.2.4.3 Expressivity

This describes differences in the severity of a disorder in individuals who have inherited the same disease alleles. In the case of sickle cell anaemia, in which all cases result from mutation affecting the same amino acid, symptoms may be sufficiently severe to cause childhood death or be so mild that the disease remains undiagnosed until middle age. One possible cause of variation is that the effect of the sickle cell mutation is modified by other genes. In the case of sickle cell anaemia, one such modifying gene has been mapped to the X chromosome.

1.2.4.4 Mosaicism

Mosaicism arises when not all cells in the body are genetically identical. This may come about through a mutation in early development and may result in either the **germline** or somatic cells being affected. If the germline is mosaic, gametes may arise from progenitor cells of different genetic constitutions. For example, in one line the progenitor cell may be heterozygous for an autosomal dominant mutation, while in another the progenitor cell may be completely normal. This will clearly disturb the proportion of gametes affected, expected to be 50% if the parent is heterozygous for an autosomal dominant allele. Note that the somatic cells of females heterozygous for a sex-linked mutation will necessarily be mosaic as a result of X-chromosome inactivation.

1.2.4.5 Phenocopy

Sometimes an environmental factor may result in a disorder with the same symptoms as an inherited disorder. For example, the infection of a mother during pregnancy with the rubella virus may result in a profoundly deaf baby, a condition that can also be caused by a number of different single-gene defects. This is known as **phenocopy**.

1.2.4.6 Environmental effects

Environmental factors may influence the penetrance or expressivity of a disease allele. A clear example of this is phenylketonuria, a disease that may be largely prevented by a low-phenylalanine diet.

Table 1.5 Some examples of genetic heterogeneity.

Allelic	Non-allelic
Muscular dystrophy	Profound deafness
Becker	Retinitis pigmentosa
Duchenne	Autosomal dominant
Cystic fibrosis	Autosomal recessive
Pancreatic sufficiency/deficiency	X-linked
CBAVD	Polycystic kidney disease

1.2.4.7 Anticipation

There are a number of diseases whose severity apparently increases with each succeeding generation. MD and fragile-X syndrome are two well-known examples. Often the disorder may be so mild in the parent that it has not been diagnosed prior to its occurrence in a child. This phenomenon is known as **anticipation**.

1.2.4.8 Genomic imprinting

Genomic imprinting is said to occur when the expression of an allele depends on the parent from which it was inherited. This has been revealed in a number of different situations. An example is the effect of parental origin of a deletion of chromosomal band 15q12. Individuals heterozygous for this deletion suffer from a different disease according to the parent from which it was inherited. If the deletion was inherited from the father Prader–Willi syndrome results, characterised by mental retardation, hypotonia, gross obesity and hypogenitalism. If the deletion was inherited from the mother Angelman syndrome results, characterised by mental and growth retardation, hyperactivity and inappropriate laughter. This suggests that the function of the remaining allele depends on its parental origin. The mechanism is thought to involve DNA methylation, but the details are currently unclear.

1.3 Multifactorial or complex disorders and traits

It is a common observation that members of the same family are likely to resemble each other more than they resemble the general population and are more likely to suffer from the same diseases. Because family members are likely to share the same environment as well as the same genes, it is not easy to disentangle the relative contributions of genetic and environmental factors that may determine what we are like and what diseases we are likely to suffer from. For the most part, susceptibility to common diseases and phenotypic characteristics do not follow simple patterns of inheritance, so neither is likely to be determined monogenically. Nevertheless, family, adoption and twin studies clearly show a strong genetic component. A list of common disorders in which a genetic component has been demonstrated is given in Table 1.6. This list encompasses such a broad range of diseases that a patient's genetic constitution may be considered a factor in the majority of medical conditions.

These diseases are often referred to as **multifactorial diseases** because there are both genetic and environmental factors responsible for the onset of the disease. The ways this may occur are discussed in more detail in Chapter 6. At this point it is sufficient to note that the interactions are likely to be complex because they involve the interactions of several genes (polygenes or oligogenes) with each other and with the environment. For this reason they are commonly referred to as **complex diseases**. Because of this complexity, it is unlikely an individual's health prospects can ever be predicted in a deterministic fashion from genetic tests. However, genes that act as risk factors are

Table 1.6 Examples of common disorders for which there is evidence of a genetic component.

Congenital disorders	Common diseases of later life	Psychiatric disorders
Neural tube	Rheumatoid arthritis	Manic depression
Spina bifida	Various cancers	Alcoholism
Anencephaly	Epilepsy	Schizophrenia
Congenital heart disease	Multiple sclerosis	Tourette's syndrome
Cleft lip and palate	Insulin-dependent	Dyslexia
Mental retardation	diabetes mellitus	Alzheimer's disease
	Non-insulin-dependent	
	diabetes mellitus	
	Peptic ulcer	
	Ischaemic heart disease	
	Hyperthyroidism	
	Gallstones	
	Migraine	
	Asthma and other allergies	

being identified for a wide range of common diseases (see Chapter 6). These may well serve as the basis for tests to determine an individual's predisposition or susceptibility to common disorders, allowing early diagnosis and consequently more successful treatment if the disease should occur. Moreover, defining genetic variables simplifies the analysis of environmental factors. Thus genetic tests for predisposition to common diseases may allow at-risk individuals to change their lifestyle to avoid contracting the disease.

1.4 Chromosomal mutations

During gametogenesis the diploid complement of chromosomes is reduced through meiosis, so that each gamete receives one member of the 22 pairs of autosomes and one sex chromosome. During this process recombination takes place, resulting in the exchange of information on homologous chromosomes. Failure of this process results in gametes that do not have the correct complement of chromosomes; when such a gamete fuses with another the resulting zygote will have a chromosome imbalance. If the failure results in a whole chromosome failing to segregate, the event is called a **non-disjunction**. Non-disjunction can result in a gamete with an extra chromosome, in which case the zygote is said to be **trisomic**, or a gamete in which a chromosome may be missing, in which case the zygote will be **monosomic**.

Accidents at meiosis can also result in chromosome rearrangements, duplications and deletions. A reciprocal **translocation** is an exchange of segments between non-homologous chromosomes (Figure 1.5). If the pair of translocated chromosomes are inherited together there will be no change in the information content, apart from the breakpoint, which, if it occurs within a gene, may result in that gene being damaged. Such translocations

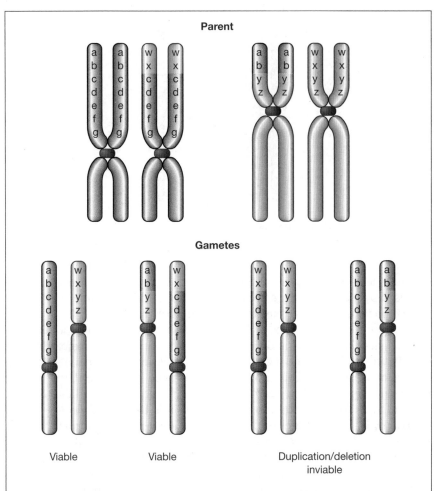

Figure 1.5 A balanced translocation between non-homologous chromosomes. Each gamete receives one copy of each non-homologous chromosome. Only two of the combinations will result in the normal gene complement. The other two combinations result in duplication and deletions of genes, resulting in inviability or a mutant phenotype.

have been very valuable in mapping disease genes, because the position of the breakpoint may be mapped cytogenetically by examining banded chromosomes during metaphase in an affected individual (see Chapter 5). If the pair of translocated chromosomes are not inherited together in a gamete, then in the resulting zygote some segments of chromosomes may be missing and some present in triplicate. Such gametes are not likely to be viable, so normally both members of a pair of reciprocally translocated chromosomes must be inherited together. Such a situation is described as a **balanced translocation** (Figure 1.5). An **unbalanced translocation** is the non-reciprocal duplication of a chromosome segment (Figure 1.6). Inheritance of such a chromosome may lead to a chromosome imbalance, for example a minority of Down's syndrome cases are familial. They result from trisomy for part of

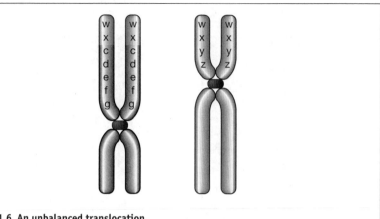

Figure 1.6 An unbalanced translocation.

chromosome 21 caused by an unbalanced translocation. Deletions of chromosomes lead to monosomy of chromosomal regions. *Cri-du-chat* syndrome is caused by a deletion of part of chromosome 5. This syndrome is identified by a characteristic cat-like mewing and mental retardation.

Chromosomal mutations are very common. One survey of spontaneous miscarriages showed that in 50% the foetus was affected by a major chromosomal abnormality. Some foetuses carrying such mutations survive to term but are consequently affected by the chromosome imbalance. These form a major category of genetic disease. Some examples are shown in Table 1.7.

The most common chromosomal mutation is Down's syndrome caused by trisomy 21. The features of this syndrome are mental subnormality, characteristic broad facial features and short stature. This is usually a result of an extra copy of the whole chromosome 21, although it can be caused by a partial trisomy caused by a parent with a balanced translocation. The former result from accidents at meiosis and are therefore sporadic. Nevertheless, the frequency increases with maternal age. A young mother age 21 has only a 1 in 2000 chance of giving birth to a child affected by Down's syndrome, whereas the risk increases to 1 in 45 for mothers who are 45 years old. This is thought to be due to the fact that the mother's eggs

Table 1.7 Some examples of chromosomal mutations.

Condition	Frequency
Sex chromosomes	
45X: Turner's syndrome	1 in 5000
47XXY: Klinefelter's syndrome	1 in 1000
Autosomes	
Trisomy 21: Down's syndrome	1 in 800 (maternal age dependent)
Trisomy 18: Edward's syndrome	1 in 10 000
Balanced translocations	1 in 500

start to develop before birth and become arrested at prophase of meiosis I. Thus, at the time of ovulation an egg may be over 45 years old and the consequent deterioration results in the decreased fidelity of the meiotic process.

Another very common class of chromosomal mutations are those involving sex chromosomes. Monosomy for the X chromosome (45X) results in Turner's syndrome, characterised by a sterile female phenotype with short stature and a web of skin between the neck and shoulders. Klinefelter's syndrome results from an XXY chromosome composition. This syndrome is characterised by a sterile male phenotype, and a tall and thin body form with breast development.

1.5 Mitochondrial mutations

Mitochondria provide 90% of cellular energy and thus the energy needed by organs, tissues and the body as a whole. Energy is generated in the mitochondria by the respiratory chain in a process known as **oxidative phosphorylation**. The respiratory chain consists of five protein complexes (complexes I–V) involving a total of 90 separate proteins. During oxidative phosphorylation electrons are passed along the chain from one complex to another. At the same time protons are pumped out of the mitochondrial matrix, generating a gradient across the inner mitochondrial membrane. The protons flow back into the matrix through complex V (ATP) synthase, which provides the energy for ATP production.

Each cell contains hundreds of mitochondria. Each contains between two and ten copies of a 16.6-kb circular DNA genome (mtDNA; Figure 1.7). In terms of size this corresponds to 0.0006% of the nuclear genome, but because there are approximately 10 000 mtDNA molecules per cell it amounts to about 1% of the total mass of cellular DNA. mtDNA encodes a number of essential functions, which are translated within the mitochondria using a mitochondrial-specific protein synthesis apparatus. The functions encoded by mtDNA are as follows:

- 13 respiratory chain subunits
  - seven subunits of complex I (NADH dehydrogenase)
  - three subunits of complex IV (cytochrome *c* oxidase)
  - two subunits of complex V (ATP synthase)
  - cytochrome *b* (a subunit of complex III)
- tRNA for each amino acid
- 12S and 16S rRNA for mitochondrial ribosomes.

The remaining functions are encoded by nuclear genes and synthesised in the cytoplasm before import into mitochondria.

Impairment of mitochondrial function leads to a decline in energy availability and results in clinical disorders. In principle, mutations to genes in both the nuclear and mitochondrial genomes could bring this about. In the last 10 years mutations to mtDNA have been shown to be the cause of

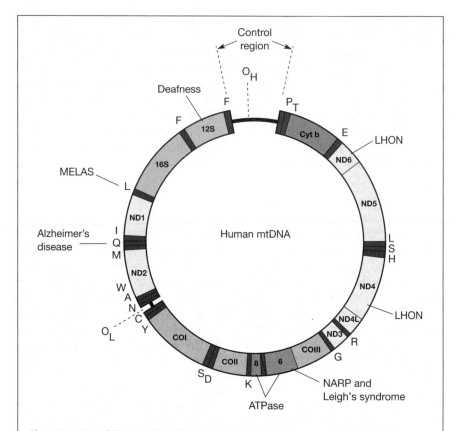

Figure 1.7 Map of the mitochondrial genome. The 22 tRNA genes are denoted by the single-letter amino acid code. 12S and 16S, 12S and 16S rRNA genes; COI–COIII, subunits of cytochrome *c* oxidase (complex IV); Cyt b, cytochrome *b*; ND1–ND7, subunits of NADH dehydrogenase; ATP6 and ATP8, subunits of ATP synthase (complex V); O_H and O_L, origins of replication for the heavy and light strand respectively. The sequence of the control region is variable. It contains the D-loop, required for DNA replication.

a number of disorders. In general these are multisystem disorders affecting the central nervous system, sight, hearing, heart and skeletal muscles, the kidneys and endocrine glands (Table 1.8).

The part of the central nervous system most often affected is a region of the brain known as the basal ganglia. The basal ganglia is important for coordinated motion and its dysfunction results in ataxia. The effect of mitochondrial dysfunction on muscles is to cause mitochondrial myopathy, a general term describing the degeneration and loss of function of muscles associated with the presence of ragged red fibres: degenerating muscle fibres containing defective mitochondria that turn red in the presence of a specific stain. The exocrine gland most commonly affected is the pancreas, resulting in diabetes mellitus due to lack of insulin and pancreatic exocrine deficiency, which is a failure to secrete the digestive enzymes that originate in the pancreas.

Table 1.8 Diseases caused by mitochondrial mutations.

19

Disease	Acronym	Symptoms	Gene affected
Chronic progressive external ophthalomoplegia	CPEO	Mitochondrial myopathy and paralysis of eye muscles	Multiple gene loss because of deletion
Kearns–Sayre syndrome		CPEO plus ataxia, retinal deterioration, heart disease, hearing loss, diabetes and kidney failure	Multiple gene loss because of deletion
Leber's hereditary optic neuropathy	LHON	Blindness caused by damage to the optic nerve	Mutations in subunits of NADH dehydrogenase
Leigh's syndrome		Degeneration of basal ganglia leading to loss of motor and verbal skills	ATP synthase
Mitochondrial encephalomyopathy, lactic acidosis and stroke-like episodes	MELAS	Dysfunction of brain tissue causing dementia and seizures, mitochondrial myopathy and lactic acidosis	tRNA^{Leu}
Neurogenic muscle weakness, ataxia and retinitis pigmentosa	NARP	Muscle weakness, ataxia and blindness	Subunits of ATP synthase
Pearson's syndrome		Childhood bone marrow dysfunction, leading to multiple blood disorders and pancreatic failure	Multiple gene loss because of deletion

Mitochondria are maternally inherited, being passed on to progeny in the cytoplasm of the egg cell. Although the sperm acrosome contains mitochondria, they are not retained after fertilisation. Normally the 10 000 copies of the mitochondrial genome in each cell are identical and the cell is said to be **homoplasmic**. When a mutation occurs in one of the mitochondrial genomes there will be a mixture of different genome types in the cell and it is said to be **heteroplasmic**. The heteroplasmic state is short-lived and the descendants of a heteroplasmic individual become homoplasmic in a few generations. This is thought to occur because the oogonia contain a much smaller number of mitochondria (about 200) than other cell types, creating a situation similar to a **population bottleneck,** which reduces genetic diversity (see Section 9.2.4).

Mitochondrial disorders do not show regular Mendelian patterns of inheritance. In fact the pattern of inheritance can be complex. There are a number of interrelated reasons for this.

1. Mitochondria are maternally inherited so mitochondrial disorders can only be inherited from an individual's mother.
2. If a mother is heteroplasmic each of her children may receive different proportions of affected mitochondria. Therefore there will be a wide variety in the type and severity of symptoms in each child.
3. During development and the lifetime of an individual derived from a heteroplasmic zygote, the proportion of defective genomes may change. This may occur through chance sampling effects during segregation of mitochondria to daughter cells after cell division, or through the acquisition of new mutations to previously wild-type genomes. As a result, in an affected individual, there will be variation in the proportion of defective genomes, both spatially in different tissues and temporally throughout the individual's lifetime. The effect of mutation will lead to an increase with time in the proportion of defective genomes. This is probably the reason why mitochondrial disorders often only become manifest after a delay of several years and why some disorders become progressively more severe with age. A factor that will exacerbate this process is that interruption of the electron transport chain leads to the production of highly reactive free oxygen radicals, which are mutagenic. Thus a slight impairment of mitochondrial function, which initially is not severe enough to cause clinical symptoms, may nevertheless interrupt electron transport sufficiently to trigger this process and eventually result in the onset of a disorder.

Both large-scale deletions and mutations to specific genes can affect the mitochondrial genome.

Point mutations affecting NADH dehydrogenase cause LHON. This disease predominantly affects young men, resulting in blindness as a result of damage to the optic nerve and sometimes heart and neurological abnormalities.

Mutations to ATP synthase result in variable phenotypes. Severely affected individuals suffer from Leigh's syndrome, a devastating and often lethal childhood disorder in which the basal ganglia of the brain degenerate. Less severely affected individuals suffer from NARP, characterised by muscle weakness, ataxia and a form of blindness called retinitis pigmentosa (which can also be caused by a variety of chromosomal mutations). The difference in severity in different individuals with the same mutation reflects the proportion and tissue distribution of mutant mtDNAs in affected individuals. Individuals with Leigh's syndrome have high levels of mutant mtDNA in multiple tissues.

Point mutations affecting tRNA or rRNA have severe effects because they simultaneously reduce the ability to make different mitochondrial proteins. MELAS is caused by a mutation in the tRNA^{Leu} gene. Other mutations of this gene have been shown to result in a variety of severe symptoms, including mitochondrial myopathy, heart disease and diabetes mellitus. Indeed, mutations to tRNA^{Leu} are responsible for 1.5% of all incidences of diabetes mellitus. Mutations to tRNA^{Glu} have been found to be associated with 5% of patients with late-onset Alzheimer's disease. Mutations in the 12S rRNA gene also generally affect protein synthesis. They have been shown to be responsible for congenital deafness.

The effect of deletion in mtDNA (ΔmtDNA) is different in adults and children. Adults generally suffer from slowly progressive neurodegenerative disorders such as CPEO and Kearns–Sayre syndrome. Children suffer from severe disorders, such as Pearson's syndrome, that progress rapidly and affect multiple organs. CPEO is characterised by mitochondrial myopathy and paralysis of the eye muscles. Kearns–Sayre syndrome is characterised by CPEO plus retinal degeneration, heart disease, ataxia, deafness, diabetes mellitus and kidney failure. Pearson's syndrome causes a failure to make blood cells, resulting in severe anaemia and other blood disorders, pancreatic failure leading to diabetes mellitus and pancreatic exocrine dysfunction. Children who suffer from Pearson's syndrome generally die in the first few years of life. A few children who do survive generally develop Kearns–Sayre syndrome as they grow older.

The differences in clinical phenotypes between children and adults may result from differences in the proportion and tissue distributions of ΔmtDNA. A more widespread distribution with a higher proportion of ΔmtDNA will result in earlier onset with more severe symptoms. The tendency of mitochondrial deletions to worsen over time is probably caused by preferential replication of ΔmtDNA in non-dividing cells. The reasons for this are not clear at present. The origin of the deletions is also unclear, because they are rarely passed on from a mother with ΔmtDNA, presumably because an egg cell containing ΔmtDNAs could not survive. mtDNA with duplicated segments is often found to be associated with ΔmtDNA. A duplicated genome would not result in any impairment since all functions are still present. However, they may give rise to deletions as a result of complex intramolecular recombination events between the duplicated segments.

1.5.1 Mitochondria and ageing

Common symptoms of mitochondrial disorders, such as diabetes, dementia, muscle weakness, loss of sight and hearing, ataxia, etc., are also characteristic of the normal ageing process. This has led to the suggestion that accumulation of somatic mutations in mtDNA may contribute to the deterioration of body function with age. Many environmental toxins inhibit mitochondria and so may be a contributory factor. Evidence is accumulating in support of this hypothesis.

- The performance of the respiratory chain complex deteriorates with age in the brain, skeletal muscle, heart and liver.

- Rearrangements in mtDNA accumulate with age in these same tissues. Δ mtDNAs accumulate in skeletal muscle after age 40, consistent with other observations that show that Δ mtDNA is preferentially replicated in non-dividing tissue.

- Genetic predisposition to type II diabetes often tends to be maternally inherited, consistent with a mitochondrial component to the aetiology.

- Animals raised on restricted-calorie diets have longer lifespans. Animals fed on such diets produce fewer free oxygen radicals and accumulate less damage to mtDNA.

- Individuals who suffer from ischaemic heart disease experience temporary interruptions to the blood flow to the heart caused by atherosclerotic plaques. During the resultant anoxia in the heart muscle, the respiratory chain is blocked; upon restoration of the blood supply (reperfusion) there is a burst of free oxygen radical production. The mitochondria in heart muscle of such patients contain a high level of mutant mtDNA, which may accelerate the onset of heart failure.

One intriguing possibility is that dietary supplements of antioxidants such as vitamin C may help limit the production of free radicals in the mitochondria and thus delay the ageing process.

1.6 The Human Genome Project

In the last two decades most of the genes involved in common monogenic diseases have been cloned and the cDNAs sequenced. This has led to enormous advances in understanding the molecular nature of the diseases and in developing robust diagnostic tests for their detection. The genes involved in monogenic diseases will be discussed in Chapter 5 and the tests used in their detection will be described in Chapter 8. The focus of human genetics has now shifted to understanding complex or multifactorial diseases. As we shall see in Chapter 6 this is a very difficult task, undoubtedly far more difficult than for monogenic diseases. Even cloning the genes involved in monogenic diseases proved very difficult and required large teams of scientists and large-scale international collaboration. From the time of the first efforts to clone genes involved in monogenic diseases it was appreciated that there was a need to map and then sequence the human genome. This led to a formal international collaboration called the Human Genome Project. Planning started in the late 1980s and work formally began in 1990. The impetus for the project initially came from the DOE and NIH in the United States. However, large-scale and important contributions came from laboratories all over the world, particularly the Genethon laboratory in Paris, funded by the French Muscular Dystrophy Association, and the

Biomedical Research.

The work was carefully planned at a strategic level, with a number of goals established in 1990. These were formally revised in 1993 and 1998. Broadly speaking, the project was divided into two phases. The first, which lasted from 1990 to 1998, was aimed at producing different types of map of the human genome. The second, from 1998, was to derive the sequence itself. The reason for planning the work like this was three-fold. Firstly, the maps would be needed before sequencing could begin. Secondly, in 1990 the technology for sequencing was inadequate for the task and would need to be improved before sequencing the genome could be undertaken. Thirdly, maps of the human genome would be tremendously valuable in their own right for identifying and isolating genes involved in disease. Moreover, the technology for their construction was either available in 1990 or at least imminent. The Human Genome Project is now drawing to a close. The maps were completed on schedule by 1998. A draft sequence of the genome was announced in 2000 and published in 2001. The finished sequence is on schedule to be completed by 2003.

The physical and genetic characterisation of the human genome will occupy the first part of this book. Chapter 2 will introduce the basic structure of the complex human genome. Chapter 3 will describe the different types of map and how they are used. Chapter 4 will describe how the sequencing was carried out and how it will be used.

1.7 Summary

- Genetic factors play a major role in determining lifetime health. Approximately 5% of babies born alive will suffer from a significant medical condition for which there is a genetic component. Susceptibility to many of the common diseases of later life is also influenced by our genetic constitution.

- There are four main classes of genetic diseases: (i) monogenic defects; (ii) multifactorial or complex disorders; (iii) chromosomal imbalances; and (iv) mitochondrial mutations.

- Monogenic disorders follow Mendelian patterns of inheritance, and are subdivided into autosomal recessive, autosomal dominant and sex linked.

- In monogenic disorders, the pattern of inheritance and the manifestation of the symptoms may be complicated by a variety of factors, such as genetic heterogeneity, penetrance, varying expressivity, mosaicism, anticipation, genomic imprinting and phenocopy.

- Multifactorial or complex disorders come about through the interaction between polygenic or oligogenic determinants and the environment.

- Chromosomal mutations arise through non-disjunction at meiosis, producing aneuploids. An extra copy of a chromosome results in trisomy, while loss of a chromosome results in monosomy. Trisomy 21 is responsible for Down's syndrome, one of the commonest genetic conditions. Other common chromosome abnormalities are balanced and unbalanced translocations, and deletions.

- Mitochondrial disorders are maternally inherited. A variety of factors lead to great complexity in the age of onset and severity of symptoms. Mitochondrial disorders affect many different organs simultaneously, including the central nervous system, skeletal muscle, heart, sight, hearing and endocrine function. Such disorders tend to become progressively worse as further mitochondrial mutations accumulate.

Further reading

General

House of Commons Science and Technology Committee (1995) *Third Report – Human Genetics: the Science and its Consequences. Volume 1. Report and Minutes of Proceedings*. HMSO, London.

Provides a general introduction to the ground covered by this book. As well as giving data for the incidence of various diseases, it provides an account of the views of leading clinicians, scientists and industrialists as to the influence of genetics on human health and the possibilities of the new developments to ameliorate the consequences of genetic defects. Finally, it provides a commentary on the legal, ethical and social impact of human genetics along with recommendations designed to ensure that scientific developments do not have a negative impact, issues discussed in Chapter 11 of this book.

WEATHERALL, D.J. (1991) *The New Genetics and Clinical Practice*, 3rd edn. Oxford Medical Publications, Oxford.

Chapter 1 provides a detailed review of the influence of genetic factors on human health. It is notable for the carefully researched data on the frequency of different genetic diseases and the extent of ethnic variation.

Monogenic disorders

Online Mendelian Inheritance in Man (OMIM)
http://www.ncbi.nlm.nih. gov/entrez/query.fcgi?db=OMIM

This database started life as a catalogue assembled by Victor McCusick, one of the leading figures in the field of human genetics. It is now a website maintained by the National Center for Biotechnology Information (NCBI) in the United States. It contains a total of 9608 gene loci (as of October 2001) that are different from each other and where the mode of inheritance is judged to have been proved. Each entry contains a textual review of the locus and, where available, links to pictures.

Complex diseases
GHOSH, S. and COLLINS, F.S. (1996) The geneticist's approach to complex disease. *Annual Review of Medicine*, **47**, 333–353.

LANDER, E.S. and SCHORK, N.J. (1994) Genetic dissection of complex traits. *Science*, **265**, 2037–2048.

Inheritance of psychiatric disorders and personality traits
BOUCHARD, T.J. (1994) Genes, environment and personality. *Science*, **264**, 1700–1701.

MANN, C.C. (1994) Behavioural traits in transition. *Science*, **264**, 1686–1689.

PLOMIN, R., OWEN, M.J and McGUFFIN, P. (1994) The genetic basis of complex human behaviours. *Science*, **264**, 1733–1739.

ROSE, S.J., LEWONTIN, R.C. and KAMIN, L.J. (1990) *Not in our Genes: Biology, Ideology and Human Nature*. Penguin, Harmondsworth.

An outspoken critique of the evidence that underpins the claims that genetic factors control behaviour and the incidence of psychiatric disorders.

Mitochondrial disorders
LARSSON, N.G. and CLAYTON, D.A. (1995) Molecular aspects of human mitochondrial disorders. *Annual Review of Genetics*, **29**, 151–178.

WALLACE, D.C. (1997) Mitochondrial DNA in aging and disease. *Scientific American*, **276**, 22–29.

CHAPTER TWO

An introduction to the structure of the human genome

Key topics

- Amount of DNA in the human genome
- Genes
  - Gene expression
  - Gene families
  - Pseudogenes
- Tandem repeat arrays of rRNA, tRNA and histone genes
- Intermediate repeated DNA
  - LINEs: L1 and retrotransposition
  - SINEs: *Alu* family
  - Processed pseudogenes
  - Selfish DNA
- Highly repetitive DNA
  - Telomeres
  - Tandem repeat arrays at centromeres
  - Minisatellites
  - Microsatellites
- Human karyotype
- Packaging DNA into chromosomes
  - Nucleosomes
  - 30-nm fibre
  - 700-nm fibre

2.1 Introduction

The human **genome** is the term used to describe the sum total of DNA molecules found within every cell of the human body except red blood cells. Nearly all the genome is found in the nucleus; however, the mitochondria also contain essential genetic information. The haploid genome, the DNA found in a gamete, is half the complement of a somatic cell. The human genome consists of 3000 million base pairs of DNA (3.2×10^9 bp or 3200 Mb).

This DNA contains all the information required for a zygote to develop into an adult human. The information is in the form of genes. Each gene consists of a length of DNA that performs some function, usually specifying the amino acid sequence of a protein. A surprising aspect of the human genome is that coding sequences form only 1.5% of the total. Some of the rest of the DNA is non-coding, but has a functional role in regulating and promoting gene expression or a structural role in chromosome integrity and segregation of chromosomes at nuclear division. A large fraction of the genome does not appear to have a function; if it does, it is a function that does not depend on the sequence of bases within the DNA. Much of this DNA consists of a menagerie of repetitive and mobile elements that may be parasitic or 'selfish' in origin. The genes themselves are not continuous stretches of DNA: the **coding sequences (exons)** are interrupted by non-coding sequences (**introns**). Sometimes the exons form only small patches in long stretches of introns. Indeed, as discussed below, in some cases only 1% or less of the total DNA that forms a gene may be coding sequence.

Thus the human genome is large, and complicated in its organisation. Before gene cloning techniques were developed, it was impossible to study individual genes. Even with gene cloning, it is still often very difficult to find and clone human genes. The small fraction of the genome that forms genes and the technical complications introduced by the repetitive and mobile elements exacerbates the problem further.

In the last few years there has been spectacular progress in characterising the genome, including mapping physical and genetic landmarks and determining the entire nucleotide sequence. This book is an account of how the maps and sequence of the human genome were obtained and how they will be exploited in research into basic human biology and in medicine. Before this account can begin, it is necessary to describe the general features of the genome and how the DNA is packaged into chromosomes. This account is only intended to be a summary of material treated in much more detail elsewhere. You should refer to one of several excellent text books listed at the end of the chapter for more detail. The facts and figures given in this chapter come mostly from the analysis of the draft human genome sequence, considered in detail in Chapter 4.

2.2 Sequence architecture of the human genome

In the 1960s a simple experiment measuring the rate of reassociation of denatured DNA led to the surprising conclusion that a large fraction of mammalian DNA consisted of repetitive sequences (see Further reading). As a result of these experiments, human DNA may be categorised as follows.

1. **Single sequence DNA** or low-copy DNA (about 45% of the total). This class contains the sequences that form genes, as would be expected from the Mendelian laws of genetics. However, coding sequences only

account for about 1.5% of the total genome. The remainder of this class is DNA from introns or from spacer DNA, sequences of no apparent function that separate genes.

2. **Intermediate repeated DNA,** present at between 10^2 and 10^5 copies per genome (about 45% of the total).
3. **Highly repetitive DNA,** present at up to 10^6 copies per genome (about 10% of the total).

2.2.1 Genes

A functional definition of a **gene** is a DNA sequence that contributes to the phenotype of an organism in a way that depends on its sequence. A change in the sequence may affect its function and may have phenotypic consequences. Most genes encode proteins. However, sequences that encode functional RNA molecules are also classified as genes, as their function depends on their sequence.

2.2.1.1 Gene expression

Transcription in humans, as in all other eukaryote cells, is catalysed by three types of RNA polymerase, designated RNA polymerase I, II and III (or pol I, pol II and pol III respectively). RNA polymerase I transcribes rRNA. RNA polymerase II transcribes nuclear genes that encode proteins. RNA polymerase III transcribes tRNA genes, and a small number of functional RNA molecules such as 5S RNA and nuclear RNAs that mediate splicing (see below). All three RNA polymerases require the action of transcription factors to bind DNA and promote transcription.

The structure of the **upstream** region of a typical gene transcribed by pol II is shown in Figure 2.1. A typical gene contains a **promoter** region, which extends for about 200 bp upstream of the transcription start site at nucleotide +1. In addition it may contain one or more of the following elements: **enhancers**, **response elements** and **silencers**. The promoter contains all of the elements required for a basal level of transcription. These elements are short conserved sequences that bind different **transcription factors**. Transcription factors are said to be *trans*-acting because they are encoded by another gene that must be translated in the cytoplasm and imported into the nucleus. The sequence elements on the DNA to which they bind are said to be *cis*-acting because they control the transcription of a gene that lies immediately adjacent on the same DNA molecule.

Most human promoters contain an element called a TATA box centred around position –30 relative to the transcription start site. The TATA box binds **general transcription factors**, designated TFIIX, where the roman numeral denotes the RNA polymerase involved and X denotes the particular protein. In pol II transcription the process is initiated by the binding of TFIID mediated by another protein called **TATA-binding protein** (TBP). This is followed by the assembly of an initiation complex with pol II and the other general transcription factors.

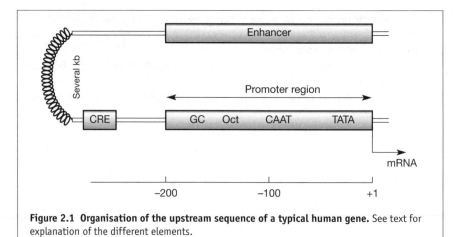

Figure 2.1 Organisation of the upstream sequence of a typical human gene. See text for explanation of the different elements.

The TATA box is responsible for correctly locating the site at which transcription starts. Three other commonly found elements within the promoter affect the efficiency of transcription.

- **GC box** (GGGCGG), which binds the transcription factor SP1.

- **CAAT box** (GGCCAATCT), which binds the transcription factors CTF and NF1.

- **Oct box** (ATTTGCAT), which binds the transcription factors Oct-1 and Oct-2.

The CAAT box is normally found around position –80 relative to the transcription start site. The spacing of the GC box and Oct box may vary. Active promoters contain at least one of these elements, but do not necessarily contain all three. As well as these elements that bind ubiquitous transcription factors, tissue-specific elements may also be found within the promoter region.

Enhancers are regions of DNA that operate to stimulate the basal level of transcription from its promoter. They are operationally distinguished from promoters by three criteria:

- they may be located a considerable distance from the transcription start site;

- their action is not dependent on their location: they may be located **upstream** or **downstream** of the gene they control;

- their action is not dependent on orientation.

Enhancers usually contain multiple sites where transcription factors bind to stimulate transcription. These may include sites for the same ubiquitous transcription factors found within the promoter as well as sites for the binding of specific transcription factors. Enhancers probably act at a distance through the formation of DNA loops, which bring the transcription factors bound to the enhancer region into close proximity with the basal transcription complex.

Response elements induce genes in response to particular signals. Examples include cyclic AMP response element (CRE), serum response element (SRE) and heat-shock response element (HRE). Response elements may be found within the promoter region, closely upstream of the promoter or in more distant enhancers.

Transcription control is generally positive in eukaryotes: transcription is stimulated by the presence of transcription factors. However, there are examples of negative control where genes are turned off by the binding of proteins to elements known as **silencers**. In yeast, where this phenomenon has been well documented, the same transcription factor can stimulate transcription at one gene and silence transcription at another. The action of silencers is less well characterised in humans.

The overall level of transcription of a gene is the outcome of the different influences exerted by the promoter and enhancers. Some genes, termed **housekeeping genes**, are expressed in most tissues most of the time, and are responsible for functions likely to be necessary in any cell. Such genes may be transcribed from a promoter alone if it contains sufficient elements for the binding of ubiquitous transcription factors such as SPI1. Other genes, whose transcription is tissue specific or depends on the presence of a specific signal, require the action of enhancers and response elements to stimulate the basal transcription from their promoters. There is great scope for sophisticated fine tuning of transcription arising from the interaction of many transcription factors to sites within the promoter and enhancers.

As soon as transcription starts, the 5' end of the nascent mRNA molecule is covalently modified to form a structure known as a **cap**. This is formed by the addition of 7-methylguanosine to the 5' nucleotide of the nascent mRNA in a 5'–5' triphosphate linkage. This effectively blocks the 5' end of the molecule because the 7-methylguanosine is in the opposite orientation to the rest of the molecule. The cap is essential for initiation of translation at the first AUG codon in the mRNA. The appearance of sequence AAUAAA in the nascent RNA molecule causes transcription to terminate a few hundred base pairs downstream; the RNA molecule is then cleaved about 20 bp downstream of the AAUAAA signal. The 3' end is also covalently modified by the addition of 100–200 adenine residues to form a structure known as the poly-A tail.

The part of the gene that encodes the protein contains sequences called **introns**, which interrupt the protein-coding regions known as **exons** (see Figure 2.2). The **primary RNA transcript** contains sequences derived from the introns. These are removed by a process known as **splicing** before the transcript leaves the nucleus. Splicing occurs through a cyclical process that occurs in a structure called a spliceosome. The spliceosome consists of small nuclear RNA molecules complexed to proteins called snRNPs (small nuclear ribonucleoprotein particles). These are sometimes colloquially known as 'snurps'. Splicing requires the concerted action of snRNPs and specific sequences at the junctions of introns and exons called **splice donor** and **splice acceptor** sites. The splice donor site is sometimes known as the 5' splice site; it consists of the first two nucleotides of the intron, which is always the dinucleotide GT. The splice acceptor site is sometimes known as the 3' acceptor site; it consists of the last two nucleotides of the intron, which is always the dinucleotide AG. Alteration of these sequences will prevent splicing and will have phenotypic consequences. In addition, there is a sequence in the intron known as the branch site, 18–40 nucleotides upstream of the splice acceptor site, where a splicing intermediate called a **lariat** is formed. However, in humans this sequence is not highly conserved and, if altered by mutation, splicing can proceed using other related sequences in the vicinity.

RNA processing is a point at which gene expression can be modified. There are now many examples of genes that are differentially spliced and give rise to sets of different polypeptides that originate from the same primary transcript.

2.2.1.2 CpG islands

DNA can be modified by methylation of cytosine to form 5-methylcytosine. This normally occurs at the dinucleotide CpG, i.e. a C residue followed in the 3' direction by a G residue on the same strand of DNA. The dinucleotide CpG occurs less frequently than would be expected by chance from the base composition of DNA. This is a consequence of DNA repair mechanisms. Accidental deamination of cytosine produces uracil, whereas deamination of 5-methylcytosine produces thymine. Because uracil is not found in DNA, it is efficiently excised and replaced with cytosine by an enzyme called glycolase. However, thymine, a natural constituent of DNA, is now mispaired with a G residue. Although this will also be repaired, the process is not perfectly efficient and over a long period has led to a gradual reduction in the frequency of the CpG dinucleotide. DNA methylation is generally associated with repression of transcription.

So CpG is relatively rare in the human genome and where it occurs it will normally be methylated. However, so-called **CpG islands** exist where the frequency of CpG is greatly elevated, approaching that expected from the percentage (G+C) base composition of the human genome, and the CpG dinucleotides are hypomethylated or not methylated at all. These CpG islands are about 1 kb in length and tend to extend over promoters of expressed genes; about 56% of human genes are estimated to be associated with such sequences. They can be recognised by restriction enzymes that contain CpG in their target sequence and which will thus cut less frequently than expected because of the relative rarity of the CpG dinucleotide and because methylation inhibits the action of the restriction enzyme. Sites where these enzymes cut DNA will be clustered in CpG islands.

2.2.1.3 The number, size and spacing of human genes

A question that has provoked much interest over the years is how many genes there are in total in the human genome. Although, as we shall see in Chapter 4, there is still some uncertainty about the final number, the completion of the draft human genome sequence has led to the conclusion that there are about 30–35 000 genes in total. This figure is lower than previous estimates and was one of the most surprising conclusions of the genome project. The draft genome sequence allows some of the properties of genes to be enumerated (these are summarised in Table 4.3 on page 105). Human genes show enormous variation in overall size and the size and number of introns. Some genes, such as the histone genes, do not contain introns. The α-globin gene is 0.8 kb in size and contains three introns, which account for 30% of the total genomic locus. Others genes are extremely large. For example, the structures of the α-globin and the blood clotting factor VIII genes are compared in Figure 2.2. The dystrophin gene (the gene defective in Duchenne muscular dystrophy) is 2.4 Mb in size and contains 79 introns, which account for 99.4% of the total genomic locus. The mean size of genes is 27 kb, including introns and exons. The whole of each gene is transcribed, but the resulting RNA molecule is processed in the nucleus to produce mRNA that is translated into protein in the cytoplasm. Only that portion of the gene that remains in the translated portions of its mRNA can be counted as coding sequence, the mean size of which is 1.34 kb. Thus on average only 5% of each gene is coding sequence. Of the total genome, 33% is transcribed but only 1.5% is coding sequence. The density of genes varies in each chromosome from 9 to 14 genes/Mb or one gene every 71–111 kb.

2.2.1.4 Gene families
Between 25 and 50% of protein-coding sequences in the genome are unique. The remainder belong to families of similar or related genes. Some of these genes are dispersed through the genome, while others are present in clusters of related genes.

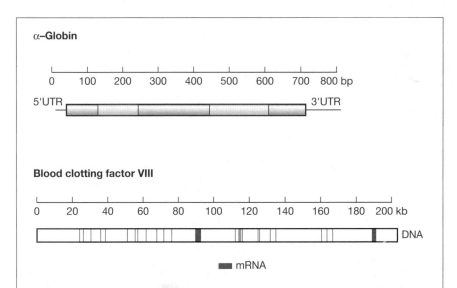

Figure 2.2 Human genes vary in size and intron content. The figure compares the size and intron content of the genes for α-globin and blood clotting factor VIII. In each case the coding regions are shown in solid blue boxes and the introns as open boxes. Note the difference in the scale shown above each gene. In the case of α-globin the 5' and 3' untranslated regions of the mRNA are shown (5' UTR and 3' UTR); these are normally classed as exons. The blood clotting factor VIII gene occupies 186 kb of the human X chromosome. It contains 26 exons ranging in size from 69 to 3106 bp and introns as large as 32 kb. The mRNA is only 9 kb in size and is shown to scale. It comprises a 7053-nucleotide coding sequence and a 3' UTR of 1806 nucleotides.

Gene families are thought to have arisen by a process of gene duplication and divergence from an ancestral gene. Figure 2.3 shows how the evolution of the different proteins that make up haemoglobin is thought to have occurred. Haemoglobin is the oxygen-carrying molecule in blood, comprising two β-globin and two α-globin protein subunits, each complexed to a molecule of haem. The different members of the globin gene families are expressed at different times during foetal, embryonic and adult stages of development (see Chapter 5).

As well as whole genes it is very common to find proteins of different overall function containing similar protein **domains**. For example, many different proteins contain nucleotide-binding domains but show no similarity throughout the rest of the protein.

Because there are multiple copies of genes in a family, sometimes loss of function in one of them may be tolerated. This has led to the evolution of **pseudogenes**, which are sequences recognisably similar to functional genes but which have accumulated nonsense and frameshift mutations that will prevent them from functioning. They may be viewed as the decaying relics of genes that were once functional. For example, the β-globin gene cluster contains two pseudogenes (Figure 2.3).

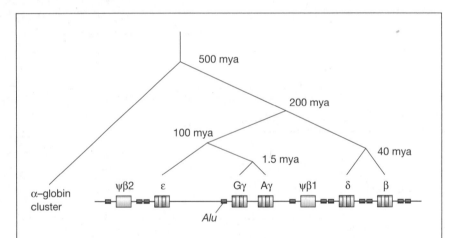

Figure 2.3 Evolution and structure of the β-globin gene family. About 600 million years ago (mya) gene duplication of a single globin-like gene produced the two genes that diverged to give rise to the ancestral genes of the α- and β-globin gene clusters. Initially the α and β genes were found as tandem repeats, as is still the case in modern amphibians. A translocation event separated the two genes to different chromosomes and this was followed by further rounds of gene duplication and divergence. ψβ1 and ψβ2 are pseudogenes. The positions of eleven *Alu* elements are shown by the solid boxes.

2.2.1.5 Non-protein-coding genes

Not all genes encode proteins. The bulk of cellular RNA consists of rRNA and tRNA required for protein translation. There are also a number of other RNA molecules that function directly within the cell. The 7SL RNA molecule forms part of the signal recognition particle required for the translocation of proteins across the endoplasmic reticulum. This is the first step in the secretory pathway that will result in the protein being exported from the cell. Small nuclear RNA molecules form complexes with protein to mediate the splicing of protein-coding RNA transcripts (see above).

2.2.1.6 Tandem repeat arrays

The rRNA molecules consist of three separate species: 28S, 18S and 5.8S. They are transcribed from a single transcription unit by a dedicated RNA polymerase known as RNA polymerase I. A single 45S transcript is produced that is processed to form the separate species. tRNA genes are also transcribed by a dedicated polymerase known as RNA polymerase III. (Protein-coding genes are transcribed by RNA polymerase II.)

   The transcription units for rRNA are repeated 150–200 times, so that each copy lies immediately adjacent to the next and are arranged so that the end of one unit abuts the start of the next. Such an arrangement is known as a **tandem repeat array** (Figure 2.4). The genes encoding the five histone proteins (see below) are found in similar tandem repeat arrays. In contrast to gene families, the sequences of these genes in tandem repeat arrays are all identical or nearly identical. There are 500 tRNA genes dispersed around the genome.

   rRNA and tRNA constitute about 90% of the total RNA within the cell. Histones are also present in large numbers as they are bound to DNA in a stoichiometric fashion (see below). When the DNA is replicated the number of histones must also double, so large quantities have to be synthesised in a short period of the cell cycle. A single copy of the histone, rRNA and tRNA genes would not be capable of synthesising enough product. Only by having several hundred copies can the rate of synthesis keep up with demand.

2.2.1.7 Spacer regions

Genes are separated from each other by long tracts of DNA known as **spacer regions**. The sequence of spacer regions changes much more rapidly in evolution than the sequence of genes, indicating that there is less selection

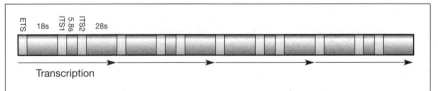

Figure 2.4 Tandem repeat array rRNA transcription units. Each unit is transcribed to produce a 45S transcript, which contains the ETS (external transcribed spacer unit), 18S, ITS1 (internal transcribed spacer 1), 5.8S, ITS2 and 28S sequences. This transcript is processed in a number of sequential steps to release the 18S, 5.8S and 28S RNA molecules.

against changes in its sequence. Thus it is thought that spacer DNA does not have a function that depends on its sequence. It is possible that the physical separation of genes is important in some way, so spacer DNA may have a sequence-independent function. The long tracts of DNA found in introns may be regarded as a particular form of spacer DNA. An example of the physical separation of genes may be seen in the region that contains the five members of the β-globin gene cluster. Each gene is 1.6 kb in size including two introns, yet the cluster is spread over 60 kb of chromosome 11 (Figure 2.3).

2.2.2 Intermediate repeated sequences

Much of the intermediate class of DNA consists of a small number of families, each consisting of sequences that are similar but not identical to each other. Most of these are **transposons**, so called because new copies are generated in a new location by a process of **transposition**. When this happens, one copy is found at the original location, and one copy at the new location, thus duplicating the original DNA sequence. Because the removal of the extra copies is a slow process on an evolutionary time-scale, they will increase in number. In humans they have come to occupy 45% of the genome. This has led to the concept of **selfish** or **parasitic DNA** because they do not contribute to the phenotype of the organism but evolve only to increase in number. However, some of these **mobile genetic elements** have been co-opted to have a role (see Section 4.8.1.3). The signature of mobile genetic elements is a short direct repeat sequence either side of the point of insertion into the host chromosome. There are four different classes of repeated DNA elements:

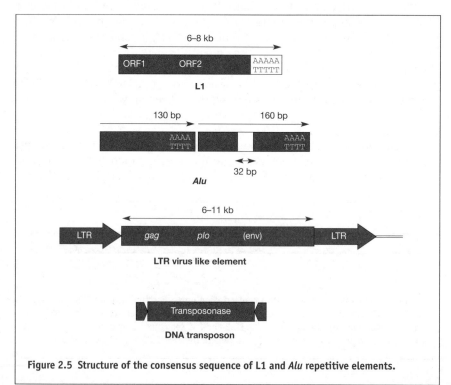

Figure 2.5 Structure of the consensus sequence of L1 and *Alu* repetitive elements.

- long interspersed elements (LINEs)
- short interspersed elements (SINEs)
- LTR retrovirus-like elements
- DNA transposons

The structures of these elements are illustrated in Figure 2.5.

2.2.2.1 LINEs

Complete **long interspersed elements** (LINEs) are about 6–8 kb in size, and contain a promoter for RNA polymerase II and two **open reading frames** (ORFs). One of the ORFs encodes a protein with similarities to **reverse transcriptase** found in **retroviruses** that synthesise a DNA copy of an RNA template. The other ORF encodes an endonuclease. An AT-rich region is located near the 3' end of the element. The propagation of LINE elements occurs by a process called **retrotransposition**, in which the LINE mRNA molecule serves as a template for the reverse transcriptase it encodes and the resulting DNA copy is inserted into a new chromosomal site (Figure 2.6). The reverse transcription of LINE elements usually fails to proceed to completion. Because the LINE mRNA template is copied from its 3' end, most LINEs are truncated at the 5' end and average only 1 kb in size. There are four distantly related LINE families in the genome (LINE1–4), together occupying over 20% of the genome. The most common is LINE1, with over 500 000 copies, representing nearly 17% of the total sequence.

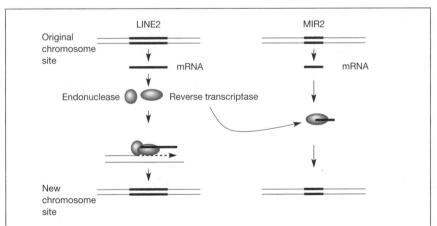

Figure 2.6 Propagation of LINEs and SINEs by retrotransposition. The retrotransposition of a LINE element is shown on the left. After transcription and translation, the endonuclease and reverse transcriptase form a complex with the LINE mRNA, which migrates to the nucleus. The endonuclease nicks chromosomal DNA and the free 3' OH end generated is extended by the reverse transcriptase, using the LINE mRNA as a template. As a result, a new copy of the LINE mRNA is inserted into the genome. The reverse transcriptase of each LINE element binds to a sequence, particular to each element, at the 3' end of the LINE mRNA. SINEs are thought to propagate by also having the recognition sequence of a reverse transcriptase encoded by a LINE in the same genome. In this case, the human MIR element has the recognition sequence for the LINE2 reverse transcriptase, which is thus able to copy it into DNA. Priming of the reverse transcription may occur by a hairpin loop formed by the polyA tail of the mRNA and the polyT region in the second repeat (see Figure 2.5).

2.2.2.2 SINEs

Short interspersed elements **(SINEs)** are 300–400 bp in size and are found in most metazoan organisms. SINE elements do not encode any proteins and are not able to transpose autonomously. Most SINE elements are similar at their 3' end to the sequence recognised by the reverse transcriptase encoded by a LINE in the same genome. This allows them to propagate by retrotranspositon (Figure 2.6). SINEs were originally derived from genes that are transcribed by RNA polymerase III (Pol III), normally tRNA genes. The Pol III promoter is retained in SINEs and is responsible for their transcription.

There are three SINE families in the human genome – *Alu*, MIR and MIR3. Together they occupy 13% of the genome. The ***Alu*** element derives its name from the restriction enzyme *Alu*1, as most copies contain a target sequence for this enzyme. It is the most common SINE in the human genome, with about 10^6 copies, representing 10% of the genome. This means that on average there will be an *Alu* element every 3 kb, although as we shall see in Chapter 4, the density is not uniform. It is the only SINE that is still active in the genome. The *Alu* element is unusual because its sequence shows that it was not derived from a tRNA gene, but from another Pol III-transcribed gene, 7SL RNA, which forms part of the signal recognition complex required for the translocation of proteins into the endoplasmic reticulum during secretion. Moreover, unlike most SINEs, its 3' end is not similar to the recognition sequence of a reverse transcriptase encoded by a LINE, so which reverse transcriptase is responsible for its propagation is still uncertain. The *Alu* elements are confined to the primate line, but there are other mammalian SINEs derived from 7SL RNA, such as the rodent B1 element. In contrast MIRs are found in all mammalian lines, hence their name: mammalian-wide interspersed elements. The 3' end of the *MIR3* transcript contains the 50 bp recognition sequence of the reverse transcriptase encoded by LINE2. Neither LINE2 nor MIR3 are still active. It is thought that when the last active LINE2 element disappeared from the genome 80–100 million years ago, transposition of MIR3 also ceased (see Section 4.8.1.3).

2.2.2.3 LTR retrotransposons

LTR **retrotransposons** (LTRs) derive their name from **long terminal repeats** at each end of the element. Between the LTRs are *gag* and *pol* genes that encode a reverse transcriptase, protease, RNAse H and integrase. These proteins are sufficient to programme the autonomous transposition of the element. The LTRs contain all the necessary regulatory elements to ensure transcription of the *gag* and *pol* genes. The resulting transcript is reverse transcribed in the cytoplasm using a tRNA primer. LTRs are very similar to retroviruses, and it is thought that retroviruses arose from LTRs by the acquisition of an *env* gene. Recombination often occurs between the LTRs, resulting in the excision of the intervening sequence and leaving a single inactive LTR in the genome. A variety of LTR elements are found in eukaryotes. Only endogenous retroviruses (ERV) are found in the mammalian genome, falling into four classes (ERVI–III and MalR). Together LTR-derived sequences occupy 8% of the genome, but 85% of this consists of isolated LTRs. Nearly all LTR transposons in the mammalian genome are now inactive (see Section 4.8.1.3).

2.2.2.4 DNA transposons

DNA transposons resemble bacterial transposons. They have terminal inverted repeats and encode a transposonase. Only an intact element can encode an active transposonase. However, the transposonase can act in the nucleus on deleted and mutated copies of the element. This results in the gradual accumulation of inactive elements and erodes the efficiency of the transposition process. For this reason DNA transposons tend to be relatively short-lived unless they can move by horizontal transfer to virgin genomes, which does happen. There are seven major classes of DNA transposons in the human genome, occupying less than 3% of the total sequence. All are thought to have been inactive for at least 50 million years (see Chapter 4).

2.2.2.5 Retrotransposons

The reverse transcriptases of the mobile elements described above may sometimes act on the mRNA sequences of other genes, resulting in the transposition of a copy to a new location. Because the reverse transcriptase uses mRNA as a template, which has been processed to remove introns, the new copy contains only the exonic sequences from the source gene. Usually these copies become inactivated by mutation and are known as **processed pseudogenes**. However, the draft sequence of the human genome shows that there are about 300 functioning genes that probably originated by retrotransposition (see Section 4.8.1.3).

2.2.3 Highly repetitive DNA

Highly repetitive DNA consists of short sequences repeated up to a million times in the genome. Because the average base composition of the repeated sequence may be different from the average of the rest of the genome, it often has a different density. Thus when total human DNA is fractionated by buoyant density ultracentrifugation, the highly repetitive DNA may form separate or satellite bands to the main peak. For this reason highly repetitive sequences are often known as **satellite DNA**. Most satellite DNA is found in long tandem arrays around the centromere, in subtelomeric regions and in the heterochromatic short arms of acrocentric chromosomes and most of the Y chromosome (see below for a description of chromosome structure). An important example of a satellite DNA sequence is α satellite DNA. The α satellite is a highly repetitive sequence that has an essential function in the centromeres of chromosomes (Box 2.1). Centromeres are the point where chromosomes attach to the spindle at nuclear division, ensuring the proper segregation of chromosomes to daughter cells.

BOX 2.1: α SATELLITE DNA

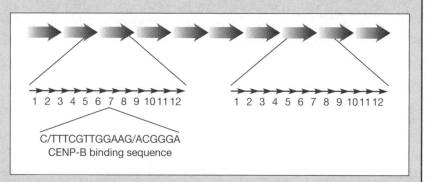

C/TTTCGTTGGAAG/ACGGGA
CENP-B binding sequence

Each centromere contains a tandem array of α satellite repeats that extend for millions of base pairs, uninterrupted by any other sequence. Each array is organised in a hierarchy of higher-order repeats. These higher-order repeats are represented by the large arrows in the figure. Each higher-order repeat contains a number of monomers, which varies between 4 and 32 depending on the particular chromosome; these are represented by the smaller arrows numbered 1–12 in the figure. Some, but not all, of these monomers contain a 17-bp binding site for the centromere-specific DNA-binding protein CENP-B. The monomers are not identical to each other; indeed they can show up to 20% sequence divergence. They may be no more similar to each other than they are to the α satellite in other primates. In any one array, each of the higher-order repeats (large arrows) are virtually identical (<2% sequence divergence). They contain the same monomer subunits repeated in the same order. The organisation and sequence of the monomers is particular to each chromosome. The number of higher-order repeats varies from 100 to 5000 on different chromosomes, giving a range of overall array sizes from 0.2 to 10 Mb. Altogether, the α satellites represent several per cent of the total genome.

The repeat structure of the α-satellite sequences makes them extremely difficult to clone in bacteria, because they are continually rearranged by recombination. Nevertheless this has recently been achieved, and it has been shown that when introduced into cells in culture they will function as centromeres. They have been used to construct artificial human chromosomes.

2.2.3.1 Minisatellites and microsatellites

Some types of satellite DNA are interspersed throughout the whole of the genome. There are two such classes, known as **minisatellites** and **microsatellites**. Both types have proved extremely important in the construction of maps of the human genome, which is discussed in Chapter 3. Minisatellites were also the original basis of **genetic fingerprinting** used in forensic science.

Minisatellites consist of sequences between 10 and 100 bp long repeated in tandem arrays that vary in size from 0.5 to 40 kb. They tend to occur near telomeres although they have been found elsewhere. Some, but not all, minisatellites show variation in repeat number so they are sometimes referred to as **variable number tandem repeats** (VNTRs). An individual who carries two different alleles that vary in size is said to be **heterozygous**. A locus that commonly exists in different forms is said to be **polymorphic**. As we shall see in Chapter 3, polymorphic loci such as minisatellites may be used as markers to construct genetic maps, although their tendency to be clustered near telomeres limits their use as a genomic mapping marker. Some minisatellite loci are hypervariable, for example alleles of a locus called D1S8 contain between 120 and 1000 repeats of a 29 bp sequence, resulting in a variation in size from 3.3 to 29 kb. As well as variation in repeat number, the sequence of the repeat unit itself can vary in different members of the repeat array, so that loci that are **monomorphic** for length may still be highly polymorphic in structure. The variability of minisatellite loci forms the basis of genetic fingerprinting used in forensic science and paternity testing (see Chapter 10).

Microsatellites consist of tandem repeats of units two to four nucleotides in length. They are also known as simple sequence tandem repeats (SSTRs). They are found at all locations within the genome, even within protein-coding sequences. Like minisatellites, the number of repeats in each microsatellite varies, changing its overall size. Thus microsatellites are polymorphic and they have proved to be valuable genetic markers. They have been more useful in this respect than minisatellites because their genomic distribution is more uniform. The most commonly used microsatellite for this purpose has the structure $(CA)_n$. Tetrameric STRs form the basis of the current method of DNA profiling for forensic casework (see Chapter 10).

2.2.4 Telomeres

At the end of each chromatid is a structure known as the **telomere.** The telomere consists of a large and variable number of tandem repeats of the sequence TTAGGG and has a protruding 3' end. This 3' end does not have the usual properties of single-stranded DNA. One possibility is that the free 3' end folds back on itself to form a hairpin structure through unusual base-pairing G residues. Other more elaborate structures have been proposed. The telomere is synthesised by an RNA-containing enzyme called **telomerase**. The RNA molecule in telomerase is used as a template to extend the free 3-OH end of the chromosome.

The telomere serves two essential functions.

1. Free ends of chromosomes produced by breakage are highly unstable and fuse with a high frequency to other chromosome ends. The telomere stops this happening to the ends of normal chromosomes by binding specific proteins to form a protective cap.
2. Telomeres solve the problem of replicating linear DNA molecules. As explained in Figure 2.7, the properties of the replication fork mean that the 3' end of a DNA molecule will be successively eroded with each round of replication. If the 3' end consists of a stretch of telomeric repeats, the erosion will affect these repeats rather than any essential functions in chromosomal DNA.

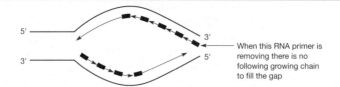

Figure 2.7 The problem of DNA replication at a chromosome end. A bidirectional replication origin near the end of a DNA molecule. Synthesis of each new DNA strand (shown in blue) proceeds in a 5' to 3' direction, extending an RNA primer (shown as a box) that is subsequently degraded. The resulting gap is filled by extension of the following chain. One strand of DNA can be synthesised continuously. The other strand, called the lagging strand, must be replicated in short sections, called Okazaki fragments, as the replication bubble grows. When the bubble reaches the end of the chromosome an RNA primer can be made for lagging strand synthesis, but when it is degraded, there is no following strand to fill the gap. Thus, that strand of the DNA molecule will be shortened by the length of the RNA primer.

The telomere may play an important part in both the ageing process and cancer. Human cells are **mortal** – in tissue culture they can only go through about 50 divisions before they stop dividing and die. This number of generations is approximately the maximum number of generations in any cell lineage from fertilisation to ageing and death, so cell mortality *in vitro* probably reflects the normal ageing process. As a cell ages its telomere becomes shorter. It is possible that cell mortality occurs when the telomere becomes too short to function efficiently and vital sequences in the chromosome are damaged. Telomerase is active in early embryonic cells but is turned off at later times and normal mortal cells do not have telomerase activity. Thus there is an in-built clock that counts cell generations and will eventually lead to cell death. This could be responsible for the normal ageing process. It is also a defence against cancers, because if cells escape other mechanisms controlling their proliferation, eroded telomeres will eventually cause their death. However, just as cancer cells have escaped other controls over their proliferation so they also eventually escape this control. One characteristic of cancer cells is that they are **immortal** and their telomeres do not shorten. Telomerase is not active in normal mortal cells, but it is active in at least some cancer cells. Forcing the expression of telomerase in human tissue culture cells that are mortal greatly increases the number of generations before division stops. Telomerase thus presents a target for anti-cancer drugs. Conversely, forced telomerase expression or delivery of telomerase to cells could theoretically provide a way of overcoming ageing. However, even if this were technically possible, and apart from the obvious ethical considerations, it would remove an important natural defence against cancer.

2.3 Structure of chromosomes

The 3200 Mb of DNA that constitutes the human genome is divided between 23 pairs of chromosomes; 22 of these chromosomes are found in both males and females and are known as **autosomes**. One pair of chromosomes is different in each sex and these are known as the sex chromosomes. Females have two copies of the **X chromosome**, while males have one X chromosome and one **Y chromosome**.

Further detail may be revealed by **banding**. In this process the metaphase spread is subject to light digestion with an enzyme such as trypsin, which breaks down proteins. It is then treated with **Giemsa**, a dye that binds DNA. Giemsa binds with different intensities at different points along the chromosome, producing a series of bands called **G-bands**. G-bands correspond to regions of the chromosome that have a lower than average proportion of GC base pairs and contain fewer genes (Section 4.8.1.1). The banding pattern is reproducible and is characteristic of each chromosome. Thus each chromosome, and even each part of a chromosome, may be recognised by its banding pattern (Figure 2.8). This provides the basis of a mapping procedure and allows changes to chromosomes to be recognised (Figure 2.9). Treatment of the metaphase spread with the fluorescent dye **quinacrine** produces a different type of banding pattern known as **Q-bands**.

At metaphase the chromosomes are **bivalent**. The two **chromatids** are joined at a constriction called the **centromere**. A protein structure called the **kinetochore** is located at the centromere and is the point at which the chromosome is attached to the mitotic spindle. The centromere divides the chromosome into two arms, the larger of which is called the **q arm** and the smaller the **p arm**. If the centromere is centrally placed the chromosome is said to be **metacentric**; where it is nearer one end the chromosome is said to be **submetacentric**. Where the centromere is near one end the chromosome is said to be **acrocentric** and when it is at one end it is said to be **telocentric**.

The chromosome number, the chromosome arm and the G-bands are used together to specify the location of genetic elements. Figure 2.9 shows a map of the X chromosome by way of illustration. Careful microscopy leads

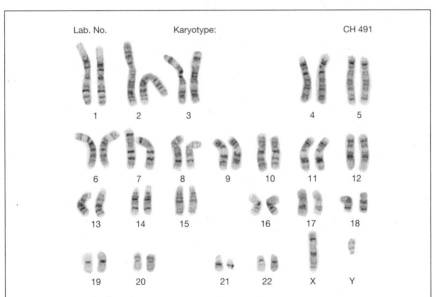

Figure 2.8 Karyotype of a human male. In the metaphase spread the different chromosomes are randomly arranged. The metaphase spread may be photographed, the individual chromosomes cut out and arranged so that each chromosome is paired with its homologue. It can be seen that chromosomes vary in size. By convention the largest pair are called chromosome 1 and the smallest chromosome 22, although it has subsequently been shown that chromosome 21 is in fact smaller than chromosome 22.

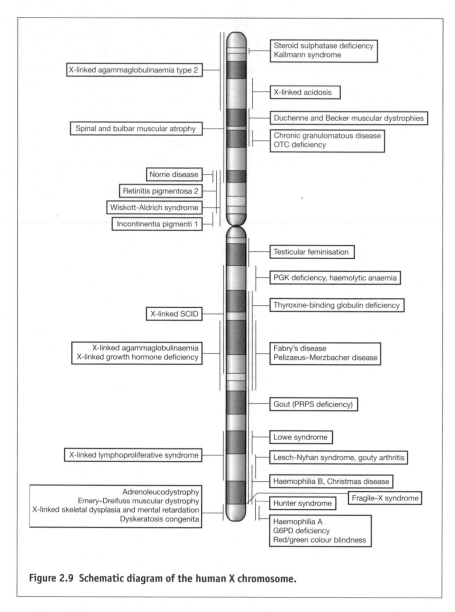

Figure 2.9 Schematic diagram of the human X chromosome.

to the subdivision bands. For example, the Duchenne muscular dystrophy locus is said to be located at Xp21.2, i.e. a subdivision of band Xp21.

2.3.1 Fluorescence *in situ* hybridisation

Chromosomes attached to a microscope slide in a metaphase spread can be hybridised with complementary sequences in probes, allowing the chromosomal origin of the probe to be identified. This process is known as ***in situ* hybridisation**. Originally it was carried out using ³H-labelled DNA. This was technically demanding and needed the use of autoradiograms, which needed long development times. The procedure was revolutionised by the use of DNA probes that were labelled using different coloured **fluorophores**. These were visualised using a fluorescence microscope that excited fluorescence using the particular excitation wavelength of the fluorophore. The use of fluorescently labelled DNA probes in this way is called **fluorescence *in situ* hybridisation** (FISH). Using different coloured fluorophores it is possible to visualise multiple regions. FISH is an important tool in physically mapping the human genome because it allows the chromosomal origin of any cloned piece of DNA to be determined. Plate 1 shows an example of this where a probe for the elastin locus hybridises to chromosome 7. The resolution of this technique is several megabases because chromosomes are in a highly condensed state in metaphase. Recently, hybridisation to interphase chromosomes or to stretched chromosome fibres has allowed mapping where the resolution is 50 kb or higher.

FISH has also proved to be a highly effective tool for karyotyping. Chromosome-specific probes can be used as 'chromosome paints' that allow the rapid identification of chromosomes without the careful characterisation of G-banded karyotypes. Plate 2 shows how this can be used for sex determination of a foetus using whole, uncultured cells from amniotic fluid. This is a powerful technique where there is a risk of a sex-linked disease. Plate 3 shows how chromosomes in a metaphase spread may be visualised using a chromosome paint. Chromosome paints are used extensively in the analysis of cancer cell karyotypes, which can show characteristic rearrangements during the course of the disease.

2.3.2 Packaging DNA into chromosomes

A single DNA molecule runs the length of each chromosome. This molecule is about 279 Mb in size in the largest of the chromosomes (chromosome 1). Physically the DNA in chromosome 1 is about 8 cm long, yet it is packaged at mitotic metaphase into a chromosome only 8 µm in length, a compaction ratio of about 10 000. How is this done?

The DNA is wound around itself in a hierarchy of coils and supercoils. At the deepest level, the DNA is wound around an octamer composed of two molecules each of the four different basic proteins called **histones**

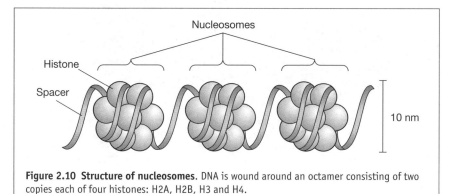

Figure 2.10 Structure of nucleosomes. DNA is wound around an octamer consisting of two copies each of four histones: H2A, H2B, H3 and H4.

H2A, H2B, H3 and H4 (Figure 2.10). This basic unit is called a **nucleosome** and in the electron microscope the DNA has the appearance of beads on a string. Each nucleosome occupies about 200 bp of DNA and is linked to the next by a short stretch of naked DNA. The register of DNA in nucleosomes may be very important in the control of gene expression.

DNA wound up in nucleosomes is coiled again to form a fibre 30 nm in diameter (Figure 2.11). The 30 nm fibre is stabilised by a fifth histone called histone H1. Finally, at mitosis, the 30 nm fibre winds around a scaffold of protein to form a fibre 700 nm in diameter (Figure 2.12). Various non-histone proteins are also associated with the DNA. The whole DNA–RNA–protein complex is known as **chromatin**.

During interphase, chromosomes are not so tightly condensed as metaphase chromosomes but the DNA must still be highly packaged. The transcription machinery must gain access to the DNA. Simple cartoons that depict transcription as an RNA polymerase moving along a DNA molecule like a train on a railway track are distorting reality to the point where it is

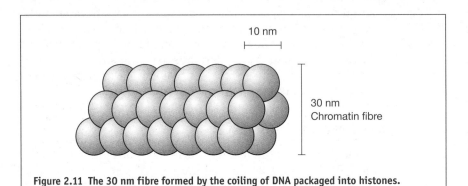

10 nm

30 nm
Chromatin fibre

Figure 2.11 The 30 nm fibre formed by the coiling of DNA packaged into histones.

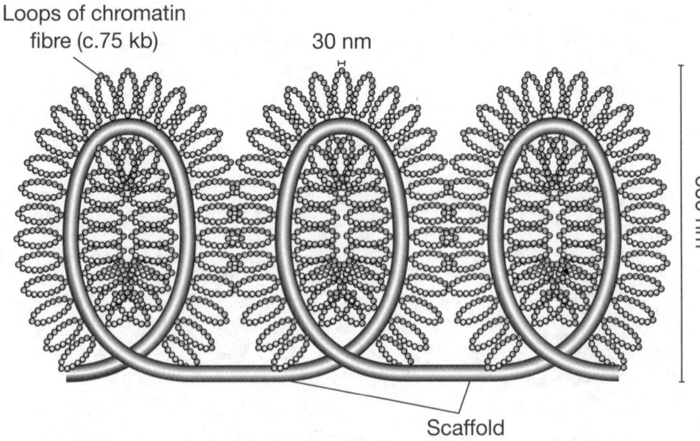

Loops of chromatin fibre (c.75 kb)

30 nm

600 nm

Scaffold

Figure 2.12 Structure of metaphase chromosomes.

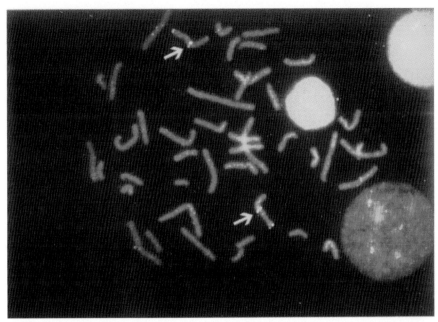

Plate 1 A gene probe for the elastin locus on the long arm of chromosome 7 is used to identify a deletion of the locus in a patient with Williams syndrome. (A) The normal karyotype. The two copies of chromosome 7 are arrowed. The signal near the telomere is a chromosome-specific probe to identify the chromosome; the signal near the centromere is the elastin locus. (B) One of the two copies of chromosome 7 (arrowed) lacks the signal near the centromere, showing the presence of a deletion. Williams syndrome affects about 1 in 10 000 of the population. It is characterised by an elfin-like appearance, hyperactive 'cocktail party' personality and multisystem disorders including multiple cardiac abnormalities, childhood hyperglycaemia, growth retardation, learning difficulties and neuromuscular disorders. It is caused by hemizygosity of the elastin locus. Elastin is a connective tissue protein that is a major component of elastic fibres in blood vessels, ligaments, etc.

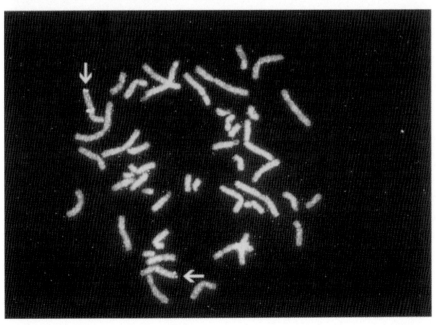

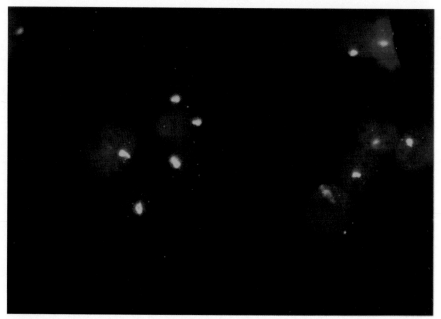

Plate 2 Probes specific for X and Y chromosones can be used to identify the sex of cells with interphase nuclei. This example shows a male foetus identified by the use of a Y-specific probe on uncultured cells from amniotic fluid.

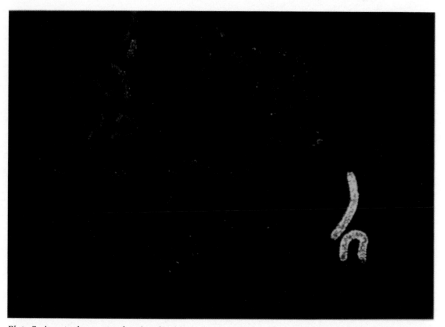

Plate 3 A metaphase spread 'painted' with a whole-chromosome paint for chromosome 2.

no longer a valid model. It is more likely that there are about 200 RNA factories within the nucleus of the human cell. The DNA is moved through these factories rather than the other way round. Perhaps a more accurate analogy is the way a tape in an audio cassette is fed through the playing head. When the tape is not being played it must be stored in a highly packaged fashion. What happens when the tape is not properly wound inside the cassette is something we are all familiar with!

2.3.3 Heterochromatin and euchromatin

Chromatin exists in two states, known as **euchromatin** and **heterochromatin**. Euchromatin is fully decondensed in interphase and is available for transcription. Heterochromatin differs in the following ways from euchromatin:

- Heterochromatin remains condensed in interphase and is stained with DNA-binding dyes such as Feulgen. In metaphase chromosomes it can be recognised by staining with a combination of distamycin and DAPI.

- It is transcriptionally inactive.

- It is late replicating.

Heterochromatin is found in the following regions:

- regions surrounding centromeres;

- subtelomeric regions;

- the short arms of acrocentric chromosomes;

- the non-expressed portion of the Y chromosome;

- short regions along the length of chromosome arms (interstitial) that are otherwise euchromatic.

These regions are always in the heterochromatic state. In addition, the inactive X chromosome in females is heterochromatic. This is known as facultative heterochromatin because the X chromosome is active in males and the choice of which is inactivated in females is random. Much of the obligate heterochromatin consists of highly repetitive simple sequence DNA or satellite DNA.

2.4 Summary

- The haploid human genome consists of 3200 million base pairs (3200 Mb) of DNA.

- It consists of different sequence classes:
  - Single sequence or low-copy DNA.
  - Intermediate repeated DNA sequences consisting of families of DNA sequences repeated between 10^2 and 10^5 times per genome.
  - Highly repetitive or satellite DNA, repeated 10^6 times per genome.

- Coding sequences represent only about 1.5% of the genome.

- Not all genes encode proteins – some encode functional RNA molecules.

- There are multiple copies of genes encoding rRNA and histones. These are arranged in long tandem repeat arrays.

- There are two different classes of intermediate repeat sequences called SINEs and LINEs. SINEs and LINEs are mobile genetic elements that may be selfish or parasitic because they undergo selection to increase in number but do not contribute to the phenotype.

- The most common LINE is called L1. It encodes a reverse transcriptase that mediates transposition to a new location by a process known as retrotransposition.

- The most common SINE is a family of sequences called *Alu* which, like the L1 family, has also spread through the genome by a process of retrotransposition.

- Most highly repetitive DNA is found in long tandem repeats at centromeres. The sequences bind proteins of the kinetechore where the spindles are attached to chromosomes at mitosis and meiosis.

- Minisatellites consist of tandem repeats of units 10–100 bp in size. The number of repeats and the sequence of the repeat units themselves are both polymorphic, making them useful genetic markers. They form the basis of genetic fingerprinting.

- Microsatellites are interspersed throughout the genome. They consist of tandem repeats of units 2–4 bp in size. They are also widely used as genetic markers.

- The number, size and appearance of chromosomes is called the karyotype. There are 22 autosomes and one pair of sex chromosomes.

- G-bands are used to construct cytogenetic maps of chromosomes, which can be used to specify the chromosomal location of genes.

- The telomere is a special structure that protects the ends of chromosomes. It may be important in cancer and ageing.

- The DNA molecules in chromosomes are physically long and so have to be tightly packaged.

- The DNA is wound around histones to form nucleosomes.

- The nucleosomes are coiled to make a fibre 30 nm in diameter.

- At mitosis and meiosis the 30 nm fibre is coiled around a scaffold of proteins to produce a 700 nm fibre.

- Chromatin exists in two states, euchromatin and heterochromatin. Euchromatin is decondensed in interphase and may be transcriptionally active. Heterochromatin is condensed in interphase and is transcriptionally inactive.

Further reading

General
GRIFFITHS, A.J.F., MILLER, J.H., SUZUKI, D.T. *et al.* (1999) *An Introduction to Genetic Analysis*, 7th edn. W.H. Freeman, New York.

LEWIN, B. (2000) *Genes VII*. Oxford University Press, Oxford.

LODISH, H., BALTIMORE, D. and DARNELL, J.E. (2000) *Molecular Cell Biology*, 4th edn. W.H. Freeman, New York.

Reassociation experiments
BRITTEN, R.J. and KOHNE, D.E. (1968) Repeated sequences in DNA. *Science*, **161**, 529–540.

Selfish DNA
DOOLITTLE, W.F. and SAPIENZA, C. (1980) Selfish genes, the phenotype paradigm and genome evolution. *Nature*, **284**, 601–603.

ORGEL, L.E. and CRICK, H.C. (1980) Selfish DNA: the ultimate parasite. *Nature*, **284**, 604–607.

Transcription
LOO, S. and RINE, J. (1995) Silencing and heritable domains of gene expression. *Annual Review of Cell Biology*, **11**, 519–548.

MANIATIS, T., GOODBOURN, S. and FISCHER, J.A. (1987) Regulation of inducible and tissue-specific gene expression. *Science*, **236**, 1237–1245.

MITCHELL, P. and TIJAN, R. (1989) Transcriptional regulation in mammalian cells by sequence specific DNA binding proteins. *Science*, **245**, 371–378.

PABO, C.T. and SAUER, R.T. (1992) Transcription factors: structural families and principles of DNA recognition. *Annual Review of Biochemistry*, **61**, 1053–1095.

ZAWEL, L. and REINBERG, D. (1993) Initiation of transcription by RNA polymerase II: a multi-step process. *Progress in Nucleic Acid Research and Molecular Biology*, **44**, 67–108.

Splicing
KRAMER, A. (1996) The structure and function of proteins involved in mammalian pre-mRNA splicing. *Annual Review of Biochemistry*, **65**, 367–410.

PADGETT, R.A., GRABOWSKI, P.J., KONARSKA, M.M. *et al.* (1986) Splicing of messenger RNA precursors. *Annual Review of Biochemistry*, **55**, 1119–1150.

α Satellite sequences, centromeres and human artificial chromosomes
HARRINGTON, J.J., VAN BOKKELEN, G., MAYS, R.W. *et al.* (1997) Formation of *de novo* centromeres and construction of first generation human artificial microchromosomes. *Nature Genetics*, **15**, 345–355.

SCHULMAN, I. and BLOOM, K.S. (1991) Centromeres: an integrated protein/DNA complex required for chromosome movement. *Annual Review of Cell Biology*, **7**, 311–336.

WILLARD, H.F. (1991) Evolution of alpha satellite. *Current Opinion in Genetics and Development*, **1**, 509–514.

LINEs
HATTORI, M., KUHARA, S., TAKENAKA, O. and SAKAKI, Y. (1986) L1 family of repetitive elements in primates may be derived from a sequence encoding a reverse-transcriptase related protein. *Nature*, **321**, 625–627.

Retrotransposons
ROGERS, J. (1986) The origin of retroposons. *Nature*, **319**, 725.

WAGNER, M.A. (1986) Consideration of the origin of processed pseudogenes. *Trends in Genetics*, **2**, 134–136.

***Alu* elements**
BRITTEN, R.J. (1996) Evolution of *Alu* retroposons. In *Human Genome Evolution*, Jackson, M., Strachan, T. and Dover G. (eds), pp. 211–228. Bios Scientific Publishers, Oxford.

OKADA, N. (1991) SINEs. *Current Opinion in Genetics and Development*, **1**, 498–504.

ULLU, E. and TSCHUDI, C. (1984) *Alu* sequences are processed 7SL RNA genes. *Nature*, **312**, 171–172.

Minisatellites
ARMOUR, J.A.L. (1996) Tandemly repeated minisatellites: generating human genetic diversity via recombinational mechanisms. In *Human Genome Evolution*, Jackson, M., Strachan, T. and Dover G. (eds), pp. 171–190. Bios Scientific Publishers, Oxford.

JEFFREYS, A.J. (1987) Highly variable minisatellites and DNA fingerprints. *Transactions of the Biochemical Society*, **15**, 309–316.

JEFFREYS A.J., WILSON, V. and THEIN, S.L. (1985) Hypervariable minisatellite regions in human DNA. *Nature*, **314**, 67–73.

Microsatellites

HANCOCK, J.M. (1996) Microsatellites and other simple sequences in the evolution of the human genome. In *Human Genome Evolution*, Jackson, M., Strachan, T. and Dover G. (eds), pp. 191–210. Bios Scientific Publishers, Oxford.

WEBER, J.L. and MAY, P.E. (1989) Abundant class of human polymorphisms that can be typed using the polymerase chain reaction. *American Journal of Human Genetics*, **44**, 388–396.

Telomeres

ZAKIAN, V. (1995) Telomeres: beginning to understand the end. *Science*, **270**, 1601–1607.

Chromatin structure

FELSENFELD, G. (1992) Chromatin as an essential part of the transcriptional mechanism. *Nature*, **355**, 219–224.

FELSENFELD, G. and MCGHEE, J.D. (1986) Structure of the 30 nm fibre. *Cell*, **44**, 375–377.

TRAVERS, A.A. and KLUG, A. (1987) The bending of DNA in nucleosomes and its wider implications. *Philosophical Transactions of the Royal Society of London Series B*, **317**, 537–561.

CHAPTER THREE

Mapping the human genome

Key topics

- Importance of genomic maps
- Sequence tagged sites
- Genetic maps
  - Key concepts, LOD scores, informative meioses, phase, haplotype
  - RFLP, minisatellite and microsatellite mapping markers
  - The Genthon map
- Physical maps
  - Clone maps
  - Radiation hybrid maps
  - STS content maps
  - FISH
  - Long-range restriction maps
- Expression maps
- Integration of genetic, physical and expression maps

3.1 Introduction

The construction of detailed maps of the human genome has arguably been the single most important development in human genetics over the last few years. The reason is simple: maps allow us to find and isolate genes when we have no other information about them apart from their location. In the last few years genes responsible for many of the major monogenic disorders have been isolated by such **positional cloning**. Furthermore, the focus of research in human genetics has now shifted to the analysis of complex diseases. Detailed maps provide the only means by which the genes that contribute to disease susceptibility can be identified and ultimately characterised.

Maps of the human genome can take different forms. The difference between them is important. Genetic maps are based on recombination frequencies between genetic markers at meiosis. Genetic markers are features of the human genome that are variable in different individuals. They may control phenotypic characteristics, which include inherited diseases, or they may be a variable feature of the genotype observed but have no phenotypic consequences. Genetic markers freely recombine at meiosis unless they are located in the same region of the genome, in which case they are said to be linked. Linked markers tend to be co-inherited; the closer they are, the higher the frequency of co-inheritance. Once a detailed genetic map has been constructed using a reference set of neutral markers, a gene responsible for a genetic disorder can be mapped by observing co-inheritance between the disease and a reference marker. Here we see the critical importance of genetic maps: they provide the only connection between biological reality and the underlying genome. In many cases, without genetic maps nothing else follows.

The location of a gene on the genetic map is only important in so far as it provides a means of identifying and isolating the DNA sequence that forms the gene. This allows us to study the nature of the protein encoded by the gene and characterise the mutations that may affect its function. Physical maps describe the location of DNA sequences. They are constructed by plotting the location of physical mapping markers based on DNA sequences. If we can align the genetic and physical maps then the genetic location of a gene directs us towards the DNA sequence that forms the gene.

Only a small fraction of DNA actually encodes proteins so often we must examine a large amount of non-coding DNA to find a gene that we are looking for. Here another form of map comes to our aid. Genes are DNA sequences that are expressed to form mRNA molecules. If we plot the origin of all mRNA molecules, then we will form a map of the DNA sequences that are expressed. Once genetic and physical maps have identified the genomic region in which a gene resides, we can restrict our attention to those DNA sequences that are expressed.

Thus genetic, physical and expression maps can be used together to find genes. Figure 3.1 illustrates the general process by which we can move from studying the segregation of a genetic disorder in a family to identifying the DNA sequences that have been altered by mutation. All of this depends on being able to align the three different map formats. Initially the different types of map were based on landmarks that were not necessarily compatible. One key advance in constructing genomic maps was the development of a marker, called a **sequence tagged site** (STS) (see below), that could be used with all three types of map. Thus when its location was defined in one map type, its location was also defined in the two other types of maps.

There is a second major reason for constructing maps. Before the human genome can be sequenced, it is necessary to assemble all the DNA into a set of ordered clones. This is essential to allow sequencing to proceed systematically instead of chaotically. As discussed below, the order of the clones itself is a form of map. Clone maps have so far been constructed in vectors called **yeast artificial chromosomes** (YACs) that carry large DNA inserts (~1 Mb).

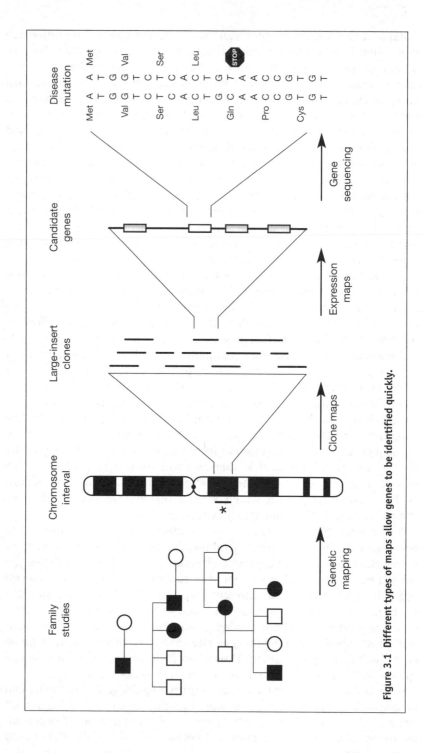

Figure 3.1 Different types of maps allow genes to be identified quickly.

This is far too large for sequencing, so it will be necessary to construct new libraries with much smaller inserts. The map of physical landmarks will allow us to quickly order these clones in preparation for sequencing. However, to do this the physical map will have to be detailed. The first goal of the Human Genome Project was to construct maps of the genome. It established targets for the resolution of these maps. These are an STS map with STS markers spaced every 100 kb and a genetic map with a resolution of 0.7 cM. As we shall see this goal for the genetic map has been achieved and the physical map is nearly complete.

3.2 Sequence tagged sites: a common currency for the different types of map

Mapping markers are the landmarks whose positions are plotted to construct the different types of map. As discussed above, it is important that the different types of map can be aligned with each other. However, markers that can be used for one type of mapping cannot necessarily be used for another. For example, genetic markers must be based on features that are naturally polymorphic. A disease segregating in a family is a genetic marker because some members are affected and others are not. However, unless the disease gene has been cloned, it cannot be used as a physical marker. A physical marker could be a sequence of DNA whose presence in a clone or location on a chromosome can be determined. However, if the sequence is the same in all individuals it cannot be used as a genetic marker.

There is clearly a need for a marker that can be used in the different types of map. The **polymerase chain reaction** (PCR) is used to generate a type of physical landmark called an STS that meets this requirement. PCR amplifies DNA using synthetic oligonucleotides as primers (Box 3.1). An STS is a stretch of DNA about 300 bp in length. As its name implies, the sequence of the STS is used to tag the larger DNA molecule from which it was derived so that the DNA region can be identified wherever it occurs. Figure 3.2 illustrates how STS markers may be generated and used to order a series of clones. The nucleotide sequence of the STS is used to specify the sequence of two synthetic oligonucleotides that will bind in opposite orientations at either end of the STS. Thus they will amplify the sequence if they are used as primers in a PCR reaction. PCR using these primers can therefore be used as a test to discover if the sequence is present in a particular DNA sample.

Originally the concept of an STS marker was developed for physical mapping because it was based on recognition of a sequence. However it was subsequently realised that an STS could be polymorphic, so long as the primers were designed to anneal to constant regions flanking some form of length polymorphism. We shall see below how microsatellites can be used to provide polymorphic STSs that are used for genetic mapping. STSs can also be generated from cDNA molecules and can be used to tag sequences that are expressed (**expressed sequence tag** or EST). The STS therefore provides a type of marker that is compatible with the three major map formats and so allows them to be aligned with respect to each other.

BOX 3.1: THE PCR REACTION

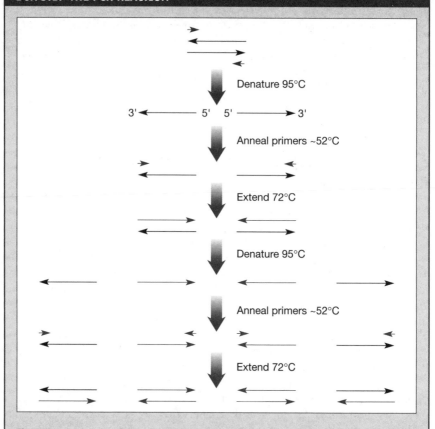

The PCR reaction is a form of *in vitro* cloning that can be used to amplify very small quantities of DNA. It depends on DNA polymerases from thermophilic bacteria, which can withstand temperatures of 95°C and have a temperature optimum of 72°C. DNA polymerases can only synthesise new DNA by extending the 3' end of a pre-existing primer. This is exploited to target new synthesis to a region between sites where two primers bind. In each cycle the template DNA is denatured by heating to 95°C, annealed to the primer at a temperature of about 52°C (see below) and then incubated at 72°C for the polymerase to extend the primer using the target DNA as a template. In the figure the target DNA is shown in black and newly synthesised DNA is shown in blue; the arrows on each DNA strand point toward the 3' end. Two complete cycles are represented, and it can be seen that in each the amount of target DNA is doubled. A total of 30 cycles means that amounts of DNA as small as a single molecule can be amplified to produce microgram quantities of DNA. The reaction is carried out in a thermocycler that can be programmed to carry out the cycles as specified by the user. The primers can be readily synthesised using automated technology. The length of the primers is usually about 20 bases; the exact length and its base composition determines the optimum annealing temperature. The annealing temperature and ionic composition of the buffer are adjusted to ensure maximum stringency.

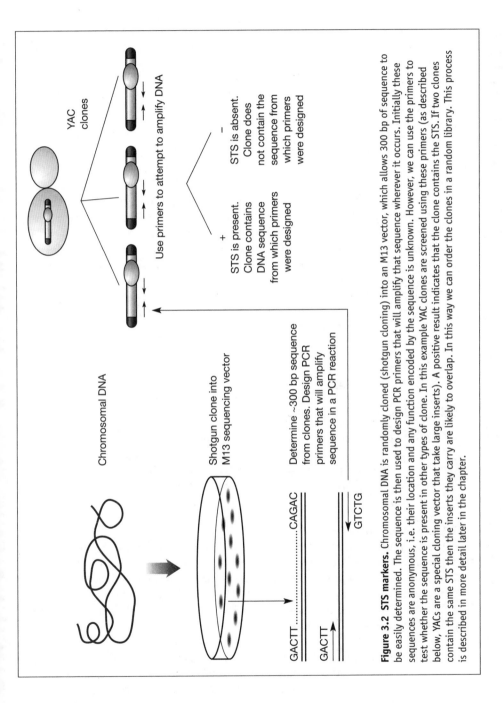

Figure 3.2 STS markers. Chromosomal DNA is randomly cloned (shotgun cloning) into an M13 vector, which allows 300 bp of sequence to be easily determined. The sequence is then used to design PCR primers that will amplify that sequence wherever it occurs. Initially these sequences are anonymous, i.e. their location and any function encoded by the sequence is unknown. However, we can use the primers to test whether the sequence is present in other types of clone. In this example YAC clones are screened using these primers (as described below, YACs are a special cloning vector that take large inserts). A positive result indicates that the clone contains the STS. If two clones contain the same STS then the inserts they carry are likely to overlap. In this way we can order the clones in a random library. This process is described in more detail later in the chapter.

Apart from the fact that they provide a common marker for different types of map, there are four other important advantages of the STS methodology.

1. They allow different sources of DNA fragments to be examined for the presence of a common sequence; for example two clones may be compared to detect any overlap between them. The sources of DNA may be totally different: we discuss below how YAC clones and radiation hybrid clones are forms of physical mapping that both use STS markers. Because of this the two types of map can be aligned with each other.
2. If a researcher in one laboratory wishes to examine whether an STS is present in clones they are working on, they do not have to physically obtain a clone from another laboratory that produced the STS. Instead the sequence of the primers is used to synthesise the oligo-nucleotides using a widely available technology for automatic synthesis. There is no need to physically maintain and distribute stocks of DNA. This solves an important logistical problem of maintaining and distributing clone libraries.
3. The PCR procedure requires both primers to bind at the correct distance apart *and* in the correct orientation. This reduces the number of false positives that can arise using other methods to compare DNA molecules, such as hybridisation.
4. The PCR test lends itself to automation, allowing very large numbers of tests to be performed. Since the human genome is so large this is an important practical consideration.

3.3 Genetic maps

3.3.1 Introduction

Genetic maps are based on the order of genetic mapping markers and the genetic distance between them measured by the recombination frequency. There are a number of concepts connected with genetic maps that we need to consider before we discuss the construction of a comprehensive map of the human genome.

3.3.1.1 Map units
All maps must have a unit of scale. The unit of human genetic maps is the **centimorgan** (cM). 1 cM is defined as a recombination fraction of 0.01, i.e. 1 in 100 gametes will be recombinant and the remaining 99 will have the parental configuration. Physically, 1 cM corresponds to between 0.7 and 1 Mb of DNA sequence, but there is no invariant relationship between physical and genetic distances (see Figure 3.9).

3.3.1.2 Informative and non-informative meioses
In order to measure the recombination frequency between two markers at meiosis it is necessary that the meiosis is **informative**, i.e. it is possible to distinguish between parental and recombinant chromosomes. This requires that both markers are heterozygous. Figure 3.3 illustrates why this is the case.

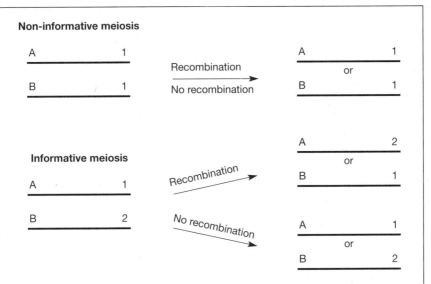

Figure 3.3 Informative and non-informative meioses. *Left*: two homologous chromosomes in a diploid cell about to undergo meiosis. Two adjacent loci, each with two alleles, are located on these chromosomes. *Right*: the structure of haploid gametes that result from the meiosis is shown. *Top*: the meiosis is uninformative because one of the loci is homozygous; after meiosis the gametes will have chromosomes that are identical whether or not a recombination event has occurred. *Bottom*: recombination results in chromosomes that can be distinguished from non-recombinant chromosomes; the meiosis is therefore said to be informative.

3.3.1.3 Phase

Phase specifies whether particular alleles at adjacent loci are on the same (*cis*) or different chromosomes (***trans***). For example, in the informative meiosis depicted in Figure 3.3, alleles A and 1 are in *cis* while alleles A and 2 are in *trans*. Clearly it is necessary to know the phase in order to know which outcomes of the meiosis represent recombinant chromosomes and which represent parental chromosomes. Figure 3.4 shows an example of how the phase can be established by examining three or more generations.

3.3.1.4 Haplotype

A **haplotype** is a set of closely linked alleles that tend to be inherited together at meiosis, i.e. not separated by recombination ('haplotype' is a contraction of 'haploid genotype'). In the previous example there are two parental haplotypes, A1 and B2. Haplotypes may consist of many more than two alleles. Generally, alleles making up a haplotype will be inherited as a block because their close proximity makes it unlikely they will be separated by recombination.

3.3.1.5 CEPH families

Recent genetic maps have been constructed using a set of reference families. These consist of samples collected from Mormon families living in Utah,

Mapping the human genome

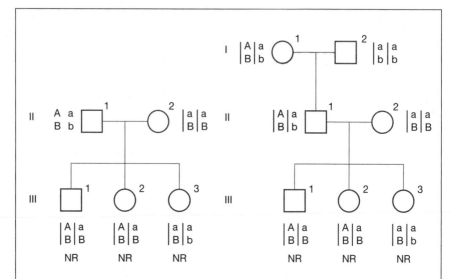

Figure 3.4 Phase can be worked out in a three-generation family. The figure shows the segregation of two linked loci each with two alleles (A/a and B/b). In the two-generation family on the left, the phase of the alleles in individual II-1 is unknown; therefore it cannot be established whether the progeny in generation III have inherited recombinant or parental chromosomes from II-1. The genotype of the grandparents allows the phase of the alleles in II-2 to be established. This reveals whether individuals in generation III have inherited parental or recombinant chromosomes. The vertical lines indicate the phase of the alleles where it is known.

USA, and French–Venezuelan families. Mormon families are typically much larger than normal and the keeping of detailed family records is part of the Mormon culture. The samples were used to establish permanent cell lines, which are maintained by the Centre d'Étude Polymorphism Humain (CEPH) in Paris. Each family consists of three generations with four grandparents, two parents and at least six children. Three generations allows the phase of markers in the parents to be established. The segregation of markers in the six children allows the results of meiosis in the parents to be determined. These cell lines provide an invaluable source of reference pedigrees, which are used in many different sorts of investigation into the human genome.

3.3.1.6 LOD score analysis

In experimental organisms, a large number of progeny from a genetic cross may be examined to accurately measure recombination frequency. Clearly this is not possible in human families. This problem is overcome by using a special form of mathematical analysis known as log ratio of odds or **LOD score** analysis (Box 3.2). The result of LOD score analysis is a numerical value that measures the ratio of odds that two loci are linked at given recombination fraction θ, compared with the chances that they are unlinked ($\theta = 0.5$). The threshold for declaring linkage is an LOD score of

An LOD score is defined as the logarithm of the ratio of odds (Z) of the observed outcome if two loci are linked with a recombination fraction θ, to the odds of the observed outcome if they are unlinked ($\theta = 0.5$). Thus

$$Z_x = \frac{\text{Odds of observed result if } \theta = x}{\text{Odds of observed result if } \theta = 0.5}$$

$$\text{LOD}_{\theta = x} = \log_{10} Z_x$$

The calculation is then repeated using different values of θ. The value of θ when LOD is at a maximum is called the MLS. The simplest application of LOD score analysis compares two markers, but it may be adapted to analyse multiple loci. This is known as multipoint LOD score analysis.

Statistical significance

Prior to the observation, the odds that any two loci are linked is about 1 in 50. This is known as the **prior probability.** The odds that they are linked based on observation is known as the **conditional probability**. This is what is measured by the LOD score. The overall odds that they are linked is the product of the prior and conditional probabilities. This is known as the **posterior probability**. Conventionally, for a result to be judged statistically significant the odds of it being true should be equal to or greater than 20:1. Therefore an LOD score of 3 is required (i.e. $Z = 1000{:}1$) for statistical significance because:

Posterior probability = prior probability x conditional probability
$$= 1{:}50 \times 1000{:}1$$
$$= 20{:}1$$

A worked example

Recall the three-generation pedigree from Figure 3.4 and consider the genotypes of the children in generation III. Are loci A and B linked? Assume that it is possible to determine the status of both alleles at both loci as shown overleaf. All the children must have inherited a chromosome with the genotype *aB* from parent II-2. The other chromosome must have come from parent II-1 and can be categorised as recombinant (R) or non-recombinant (NR). In fact, in this example, they are all non-recombinant.

$\theta = 0$ (the loci are completely linked)
For each child there is a 0.5 chance of the observed result, because the child inherits one of two chromosomes. For the whole family, the odds of the observed result is $0.5^3 = 0.125$.

If $\theta = 0.5$ (the loci are unlinked): for each child there is a 0.25 chance of the observed genotype, because there are four genotypes and each is equally likely. For the whole family, the odds of the observed result is $0.25^3 = 0.015625$. Thus

$$Z_{\theta = 0} = \frac{0.125}{0.015625} = 8$$

$$\text{LOD}_{\theta = 0} = 0.9$$

Continued ▶

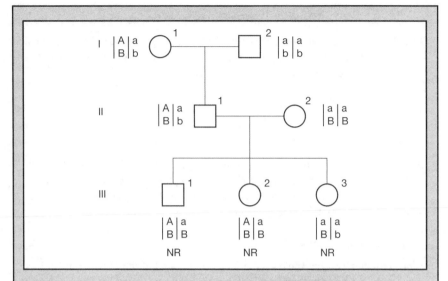

If θ = 0.25

For each child the odds of the observed result is 0.5 × 0.75 = 0.375, because the odds of inheriting a non-recombinant chromosome is 0.75 and there are two possible chromosomes that could be inherited. For the whole family, the odds of the observed result is $0.375^3 = 0.053$.

If θ = 0.5: for the whole family, the odds of the observed result = 0.015625 (as above).

$$Z_{θ=0.25} = \frac{0.053}{0.015625} = 3.39$$

$$LOD_{θ=0.25} = 0.53$$

The calculation is then repeated for other values of θ (0.1, 0.2, 0.3, etc.).

From this limited calculation we conclude that MLS = 0.9 when θ = 0, i.e. the alleles appear to be linked but the result falls short of significance.

We could now take account of results from other families. The odds of seeing an observed result in more than one family is the product of the odds in the individual families. Since LOD scores are logarithmic, LOD scores from different families can be directly added together. Thus if a similar LOD score was seen in three to four other families, the combined LOD score would become significant (four similar families would give an LOD score of 3.6).

This example is highly idealised. For example, the phase and genotypes are known. In practice, more complicated mathematical models are used and it is possible to calculate LOD scores for imperfect situations, for example if the phase is not known or penetrance of a disease is incomplete. The penalty is that the resulting LOD scores are lower so that more data need to be collected for significance. In the above example, if the phase is not known (i.e. only the two-generation family in Figure 3.4 was available) then MLS = 0.62 when θ = 0.

3 or more. It is unlikely that an LOD score of 3 will be obtained from one

63

Genetic maps

family. However, the mathematical properties of the test allow data from a number of families to be combined.

A related parameter to the LOD score is **maximum likelihood score** (MLS). This is the LOD score for the most likely of a series of alternatives. For example, an MLS score can be calculated for the most likely recombination fraction (θ) between two markers. This is done by iterating the LOD score calculation assuming a different value of θ each time. The MLS obtained gives the most likely value of θ. MLS can also be calculated when other parameters of a genetic model are varied, i.e. degree of penetrance. In practice, LOD scores and MLS values are derived by computer analysis of the observed results. One popular program for this purpose is called LIPED.

3.3.2 Mapping markers

The first maps of the human genome were produced in the 1970s and were based on linkage between markers such as enzyme polymorphisms, disease genes, blood groups and other phenotypic traits that were monogenic in origin. Linkage between such markers was rare so the resolution of the maps was low and large areas of the genome were not represented. Progress in the construction of human genetic maps was blocked because of a lack of usable markers in relation to the size of the genome.

3.3.2.1 Restriction fragment length polymorphism
The breakthrough came in 1980 with the realisation that many sequences in the genome were polymorphic in a way that did not have phenotypic effects. Such markers are called **neutral molecular polymorphisms**. The first such marker to be used is called a **restriction fragment length polymorphism** (RFLP). It is based on the presence or absence of a target for a restriction enzyme, usually due to a polymorphism at a single base pair (Figure 3.5).

RFLPs allowed the construction of a comprehensive map for the first time in 1987. This map had an enormous impact on human genetic research and was the basis for the cloning of many of the major monogenic disease loci. RFLP markers also provided an important tool for prenatal diagnosis, allowing a chromosome carrying a disease locus to be tagged and its segregation in a family to be followed.

3.3.2.2 Minisatellites
The problem with RFLP markers is that by definition there can be only two alleles at a locus, i.e. the restriction site is either present or absent. The maximum proportion of individuals in a population that are heterozygous is only 50%. Since genetic markers are only informative when they are heterozygous, on many occasions an RFLP marker will not provide useful information. In practice the situation is much worse, because usually one allele is less common and the frequency of heterozygotes is less than 50%. The proportion of meioses using a particular marker that will be informa-

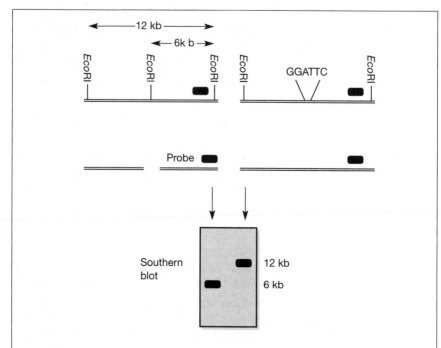

Figure 3.5 An RFLP marker. Parts of DNA molecules from two chromosomes differ from each other by a single base pair, which results in the absence of an *Eco*RI site in one of the chromosomes. Upon digestion with *Eco*RI, the chromosome without the extra *Eco*RI site produces a larger fragment than the other chromosome. The difference is recognised by Southern hybridisation using a probe (thick line) that hybridises within the region encompassed by the two flanking *Eco*RI sites present in both molecules.

tive is called its **polymorphism information content** (PIC). The maximum PIC value of a biallelic marker is 0.375 (it is less than 0.5 because some meioses are not informative even if both parents are heterozygous). Minisatellite and microsatellite markers described below typically have much higher PIC values.

A more useful type of marker is one that is naturally multiallelic, so there is a much higher chance that an individual will be heterozygous. Minisatellite loci, also known as VNTRs (see Chapter 2), meet this criteria. The number of repeats in the locus is usually determined by Southern hybridisation. Figure 3.6 illustrates how the segregation of a VNTR may be followed in a family.

3.3.2.3 Microsatellites
Minisatellites have been successfully used in the construction of genetic maps. However, their use is limited by their distribution in the genome, as they tend to be clustered near telomeres. Furthermore, the Southern blot procedure needed to score them is both laborious and expensive. PCR can be used but the large size of some minisatellites makes them difficult to amplify. A more useful type of marker is the microsatellite (see Chapter 2),

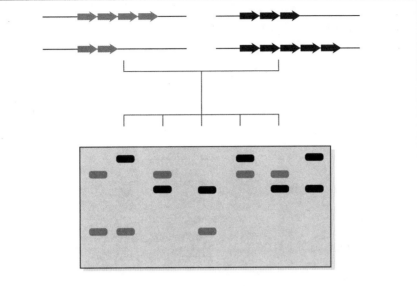

Figure 3.6 Segregation of a VNTR genetic marker in a family. The length of the VNTR depends on the number of repeats present. This can be recognised in a Southern blot using a sequence unique to the minisatellite as a probe. The genotypes of the two parents are shown in the left and right lanes respectively. Note that there are four alleles at this locus segregating in a Mendelian fashion, each child receiving one allele from each parent. This is indicated by colour coding.

which is more common and more evenly distributed than VNTRs. The microsatellite based on CA repeats has come to be the standard in the construction of genetic maps. The length of microsatellites is determined by PCR using primers designed from surrounding unique-sequence DNA (Figure 3.7). This makes them a special form of STS, which allows genetic maps based on microsatellite markers to be aligned with the different forms of physical map.

3.3.3 A comprehensive genetic map of the human genome

To be useful, a genetic map must be both comprehensive and detailed. Comprehensive means that it extends over the entire genome with no regions left unmapped. Detailed means that the distance between the markers is small, so that it is easy to detect linkage between the markers and biologically relevant loci, such as those which contribute to a familial disorder. Microsatellite loci have been used to construct a comprehensive map of the human genome, the resolution of which meets the goal set by the Human Genome Project. This map, constructed by the **Genethon** laboratory in Paris, is regarded as the definitive genetic map of the human genome.

The Genethon map is based entirely on microsatellite markers. The process of constructing the map began with identifying and sequencing 5 264 CA repeat loci. The genotype of these loci was then determined in

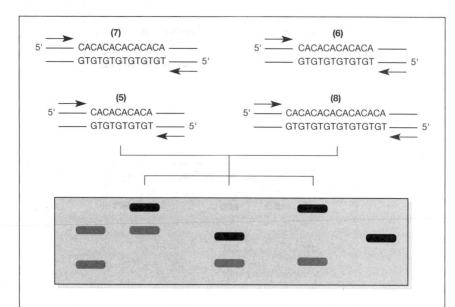

Figure 3.7 Segregation of a microsatellite locus in a family. The microsatellite locus is amplified using the PCR reaction with primers that anneal to the unique sequence either side of the CA repeats (shown by arrows). The size of the fragment produced depends on the number of CA repeats at each locus, indicated by the figure in parentheses above each DNA molecule. The products of the PCR reaction are separated by polyacrylamide gel electrophoresis, which is capable of resolving the difference in size between the different alleles. Note that there are four different alleles at the locus which show Mendelian segregation, each child receiving one allele from each parent (shown by colour coding).

members of large CEPH families. To map the autosomes a total of 134 individuals in eight families were examined, allowing the outcomes of 186 meioses to be analysed. The mapping of the X chromosome required the examination of 20 families, 304 individuals and 291 meioses.

Once the genotypes of the markers had been obtained they were analysed first to determine the order of the markers and then the genetic distance between them. This is done using specially developed computer programs. The 5264 markers analysed were found to define 2335 positions. The order of 2032 of these was determined with odds of 1000:1 against alternative orders. One of the most difficult and time-consuming steps in this process was checking for errors.

The length of the map in females is 4396 cM but only 2769 cM in males. The difference is due to an elevated rate of recombination in female meioses compared with male meioses. The sex-averaged length is 3699 cM. The length of the genetic maps for individual chromosomes generally reflects their physical size. The map of the largest, chromosome 1, is 292 cM; the smallest, chromosome 21, is 58 cM.

The average distance between markers is 1.6 cM. An important consideration is how evenly spaced these markers are. There are still some parts of the genome where there are not many markers, although these are a small minority of the total. For example, in 1% of the genome the distance between markers is over 10 cM. This actually consists of three positions where the interval between adjacent markers is 11 cM. This could be due to a large physical distance separating the markers, although it could also be due to enhanced rates of recombination in a particular region of the genome. In fact the physical map showed that at least part of the distance is due to increased recombination. In contrast, there were some places where markers showed no recombination with each other but were physically separated by several megabases of DNA.

Such observations emphasise that measurements of genetic map distances assume that the rate of recombination between two markers depends only on the physical distance that separates them. This requires that the frequency of recombination is constant throughout the genome. In practice this is not true. There are recombination hotspots where recombination is more frequent and coldspots where it is less frequent.

The completion of a detailed and comprehensive genetic map is a milestone in human genetics. It provides an essential resource that will be used for the rapid identification of monogenic disease loci, allows the mapping of genes that make a contribution to multifactorial disease and provides a framework to check the authenticity of physical maps. Panels of markers evenly spaced through the genome may now be purchased from biotechnology companies and these are routinely used to hunt for genes of interest.

A sample of a similar genetic map constructed by the Cooperative Human Linkage Centre is shown in Figure 3.8, which shows an overall representation of chromosome 18 on a web page. Greater detail for any part of the map is obtained by selecting a particular chromosome interval.

3.4 Physical maps

Physical maps plot the actual location of DNA sequences in the genome. There are a number of different ways in which this can be achieved.

- **Clone maps** consist of libraries of overlapping clones where the relationship of each clone to the other clones in the library is defined.

- **Radiation hybrid (RH) maps** are based on the frequency with which markers are found on the same fragment of DNA after the genome has been fragmented by irradiation with X-rays.

- **Expressed sequence maps** plot the sites on the genome called ESTs, which are expressed in mRNA. Essentially this type of map is based on the physical location of genes. Remember that these constitute less than 5% of the total genome.

- **STS content maps** plot the location of a physical mapping marker (an STS).

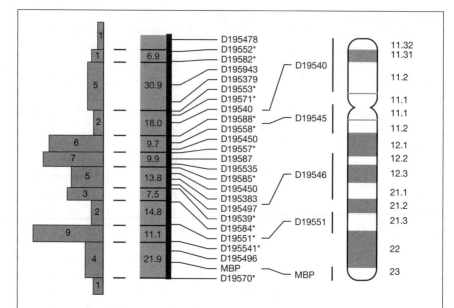

Figure 3.8 Genetic map of chromosome 18. The vertical histogram on the left shows the number of markers in each chromosome interval, shown in the centre (the figures show the size of the interval). The physical location of selected markers revealed by FISH is indicated on the cytogenetic map on the right. The map can be visited at the web site of the Cooperative Human Linkage Centre (http://www.chlc.org/HomePage.html). Clicking one of the intervals reveals more detail of the region.

- **Long-range restriction maps** are based on the position of the target sites of rare cutting enzymes determined by pulsed field gel electrophoresis.

- FISH locates the chromosomal position of cloned fragments of DNA.

In practice these overlap because clone maps, RH maps and EST maps all make use of STSs in their construction. Moreover, as we have seen the latest form of genetic map is based on polymorphic STSs. This means that we can align all of these maps, so that it is possible to move from one format to another. This is illustrated in Figure 3.9, which should be referred to throughout the following sections. Originally the term 'physical map' was most often used to describe clone maps, but is now more often used to describe a map based on the order of STS markers. This reflects the central importance of STSs in mapping. They are used to construct clone and RH maps and can be cross-referenced to expression maps based on ESTs and genetic maps based on microsatellites.

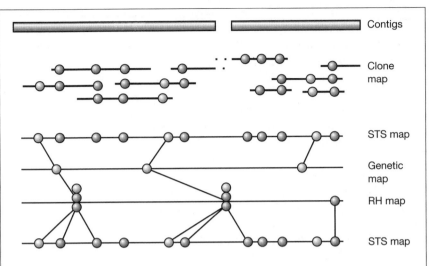

Figure 3.9 Different types of map can be aligned by STS content. Each STS is shown as a sphere; microsatellite-based STSs suitable for genetic mapping are shown in blue. The top part of the diagram shows the inserts of different clones aligned using STS content. The clones fall into two groups called contigs. Within each contig the overlap between clones may be traced through shared STSs, thus defining the spatial relationship of each clone to all the other clones in the contig. The extent of each contig is shown by the shaded bars. It is possible that the two contigs overlap (dotted line) but this has not been revealed by STS content. The overlap between the contigs allows the position of the STSs to be plotted to form the STS content map (shown twice, on the first and fourth line). Genetic and RH maps are also based on STSs, allowing these maps to be aligned with the STS content map (shown by fine lines). Note that the distortion of map intervals in the genetic map is caused by variation in recombination frequencies. RH and genetic maps can link STSs together over long distances but the order of closely linked STSs may not be resolved. The genetic and RH maps serve as a framework for the clone maps. In this example, they place the two contigs next to each other in the absence of any physical overlap being detected between them.

3.4.1 Clone maps

The original concept of a clone map was to specify the relationship of the clones in a gene library by determining how they overlapped. A group of clones whose relationship is defined in such a manner is called a **contig** (Figure 3.9). Note that the overlap between clones allows the physical relationship between two non-overlapping clones to be specified because their proximity can be traced through the overlap of intervening clones.

A perfect clone map of the human genome would consist of 24 contigs, one for each of the 22 autosomes and each of the sex chromosomes. In practice this is difficult or impossible to achieve because there are always gaps where two contigs cannot be joined. This may be because some parts of the genome are unclonable so there are no clones in the library that cover the region, or because an overlap does exist but cannot be recognised by the methods employed to detect overlaps.

This problem may be solved by using other mapping formats, such as genetic maps or RH maps, to provide an overall framework for the position of the contigs and thus detect those adjacent to each other (Figure 3.9). In order to do this, it is necessary to locate the position of STS mapping markers within the clones so that they can be anchored to the framework provided by the other mapping formats.

There are three phases in the construction of a clone map:

1. construction of the genomic library;
2. detection of overlap between the clones to produce contigs;
3. closing the gap between contigs to produce a continuous map.

3.4.1.1 Constructing a genomic library

A gene library consists of a collection of clones that together contain all the sequences present in the donor genome. Figure 3.10 illustrates the process using a YAC vector. For these to be used to construct a clone map, it is essential that two conditions are fulfilled:

1. every sequence from the donor genome is present so that the library is complete;
2. the **topology** (the geometrical arrangement) of sequences in the recombinant clones faithfully reflects that in the donor genome.

The completeness of the library is affected by two factors: the number of clones in the library and whether any sequences in the donor genome are unclonable in the host. The number of clones required is a function of the size of the insert in each clone related to the size of the total genome. Generally, the larger the insert that can be carried by the vector, the fewer clones are required for completeness. For this reason YAC vectors have been used for the first clone maps of the human genome because they can carry large inserts and thus complete libraries contain far fewer clones (~10 000) than are necessary using other types of vector (see below).

Sequences from the donor genome may be unclonable for two main reasons. Firstly, they may not be biologically neutral in the cloning host, for example they may encode a protein whose biological activity slows cell growth. Secondly, recombination may take place between direct repeats that results in deletion of the intervening sequences (Figure 3.11). Such recombination can make it very difficult to clone highly repetitive DNA sequences arranged in tandem repeats.

The major reason why the topology may be disrupted is that fragments of DNA from different parts of the genome have ligated together to form chimeric clones. When analysed, such clones will falsely indicate that the two sequences are adjacent to each other in the genome.

These considerations determine the choice of vector for the library construction. Routine cloning plasmids, such as the pUC series, cannot be used for this purpose because the size of the inserts they can carry are far too small for the scale of the task. Vectors based on phage λ (maximum insert size, 20 kb) and **cosmids** (maximum insert size, 40 kb) have been used for many

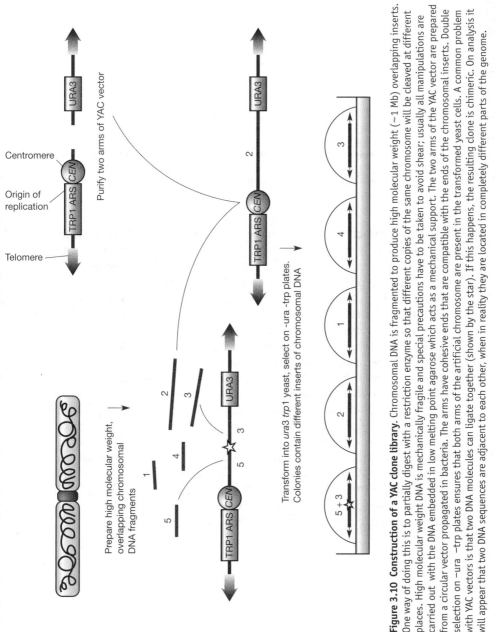

Figure 3.10 Construction of a YAC clone library. Chromosomal DNA is fragmented to produce high molecular weight (~1 Mb) overlapping inserts. One way of doing this is to partially digest with a restriction enzyme so that different copies of the same chromosome will be cleaved at different places. High molecular weight DNA is mechanically fragile and special precautions have to be taken to avoid shear; usually all manipulations are carried out with the DNA embedded in low melting point agarose which acts as a mechanical support. The two arms of the YAC vector are prepared from a circular vector propagated in bacteria. The arms have cohesive ends that are compatible with the ends of the chromosomal inserts. Double selection on −ura −trp plates ensures that both arms of the artificial chromosome are present in the transformed yeast cells. A common problem with YAC vectors is that two DNA molecules can ligate together (shown by the star). If this happens, the resulting clone is chimeric. On analysis it will appear that two DNA sequences are adjacent to each other, when in reality they are located in completely different parts of the genome.

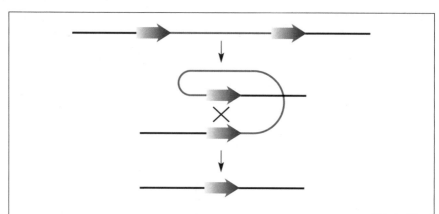

Figure 3.11 Recombination between direct repeats leads to deletions of DNA between them. Two direct repeats are shown by the arrows. DNA can loop back so that two sequences are aligned, allowing recombination to take place. As a result the intervening DNA (shown in blue) and one copy of the direct repeat is lost.

years to construct libraries. However, the maximum size of the insert they can carry is small in comparison to the size of the genome. This means that the number of clones necessary to have a reasonably complete genomic library is very large and ordering them is very difficult. Moreover the clones are not always stable and deletions and other rearrangements occur during propagation of the cloned DNA. Cosmids and λ vectors have been most useful for local genomic regions rather than whole genome libraries.

The next type of vector to be developed was the YAC (Box 3.3). YAC libraries have been made that contained very large inserts (up to 1 Mb). This meant that the whole genome could be encompassed in a few thousand colonies, allowing clone maps of the whole genome to be constructed. They have been the workhorse for the construction of the physical maps of the genome described below. However there are several major drawbacks to YAC vectors, the most important of which is a very high rate of chimeric clones. **Bacterial artificial chromosome** (BAC) and **P1-derived artificial chromosome** (PAC) vectors (see Box 3.3) accept inserts in the 100–300 kb range and maintain the topology of inserts much more faithfully than YAC vectors. An ordered library of BAC clones was used to sequence the human genome by the clone-by-clone approach (see Section 4.3.1).

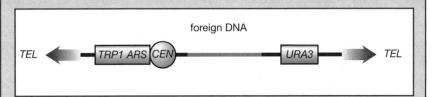

YACs contain all the elements found in a natural chromosome assembled into a vector that can propagate foreign DNA in a yeast cell. The *ARS* element is an origin of DNA replication, *CEN* is the centromere that ensures segregation to daughter cells and *TEL* is a telomere that protects the chromosome ends. *TRP1* and *URA3* encode enzymes required for the biosynthesis of tryptophan and uracil respectively and are used as selection markers (the host cell is *trp1⁻ ura3⁻*). These essential elements occupy only 6 kb in a circular vector that is maintained in *E. coli*. The two arms of the YAC are purified from this vector and ligated to high molecular weight DNA from the donor genome; the two selection markers ensure that each arm is present. Yeast chromosomes can be up to 1 Mb in size and so DNA in this size range can be carried in a YAC vector. This enormously simplifies the task of constructing and ordering a human genome library because so many fewer clones are required compared with a λ or cosmid library. However, they suffer from three major drawbacks.

1. Up to 60% of the clones are chimeric. It is not clear whether this occurs during the *in vitro* ligation process or *in vivo* by recombination between different vectors after transformation into the yeast cell. The transformation process involves the enzymic removal of the yeast cell wall. This is known to be a mutagenic process and it may stimulate mitotic recombination mechanisms. An alternative method of transformation that keeps the yeast cell intact results in a much lower level of chimeric clones.
2. Clones are unstable in the yeast host, with deletions occurring at a high frequency.
3. There is no easy way to purify the recombinant YAC vector from the remaining yeast chromosomes. It is necessary to fractionate the chromosomes by pulsed field gel electrophoresis and recover the chromosomes from blocks cut out of the agarose slab used for electrophoresis.

Phage P1 is a temperate bacteriophage. It can exist in one of two states in the host cell: in a lytic cycle where it multiplies and eventually lyses the cell, releasing progeny phage particles into the surrounding environment or in a dormant lysogenic state, where it is passively replicated along with the host chromosome. Lysogenic phage are induced to re-enter the lytic cycle by cues such as damage to the host DNA. It has been known for a long time that phage P1 mediates generalised transduction, i.e. it can mediate gene transfer from one *E. coli* cell to another. This formed the basis of one of the original methods used to map the *E. coli* genome. What happens in generalised transduction is that fragments of host

Continued ▶

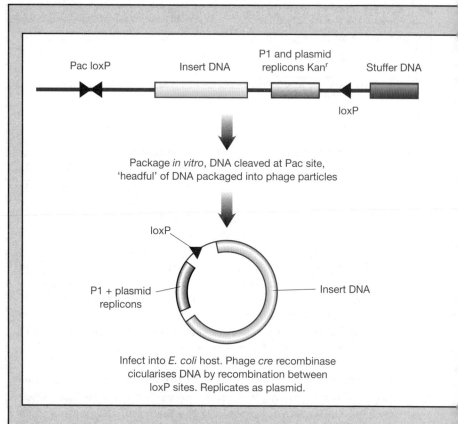

Pac loxP   Insert DNA   P1 and plasmid replicons Kan^r   Stuffer DNA

loxP

Package *in vitro*, DNA cleaved at Pac site, 'headful' of DNA packaged into phage particles

loxP

P1 + plasmid replicons   Insert DNA

Infect into *E. coli* host. Phage *cre* recombinase cicularises DNA by recombination between loxP sites. Replicates as plasmid.

DNA are occasionally mistakenly incorporated into the phage head during a lytic cycle. T̶ DNA is then injected into a new cell when the phage particle attaches to its surface. T̶ ability to transfer non-phage DNA is the basis of PAC vectors. They can accept about 100 of insert DNA and maintain it in low copy number in the host cell as a replicating plasm̶ when desired the phage lytic system can be induced to increase copy number. The leve̶ chimeric clones in libraries is less than 5%, which is much lower than YACs.

3.4.1.2 Ordering the clones

Once a complete gene library has been constructed the clones must be ordered with respect to each other. This is done by detecting sequences pre-sent in two clones which indicate that they overlap. This can be done by a variety of methods, including STS content, hybridisation and fingerprinting.

STS content mapping seeks to identify clones that both contain the same STS site, indicating that they share an overlapping sequence. The use of STS content to order clones spanning part of human chromosome 21q is illustrated in Figure 3.12.

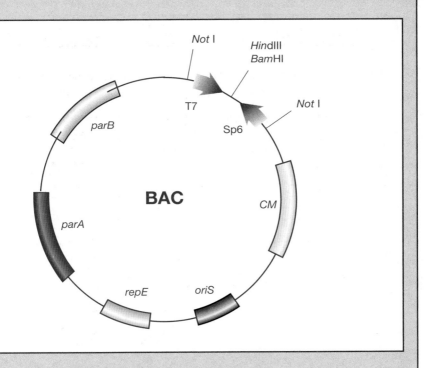

BACs are based on the F factor, the plasmid responsible for conjugation in *E. coli*. The *iS* and *repE* elements mediate replication and *parA* and *parB* maintain copy number at one two per genome. *CM*[r] (chloramphenicol resistance) provides a means of selection. Insert JA is cloned into the *Bam*HI or *Hind*III sites and excised using the flanking *Not* I sites. The serts can be transcribed *in vitro* to make RNA using the T7 and Sp6 promoters. BAC vectors n carry up to 300 bp DNA with a high level of stability and low rate of chimeric clones. One sadvantage of BAC vectors is that it is difficult to recover good yields of the recombinant JA from the host because of the low copy number.

If two clones contain sequences in common, they will hybridise to each other. In principle one clone can be used as a probe to identify overlapping clones by using the DNA from one clone as a probe to hybridise to other clones. Hybridisation is more sensitive than fingerprinting or STS content and is often used to detect overlap between adjacent contigs that have been missed by these other methods. It thus provides a method of closing the gaps between adjacent contigs. A problem encountered with hybridisation is a high rate of false-positive results. One reason for this is the presence of repetitive elements in the probe that hybridise to a large number of unrelated clones. One way to ensure that the probe carries only single sequence

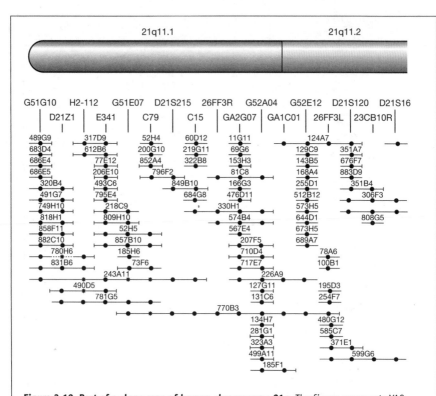

Figure 3.12 Part of a clone map of human chromosome 21q. The figure represents YAC clones, shown as lines, spanning the terminal G-band of the q arm (cytogenetic map is shown at the top). The clones are aligned by the overlap in the STS sites (shown as filled circles) in different clones. The names of the STSs are shown underneath the cytogenetic map in the order revealed by the STS content of the clones. Bars at the end of clone lines means that the clone tested negative against adjacent clones.

DNA is to use PCR to amplify the sequences between adjacent *Alu* elements (see Chapter 2) (Figure 3.13).

Fingerprinting methods aim to identify some property of a DNA sequence that allows it to be recognised where it occurs in two different clones. There are many ingenious ways in which this can be done. Figure 3.14 illustrates a simple method of how this might be achieved.

3.4.1.3 A clone map of the human genome

The first comprehensive clone map of the human genome was generated by international collaboration. The map consisted of YAC clones ordered into 225 separate contigs. The ordering was achieved by STS content, hybridisation and fingerprinting and was estimated to cover about 75% of the total genome. Only 786 independent loci were fully mapped so the resolution is much lower than the target of the Human Genome Project.

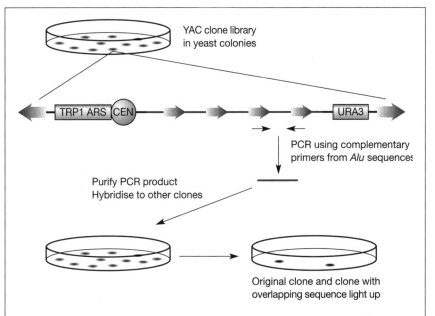

Figure 3.13 Detecting overlap between YAC colonies by colony hybridisation. Each clone contains a large insert of human DNA that has *Alu* repetitive DNA sequences along its length (large blue arrows). Complementary PCR primers (small black arrows) are designed from the *Alu* sequence and are used to amplify unique-sequence DNA located between the *Alu* sequences. The amplified products are used as a hybridisation probe to identify other clones in the library that contain the same sequence and therefore overlap with the first clone.

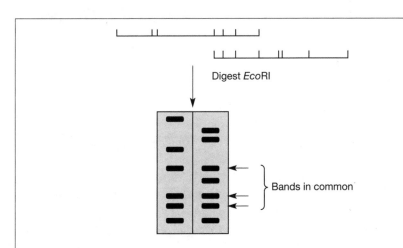

Figure 3.14 Detecting overlap between two overlapping clones by fingerprinting. The clones are digested with a restriction enzyme and the products run on a gel. Clones that overlap will display common bands. It is important to understand that there is no attempt to order the sites within the clones – overlap is indicated by the pattern of bands. Clearly many clones will share bands by chance. The number of apparently common bands expected by chance can be calculated and overlap is only declared when the number of common bands is significantly greater. In practice more complex protocols are used to derive arbitrary patterns that characterise clones.

3.4.2 Radiation hybrid maps

In RH mapping a human cell line is irradiated with X-rays before fusion to hamster cells to make lines of somatic cell hybrids (Figure 3.15). The radiation fragments the human chromosomes so that they cannot be independently maintained in the hybrid unless they become incorporated into the hamster chromosomes. Each hybrid line retains about 20% of the donor human genome in fragments of about 10 Mb. A panel of lines are examined for co-retention of different STS markers. If two markers are co-retained more frequently than would be expected by chance, then this is evidence that they are linked. The distance between markers is estimated from the frequency of co-retention. The map units are **centirays**, which are analogous to centimorgans but depend on radiation dose.

The value of RH maps is that they provide an independent method of mapping STS sites. Because it does not rely on the generation of clones it is not subject to the same systematic errors such as chimeras and deletions that affect clone maps. Moreover, RH maps can detect linkage between STSs over a much longer distance than can be detected by the overlap of clones. They therefore provide a framework to link clone contigs together.

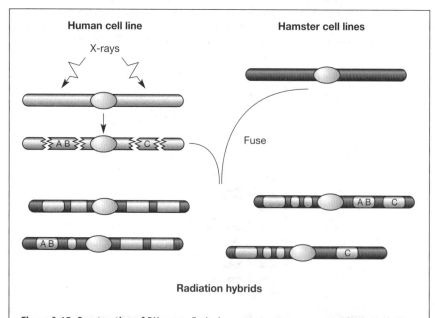

Figure 3.15 Construction of RH maps. Each chromosome represents a complete set of chromosomes in a cell line. A human cell line is irradiated with X-rays, which fragments the chromosomes. The irradiated cells are fused to a hamster cell line. The fragments of human chromosomes become incorporated into the hamster chromosomes. Markers adjacent to each other in the human genome (AB) show the same pattern of retention in the hybrid lines, i.e. they are co-retained. Marker C is distant from A and B and shows a different co-retention pattern compared with either A or B.

3.4.3 Expression maps

ESTs are a special form of STS that are generated from cDNA libraries not genomic sequences. ESTs therefore identify coding regions and may be regarded as gene-based STSs (Figure 3.16). ESTs provide a wealth of information concerning the total number of genes encoded by the human genome, the nature of the proteins encoded and the differences in expression between different tissues.

The human genome is thought to contain between 30 000 and 35 000 genes (see Chapter 2). So far 16 000 of these have been placed on the integrated physical and genetic map screening panels of YAC clones and radiation hybrids for the presence of ESTs. This represents the first step in a programme to map the majority of human genes that will complete the construction of an integrated genetic–physical–expression map.

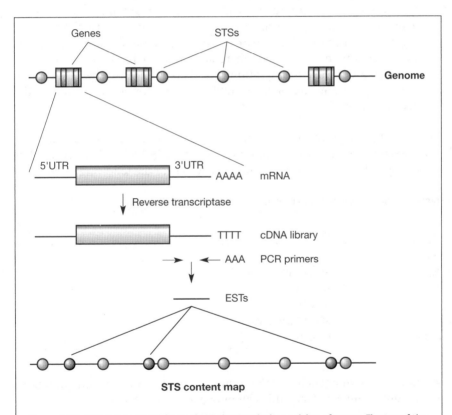

Figure 3.16 ESTs are a special form of STS that mark the position of genes. The top of the diagram shows the genome with isolated genes, each containing introns, shown as blue boxes. The STS content map is superimposed. A cDNA library is constructed from the mRNA derived from the genes. ESTs are prepared from the 3′ untranslated regions (3′ UTR). The 3′ UTR is used because it does not generally contain introns and tends to be unique, even in genes that have related coding sequences. These form STSs which mark the position of genes in the STS content map (filled blue circles).

3.4.4 Long-range restriction maps

Restriction maps have long been an important form of physical map for DNA sequences of up to a few tens of kilobases. The landmarks mapped are the sequences at which different enzymes cut the DNA molecule. Conventional methods of producing restriction maps cannot be used to generate longer maps, which might extend over hundreds of kilobases or even megabases, for two reasons.

1. Commonly used enzymes, such as *Eco*RI or *Bam*HI, cut at 6-bp target sequences. If random sequence DNA is digested with such an enzyme, the size distribution of the fragments produced would have a mode of 4096 bp (4^6). It would be completely impracticable to attempt to order the hundreds of fragments that would be produced if DNA in the megabase size range was digested with such an enzyme.

2. If a way could be found to produce larger fragments, conventional agarose gel electrophoresis (AGE) could not be used to characterise them as it becomes non-linear at large fragment sizes. As the size is increased a point is reached where additional increase in size does not result in a decrease in migration rate. This is because the DNA molecules align along the electric field and migrate end-on through the agarose matrix.

The first of these problems was solved by the discovery of so-called **rare cutting enzymes**. Some examples of these are shown in Table 3.1. These cut less frequently in human DNA than normal enzymes for the following reasons.

Table 3.1 Some examples of enzymes that cut human DNA rarely.

Enzyme	Target sequence
Mlu I	ACGCGT
Not I	GCGGCCGC
Nru I	TCGCGA
Sal I	GTCGAC
Sfi I	GGCCNNNNNGGCC
Xho I	CTCGAG

- Some enzymes such as *Not* I have an 8-bp recognition target. In random sequence DNA this will only occur every 65 kb.

- They contain the dinucleotide CpG. This has two consequences. Firstly the frequency of CpG is depressed in human DNA (see section 2.2.1.2). Secondly CpG is the target for DNA methylation, which prevents these restriction enzymes from cutting the DNA.

As a consequence the restriction enzyme *Not* I cuts human DNA on average every 500 kb, and its target sites thus provide landmarks that are conveniently spaced for the analysis of human chromosomal regions. Other enzymes, such as *Nru* I and *Sfi* I also produce fragments that are large enough that the target sites are useful landmarks for chromosomal maps. Note that enzymes such as *Xho* I and *Sal* I are comparatively rare cutters with human chromosomal DNA, even though they are commonly used in the construction of conventional restriction maps.

The problem of separating large DNA molecules was solved by a modification of AGE called **pulsed field gel electrophoresis** (PFGE). In PFGE the direction of the electric field is periodically changed. In order to make progress in the new field the DNA molecule must reorientate itself. The time taken for this reorientation is proportional to molecular weight, even with DNA molecules several megabases in size. The exact conditions of electrophoresis can be elaborate and have to be fine tuned for the particular size range employed.

Long-range restriction maps have proved very important during the course of positional cloning. They provide a framework with which to locate the position of clones derived from techniques such as chromosome walking. An example of a long-range restriction map constructed during the hunt for the cystic fibrosis gene is shown in Box 5.2. Another important practical consideration is that the sites of rare cutting enzymes will tend to be clustered in CpG islands that mark the beginning of genes. This allows the islands to be identified in long stretches of DNA.

3.5 An integrated physical and genetic map of the human genome

STSs can be used as a common marker to align genetic, RH and YAC-based clone maps of the human genome. Genetic maps can detect linkage between markers over distances of 30 Mb, while RH mapping can detect linkage over distances of 10 Mb. As a consequence, comprehensive genetic and RH maps of the genome require only a few thousand markers. However the resolution of fine structure mapping is low. In contrast, YAC-based clone maps detect linkage over the 1 Mb base range. Thus a large number of markers are required to provide adequate coverage of the genome, but the fine-structure resolution is high.

As we have seen, clone-based maps suffer from two particular problems: there are always gaps between the contigs, and false linkage between markers may be detected as a result of chimeric clones. RH and genetic maps may be used to provide a framework for clone-based maps, allowing adjacent clone contigs to be positioned with respect to each other and trapping errors that result from chimeric clones. An integrated map has been constructed that aligns the Genethon genetic map with an RH map and an STS-content map based on YAC clones (Figure 3.17).

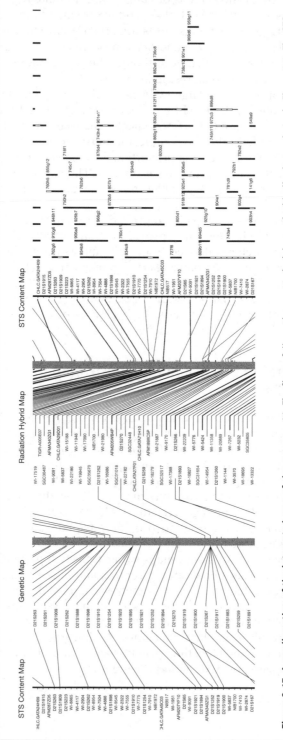

Figure 3.17 A small portion of the integrated map of chromosome 14. The figure follows the same principles as the idealised map shown in Figure 3.9. The STS content map is shown twice, flanking the genetic and RH maps. The lines connect markers found in more than one map. The bars on the right are the overlapping YAC clones which contain the STSs mapped.

3.6 Genome mapping is big science

A traditional view of the way that biological research is carried out would be of a small team of scientists working in a single laboratory. The scale of the effort required to produce whole genome maps has demanded a totally different method of working that is perhaps more familiar to the physical sciences such as particle physics. As an example, consider the following facts concerning the production of the integrated physical and genetic map (see Further reading).

- It involved 15 million separate PCR reactions.

- There were 51 authors to the paper.

- It required the construction of a robotic production line to carry out the PCR tests that was commissioned from a specialist engineering company.

- A full description of the results would have required 900 printed pages. For this reason the detailed results are published electronically and accessed via the World Wide Web (see Table 4.2 on p. 103).

These facts illustrate a number of aspects of how this type of work is carried out.

- Large numbers of personnel are involved. Often several laboratories or even institutions formally collaborate in consortia.

- Automation is essential to allow the large number of tests to be carried out. One value of PCR-based testing is that it readily lends itself to automation.

- Results are published electronically. This allows continuous publication, so that other researchers can exploit the results without the long delays of conventional means of publication. A list of websites is given in Table 4.2 on p. 103. These can be readily accessed to monitor progress of the different mapping projects and to explore the maps in detail. The reader is encouraged to visit these sites.

- PCR-based tests allow researchers to use the sequence of the PCR primers to synthesise their own primers locally and so use the maps in their own particular research.

3.7 Summary

- The human genome contains an immense amount of DNA. Maps are necessary to:
  - ○ Locate genes responsible for monogenic and multifactorial disorders.
  - ○ Order individual clones before they are sequenced.

- Maps require landmarks. The most useful form of landmark for genomic mapping is called an STS, which uses PCR to recognise a particular sequence wherever it occurs. The different types of map can all be based on STSs, allowing maps to be aligned.

- Genetic maps are based on recombination at meiosis. They often provide the only connection between biological and medical reality and the underlying genome. A special form of STS based on microsatellites is used as the marker. The definitive genetic map of the human genome has now been produced.

- Physical maps locate features to a particular part of the DNA molecules that make up the genome. There are different types of physical map:

  - ○ Clone maps use the overlap between clones in a genomic library to construct groups of clones called contigs, where the spatial relationship of each clone is defined with respect to the other clones in the contig. Comprehensive clone maps are constructed using YACs as vectors because of the large insert size they can carry.
  - ○ RH maps plot the position of STS markers by observing how often they are retained together on the same fragment when the genome is fragmented by X-rays and fused to hamster cells to make panels of radiation hybrid lines.
  - ○ Expression maps plot the position of genes using a special form of STS called an EST.

- Because the different types of map are based on forms of STS marker, they can be integrated to construct a single STS content map. This is constructed using the STS content of clones to construct contigs, then using genetic and RH maps as a framework to order the contigs and check for errors arising from chimeric clones.

- Mapping the genome is science on a big scale. It involves consortia of research groups, robotic production lines to carry out the tests and electronic publication on the World Wide Web because the data is too copious to publish in printed form.

Further reading

LOD score analysis

RISCH, R. (1992) Genetic linkage: interpreting LOD scores. *Science*, **255**, 803–804.

The genetic map

DIB, C., FAURE, S., FIZAMES, C. *et al*. (1996) A comprehensive genetic map of the human genome. *Nature*, **380**, 152–154.

Physical maps

BOUFFARD, G.G., IDOL, J.R., BRADEN, V.V. *et al*. (1997) A physical map of human chromosome 7: an integrated YAC contig map with an average STS spacing of 79 kb. *Genome Research*, **7**, 673–692.

CHUMAKOV, I., RIGAULT, P., GUILLOU, S. *et al*. (1992) Continuum of overlapping clones spanning the entire human Y chromosome 21q. *Nature*, **359**, 380–387.

COHEN, D., CHUMAKOV, I. and WEISSENBACH, J. (1993) A first generation map of the human genome. *Nature*, **366**, 698–701.

FOOTE, S., VOLLRATH, D., HILTON, A. and PAGE, D.C. (1992) The human Y chromosome: overlapping clones spanning the euchromatic region. *Science*, **258**, 60–66.

The integrated map

HUDSON, T.J., STEIN, L.D., GERETY, S.S. *et al*. (1995) An STS-based map of the human genome. *Science*, **270**, 1945–1955.

ESTs and expression maps

ADAMS, M.D., KERLAVAGE, A.R., FLEISCHMANN, R.D. *et al.* (1995) Initial assessment of human gene diversity and expression patterns based upon 83 million nucleotides of cDNA sequence. *Nature*, **377** (Suppl.), 3–17.

SCHULER, G.D., BOGUSKI, M.S., STEWART, E.A. *et al.* (1996) A gene map of the human genome. *Science*, **274**, 540–545.

Sequence tagged sites

ROBERTS, L. (1989) New game plan for genome mapping. *Science*, **245**, 1438–1440.

Radiation hybrid mapping

LEACH, R.J. and O'CONNELL, P. (1997) Mapping of mammalian genomes with radiation (Goss and Harris) hybrids. *Advances in Genetics*, **33**, 63–101.

Commentaries on genomic maps

BOGUSKI, M.S. and SCHULER, G.D. (1995) ESTablishing a human transcript map. *Nature Genetics*, **10**, 369–370.

COX, D.R. and MYERS, R.M. (1996) A map to the future. *Nature Genetics*, **12**, 117–118.

JORDAN, E. and COLLINS, F.S. (1996) A march of the genetic maps. *Nature*, **380**, 111–112.

LITTLE, P. (1995) Navigational progress. *Nature*, **377**, 286–287.

SCHMITT, K. and GOODFELLOW, P.N. (1994) Predicting the future. *Nature Genetics*, **7**, 219–220.

Long-range restriction mapping

BARLOW, D.P. and LEHRACH, H. (1987) Genetics by gel electrophoresis: the impact of pulsed field gel electrophoresis on mammalian genetics. *Trends in Genetics*, **3**, 167–171.

Websites

Visit the websites listed in Table 4.2.

CHAPTER FOUR

The sequence of the human genome

Key topics

- Technology for sequencing the human genome
- The IHGSC clone-by-clone strategy
- The Celera shotgun sequencing strategy
- Single nucleotide polymorphisms
- Sequencing the genome of model organisms
- Analysis of the structure and evolution of the human genome
- The total number of genes in the human genome
- The properties and evolution of the encoded proteins
- Oligonucleotide arrays
  - Determining the boundaries of genes
  - Expression profiling
  - Mass screening of sequence variants

4.1 Introduction

After the mapping phase of the Human Genome Project was essentially completed in 1998, attention turned to determining the nucleotide sequence of the genome. The plan was to complete the sequencing in two phases. In the first phase a draft would be produced that contained over 90% of the sequence but would contain gaps and would not have been edited to ensure high quality. The draft sequence would then be finished by closing the gaps and ensuring that the sequence contained less than one error in 10 000 nucleotides. Initially the timetable was to complete the draft sequence by the end of 2001 and improve the sequence to finished standard by 2005. However, at around this time a private biotechnology company called Celera Genomics announced that it also was going to sequence the human genome. There was great concern that the human genome sequence could be patented before it was even in the public domain. This spurred the publicly funded International Human Genome Sequencing Consortium (IHGSC) to advance their timetable, aided in the USA by greatly increased government funding totalling $86 million and in the UK by a $77 million grant from the Wellcome Trust, a medical research charity. IHGSC and Celera Genomics raced to produce the draft sequence. The race ended in an official draw when the completion of this draft sequence by both groups was announced in June 2000. The first printed accounts of the sequencing process and analysis of the draft sequence were published simultaneously in February 2001. Work is now proceeding to produce a finished sequence, which is due to be completed by 2003.

Obtaining the sequence of the human genome is an immensely difficult logistical problem. The human genome contains about 3.2×10^9 bp. The technology for deriving DNA sequence can read about 500 bp at a time from the ends of DNA fragments, which are initially randomly cloned from the intact genome. This implies that a minimum of about six million separate sequences are needed to provide one-fold **coverage**, that is on average every nucleotide will be sequenced once. In fact, the problem is very much worse than this for two reasons. First, each clone sequenced has been randomly sampled from a pool of clones. To be sure that every part of the human genome is represented in the clone collection, the laws of probability make it necessary to amass a total number greater than the number of clones necessary to provide the one-fold coverage. Secondly, to be completely sure that the sequence is accurate, each part should be sequenced on multiple occasions. Taken together these factors mean that to obtain a **draft sequence**, it was necessary to obtain four-fold coverage. Nine-fold coverage will be necessary to obtain the **finished sequence**.

When enough sequence had been obtained, fitting it together to reassemble the entire sequence of each chromosome was logistically complex. In the first place, experience of creating clone libraries shows that for various reasons there will always be some fragments that won't be present (see chapter 3). So fitting the overall sequence together was like doing a jigsaw with pieces missing. Secondly, a large part of the human genome consists of sequences that are repeated elsewhere (see chapter 3). Thus, assembling a complete sequence from the fragments is very difficult because there are apparently many possible locations for each repetitive sequence.

Celera Genomics and the IHGSC differed in the overall strategy they adopted to solve these problems. The IHGSC used a **clone-by-clone** or hierarchical approach that partitioned the genome into sections that were ordered with respect to each other. Each section was then sequenced separately before being assembled into a complete sequence for each chromosome. In contrast, Celera Genomics fragmented the genome into random short sections, which were sequenced separately.

Powerful computers and algorithms were used to assemble the complete sequence in one operation. The clone-by-clone approach was initially slower and more expensive, because of the need to order the clones before sequencing began. However, it was considered a far safer way to assemble sequence from a repeat-rich genome. Moreover, it would be far superior in the finishing phase, because the location of gaps would be precisely defined and it would be easier to retrieve the appropriate clones if further work was necessary.

4.2 Basic technology for sequencing the human genome

4.2.1 Sequencing technology

Both groups sequencing the human genome used technology that was based on the dideoxy method developed by Fred Sanger and his colleagues in Cambridge, UK, for which he received the Nobel Prize for Chemistry in 1980 (in addition to the Nobel prize he received in 1959 for developing a method to sequence proteins). The original protocol is described in Figure 4.1. This is now normally referred to as 'manual sequencing'. It is a technically demanding procedure that requires the use of radioisotopes and autoradiography, and the experimenters must read and record the results of the autoradiograph themselves.

For the amount of sequencing that is needed for the whole human genome there was a need for a procedure that did not require unstable isotopes and the attendant autoradiography, and in which the sequence data was collected automatically, not manually. This led to the development of fluorescence-based technologies described in Figure 4.2. These still made use of the basic concept of the Sanger method, but replaced labelled isotopes with four fluorescent dyes that labelled each of the four dideoxy analogues a different colour. The product of these reactions could be run on a single lane of a polyacrylamide electrophoresis gel. At the bottom of the gel a laser excites the fluorophores incorporated into the DNA molecules. The order of coloured peaks passing a detector is recorded and automatically converted into a digital file of the DNA sequence, which is ready for further analysis by programs that assemble the longer sequences from multiple gel readings.

The speed at which the products of a sequencing reaction can be separated by electrophoresis is proportional to the applied voltage. This is physically limited in slab gels by the rate at which the resultant heat can be removed. A further time-saving development was the replacement of slab gels with capillary gels with a 1 μm internal diameter. These can be cooled much more efficiently, and so the applied voltage can be increased to the extent that they can be run up to 14 times faster. They are also much more sensitive, producing a readable signal with a much smaller amount of the sequencing reaction.

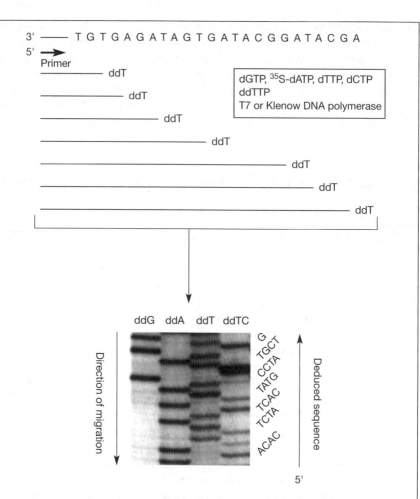

Figure 4.1 The Sanger dideoxy sequencing protocol. The DNA to be sequenced is either denatured or recovered in a single-stranded form from an M13-based vector. This is annealed to a primer that is designed to anneal to a fixed site in the cloning vector just outside the cloned fragment. The primer/template partial duplex is incubated in four parallel reactions, which each contain T7 DNA polymerase and a mixture of the four deoxynucleotides. In addition, each of the parallel reactions also contains a different 2'3' dideoxynucleotide analogue that can be incorporated into a nascent chain, but cannot serve as a substrate for the next step in the polymerisation because there is no free 3'OH group. They are therefore said to be chain terminating. In the example shown, dideoxy-TTP (ddT) is included in the reaction mix. At each A residue in the template either dTTP or ddT can be incorporated into the new chain. If dTTP is incorporated the polymerisation proceeds until the next A in the residue in the template. If ddT is incorporated the polymerisation of that particular chain is terminated. The size of the fragment upon chain determination reports the number of nucleotides from the primer to the position where ddT caused chain termination. The product is made radioactive by including ^{35}S-dATP in the reaction mix. The products of the four parallel reactions are fractionated by polyacrylamide gel electrophoresis and the molecules visualised by autoradiography. The sequence can be deduced by the order of bands in the four different lanes of the resulting sequencing ladder. On a good day about 400 bp of sequence can be read from a single gel.

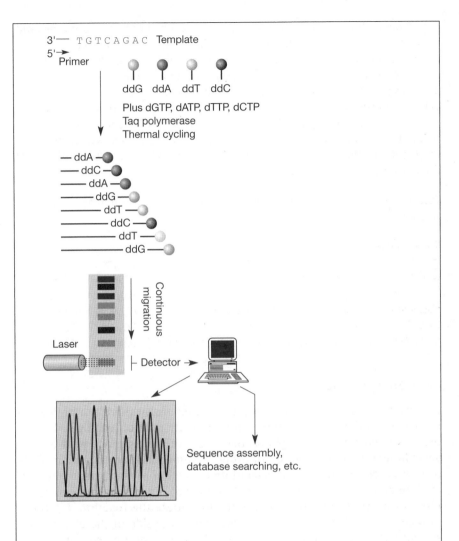

Figure 4.2 Automated sequencing using fluorescent dye terminators. The Sanger sequencing protocol described in Figure 4.1 is carried out with a different coloured fluorophore conjugated to each dideoxy analogue. A thermostable DNA polymerase is used that allows thermal cycling similar to that used in PCR. As a result the products are amplified in a linear fashion to generate enough product for the fluorophore to be detectable. The products are fractionated on a polyacrylamide gel. As they pass a detector a laser excites the fluorophore and a detector records the colour of each successive peak. The order of peaks therefore reports the DNA sequence, which is recorded by an attached computer.

4.2.2 Sequencing factories

The sequencing reaction itself was not the only aspect of the process to become automated. Library production, template preparation, setting up sequencing reactions, loading and running gels, data collection, quality assessment, storage and processing were all carried out on robotic production lines. Celera were able to process 175 000 gel reads per day operating 24 hours a day, seven days a week. At the height of the IHGSC effort, the rate of sequencing was one-fold coverage of the genome every six weeks, or 1000 bp per second. The Celera operation was confined to one location, the IHGSC consisted of a number of centres spread around the world, each of which operated as sequencing factories. Altogether 21 centres contributed to the final sequence, however, 86% of the draft sequence was contributed by just six of these, listed in Table 4.1 together with the contribution they made to the final sequence.

Table 4.1 Top six sequencing centres and their contribution to the draft and finished sequence.

Sequencing centre	Draft (kb)	Finished (kb)
Whitehead Institute, Center for Genome Research	1,196,888	46,560
The Sanger Center	970,789	284,353
Washington University Genome Sequencing Center	765, 898	175,279
US DOE Joint Genome Sequencing Center	377,998	78,486
Baylor College of Medicine, Human Genome Sequencing Center	345,125	53,418

4.2.3 Shotgun sequencing

One aspect of sequencing that has been stubbornly difficult to improve is the length of the sequence that can be reliably read off each gel. Despite the improvements in throughput, the length of the sequence read of each gel is limited to about 500 bp. In order to sequence a larger piece of DNA it must be broken down into a random collection of small fragments that are cloned into sequencing vectors (Figure 4.3). Each sequencing reaction requires a primer of known sequence to initiate the reaction. Synthetic primers are used that anneal to sequences either side of the cloning site, so that about 500 bp from either end of the fragment are determined. The sequence of the starting fragment can then be assembled by looking for overlaps between the gel readings. A longer sequence that has been reconstructed in this way is called a **contig**. The contig may not extend the full length of the starting fragment because a section may be missing, so usually some form of external **framework** or **scaffold** is needed to locate the contigs with respect to each other. One very important piece of information is that the sequence at the two ends of single clone are necessarily linked. So if the two end sequences from one clone fall into different contigs, the contigs must be adjacent to each other and the size of the gap can be estimated from the total size of the insert in the clone (Figure 4.3). The process of breaking down a larger sequence into random fragments, sequencing the smaller fragments and piecing together the

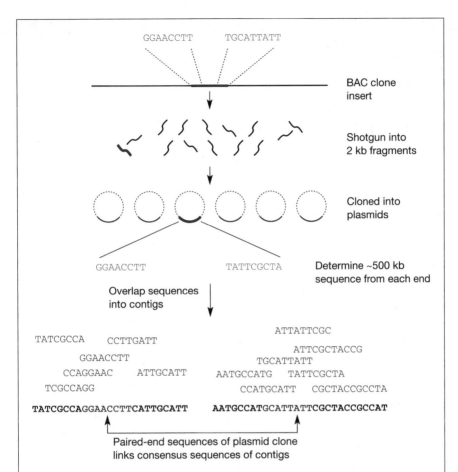

Figure 4.3 Shotgun sequencing. The insert in each BAC clone is fragmented and the pieces cloned into a sequencing vector. The method of generating the fragments and their size varied in the different sequencing centres. One common way is to sonicate DNA, which generates fragments whose size is proportional to the length and intensity of the sonication. The fragments so generated have blunt ends; they can be joined to the cloning vector by blunt end ligation. Because the breaks are introduced randomly, the breaks will occur in different places in each insert molecule; the inserts in the clones will overlap. One gel's worth of sequence is obtained from each end of each clone using primers that anneal either side of the cloning site. Sequences that overlap are arranged into contigs, from which a consensus sequence is obtained (shown in bold). Gaps between contigs are bridged by paired-end sequences that must be linked by virtue of being part of the same insert. The diagram follows the sequencing of one of the shotgun fragments (shown in blue), from its location in the original BAC clone, as an isolated fragment after sonication, as a sequence-ready clone and finally the location of the two end-paired sequences in different contigs. Only one strand of the DNA molecule is shown throughout the diagram. Often overlap will be detected between a sequence and its reverse complement that was sequenced in the opposite orientation in another clone. The computer software automatically reformats the gel readings to have the same orientation.

sequence of the original DNA is called **shotgun sequencing**. Both Celera and the IHGSC used shotgun sequencing, but differed in the size of the DNA that was sequenced by this method. Celera shotgun sequenced the whole genome, IHGSC applied this method to the 100–200 kb inserts of ordered BAC clones.

4.3 The IHGSC clone-by-clone sequencing strategy

4.3.1 Making a BAC clone map of the genome

The IHGSC project started by making genomic libraries in BAC or PAC vectors from blood and sperm samples that were collected from male volunteers. Because all identifying tags were removed and only about one-tenth of samples collected were actually used, the identity of the donors is unknown to both investigators and the donors themselves. So the published sequence comes from a mixture of different individuals whose identity is unknown. Male DNA was used because the female genome is missing the Y-chromosome sequence.

The fragments for the genomic libraries were generated by partial digestion with *Hind*III so that the inserts would be overlapping. The inserts were relatively large (100-200 kb) so that the number spanning each chromosome would be manageable. BAC and PAC vectors were used because they can carry reasonably large inserts without compromising their integrity; that is, there is a low level of deletions, rearrangements and chimeric clones (see Chapter 3).

To order the clones they were completely digested with *Hind*III and the products separated using agarose gel electrophoresis. The pattern of bands from each clone constitutes its fingerprint. Clones that overlap can be identified because their fingerprints will overlap (see Chapter 3). By overlapping the clones, they can be arranged into contigs called **fingerprint clone contigs**. Like all contigs, there will eventually be gaps because key clones will be missing or overlap between contigs exists but has not been detected. The long-range physical and genetic genomic maps described in Chapter 3 were used to position the fingerprint clone contigs along the chromosomes (Figure 4.4). As illustrated in Figure 4.4, this allowed errors in the clone map to be identified. Clones were also hybridised to whole chromosomes using **FISH** so that the map was also integrated with the cytogenetic map (Figure 4.5).

The finished clone map contains 372 000 BAC clones in 965 contigs. As well as providing the foundation of the sequencing project, the clone map provides a permanent resource that can be used to retrieve any specified region of the genome. It is tightly integrated with the genetic, STS, expression and cytogenetic maps. Thus, once a gene, disease or phenotype has been mapped by linkage to a mapping marker, the DNA surrounding the marker is physically available for study, along with its sequence and knowledge as to which parts are expressed as mRNA. The clones also provide a physical source of DNA to act as a reference sequence when searching for mutations and SNPs.

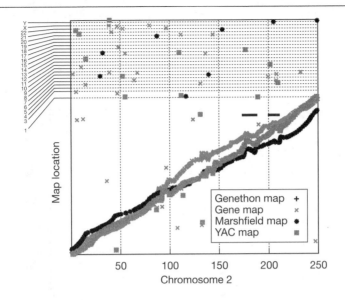

Figure 4.4 Alignment between BAC clones and other genomic maps. As the BAC clones are sequenced STS markers are revealed by **ePCR**. The position of these in the genetic (Genethon and Marshfield), gene (radiation hybrid) and STS-content (YAC) maps anchors the sequence to chromosomal position and acts to trap errors in the BAC clone map. In this example of BAC clones mapping to chromosome 2, there are two apparent discrepancies between the sequence and all three of the other types of map (shown by black bar). Inverting the segment of the BAC clone map restored consistency between the sequence and the other types of maps. A small number of markers from the other maps were found to have been misassigned to chromosome 2, their true chromosomal location revealed by sequencing (shown at the top of the figure). (Redrawn with permission from IHGSC (2001) *Nature*, **409**, 860-921.)

4.3.2 Generating and assembling the sequence

From each BAC fingerprint clone contig, clones representing the minimum overlapping set were selected for sequencing. One advantage of this scheme is that it is suitable for collaboration between different sequencing centres, because each can receive different BAC clones to sequence. The insert in these clones was then shotgun sequenced (Figure 4.3). Overall, in the draft phase the amount of sequence obtained was such that on average each nucleotide was sequenced four times, i.e. four-fold **coverage**. The gel reads were filtered for vector sequences and other extraneous DNA from the bacterial host or other non-human sources.

Sequences for each chromosome were then assembled in stages from the primary gel reads (Figure 4.6). First the gel reads were first assembled into sequence contigs for each BAC clone. These were then merged with those from adjacent BAC clones to form longer contigs. One of the guiding principles of the IHGSC was that as the primary gel readings were assembled into

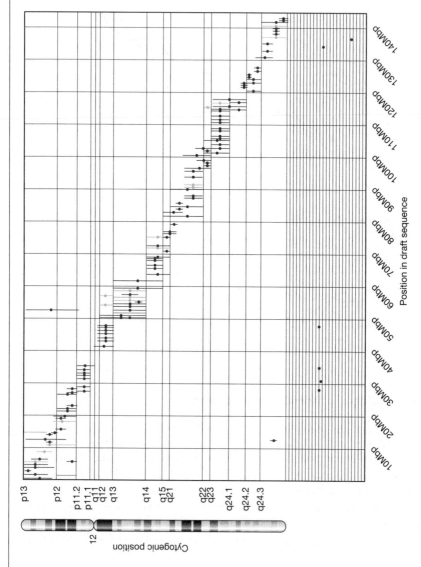

Figure 4.5 Correlation of BAC clone map with cytogenetic map. The cytogenetic map of chromosome 12 is shown on the left. The graph shows the range of bands to which each clone hybridises as determined by FISH. The position of each clone in the sequence is shown on the x-axis. A small number of clones hybridise to other chromosomes revealing errors in the original chromosome assignment. (Redrawn with permission from The BAC Consortium (2001) *Nature*, **409**, 953–958.)

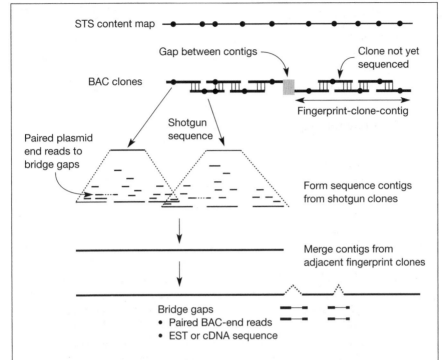

Figure 4.6. IHGSC clone-by-clone sequencing strategy. STS markers in the BAC clones, shown by solid circles are aligned with markers in the STS content map. Merged contigs are located on a scaffold. Missing sequences, caused by gaps between fingerprint clone contigs or BAC clones not yet sequenced, are bridged by paired BAC-end reads or sequence from EST or cDNA sequence already in databases. Note that cDNA and mRNA are derived from mRNA sequences from which introns have been removed by splicing, so their sequences will not be continuous in nuclear DNA and may be widely separated. They are thus a useful resource to link and order sequences that originate in different contigs.

contigs, this information was released on the internet within 24 hours. The idea was to prevent patenting of DNA sequence information by placing it in the public domain as soon as possible. Ironically, this allowed Celera to make use of the information for its own privately funded sequencing of the genome (see below). In principle, the merged sequence contigs should extend the full length of the chromosome, but in practice there were gaps in the draft sequence. However, as shown in Figure 4.6, the contigs could be placed on a scaffold using information such as paired plasmid-end reads and data already in databases from cDNA and EST sequencing.

Overall it is estimated that 88% of the human genome has been sequenced in the draft phase. The advantage of the clone-by-clone approach is that it is relatively easy to obtain most of the missing information because many of the clones that require extra sequencing are already available. Some of the sequence will be in the gap between BAC fingerprint clone contigs and this may be more difficult to retrieve, perhaps requiring the use of a different vector. The accuracy of the sequence is an important criterion. As the initial gel readings are assembled into longer consensus sequences, computer software has been developed to estimate the error rate, based on information such as the quality of the gel trace and the number of times it has been independently sequenced. In the draft sequence 91% of the sequence is judged to have an error rate of less than 1 in 10 000, and 95% is estimated to have an error rate of less than 1 in 1000.

4.3.3 Finishing the sequence

The draft nucleotide sequence provides much of the useful information to be obtained from the human genome. However, to provide a permanent and definitive resource the quality of the draft sequence will be improved to what is termed **finished** standard. In finished sequence all of the nucleotide assignments will have an error rate of less than 1 in 10 000. The gaps will be kept to a minimum, but some will remain. Some regions may be refractory to cloning for reasons discussed in Chapter 3. It may be possible to use different vectors to recover the DNA from some of these regions, but in some cases the effort required will far exceed any possible benefits. One class of DNA that will probably not be completely sequenced is **heterochromatin**. Sequences from the centromere and telomere regions that contain high copy, simple sequence repeat DNA will also be difficult to sequence. The aim is to finish the sequence of the human genome by 2003. By October 2001, 63% of the sequence had been obtained to finished standard and the project was well on schedule to achieve its target and perhaps complete the task even earlier. The portion of finished DNA includes the whole of the two smallest chromosomes, chromosomes 21 and 22. Although many of the gaps were closed in the sequence of these chromosomes, some remained. For example in chromosome 21 there were 11 gaps in the finished sequence, amounting to less than 150 kb out of a total of 33 Mb of **euchromatin** sequenced, or just 0.45% of the total. The regions containing these gaps were sequenced to 20-fold coverage, without recovering the missing information. It was concluded that for some reason the sequences were unclonable in the BAC vectors used. It is possible that some other vector would be more successful in cloning the missing sequences.

4.4 The Celera shotgun sequencing

Like the IHGSC, Celera constructed three different types of genomic libraries from anonymous volunteers. The three libraries differed in the size of their inserts, averaging 2, 10 and 50 kb, respectively. Instead of ordering the clones, Celera moved directly to shotgun sequencing using gel reads from the end of each clone to amass a total sequence amounting to five-fold coverage. This was augmented by IHGSC gel reads, which they downloaded from the internet, so that in total the amount of sequence analysed amounted to eight-fold coverage.

In order to assemble the shotgun sequence for the whole genome in one operation, a major problem must be solved – about 45% of the genome consists of repetitive elements (Table 4.2). For example, the *Alu* family (see Chapter 2) consists of 1.5 x 10^6 copies of a 300 bp sequence and occupies 13% of the genome (Table 4.2). Each copy will differ slightly, but recognisably conforms to a consensus. Deriving a longer sequence by many overlapping smaller sequences will be a hopeless undertaking, because each sequence with an *Alu* element will overlap with over a million other sequences – only one of which will be the correct match.

The strategy Celera adopted to solve this problem is shown in Figure 4.7. There were two key elements in their strategy:

- Computer software was used to mask repeated sequences in the raw sequence reads. Only the remaining unique sequence was used to construct the contigs. Once the contigs were constructed the repeated sequences were restored because their proper location had been determined.

- Three different libraries with inserts of 2, 10 and 50 kb were constructed. The paired-end reads from these libraries were used to construct a scaffold that linked the contigs together.

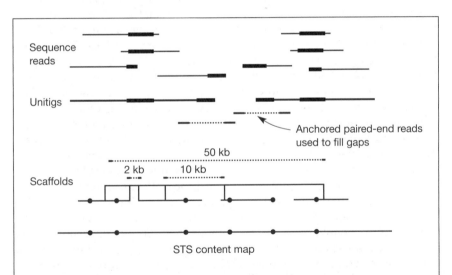

Figure 4.7 Celera whole genome shotgun sequencing strategy. Repeated sequence (blue bars) in the initial reads from the shotgun clones are masked. Unitigs are built up from overlapping the remaining unique sequence (black lines). Repeated sequence can be restored to the unitigs when its location has been unambiguously determined through overlap of the surrounding unique sequence. Unitigs ending in repeat sequence cannot be merged with the adjacent unitig because overlap of repeated sequences is disallowed. However, the unitigs can be placed on a scaffold using paired-end reads from clones from the 2, 10 and 50 kb libraries and located on the chromosomes through STS content maps. Once located on a scaffold, paired-end reads can be used to fill the gap where one end is unambiguously located in a unitig. Multiple gel readings placed in the same gap were then assembled into contigs.

Celera also independently assembled a sequence of the human genome using positional information from the BAC-clone map constructed by IHGSC. Sequences from BAC clones, made publicly available by IHGSC, were matched against Celera's own gel reads. This allowed Celera's gel reads to be placed in bins defined by the BAC clones. The whole genome sequence was then assembled using Celera's sequence and the gel reads from the IHGSC downloaded from the internet. The initial construction of contigs only allowed matches between gel readings that came from the same BAC clone, thus simplifying the problem of repetitive sequences. When compared with the sequence derived from the whole genome shot gun approach it gave a slightly more complete sequence with fewer gaps. However, the important point is that it showed it was possible to use a whole genome shotgun approach to sequence large genomes with large fractions of repetitive DNA. Because it is faster and cheaper than the clone-by-clone approach, it can be used to determine the sequence of other large genomes, such as the mouse genome, that will be very important for bio-medical research and economically important animals and plants.

4.5 Single nucleotide polymorphisms

Both the Celera and IHGSC sequences were derived from a panel of different individuals. As the consensus sequences were assembled, differences between the nucleotide sequences of different individuals became apparent. If a difference is common in a population it is defined as a polymorphism. Thus during the sequencing of the human genome **single nucleotide polymorphisms** (SNPs) were identified. Because they were identified in the course of sequencing their exact location is automatically known, so a map is generated of the common differences between individuals. As genetic differences between individuals underlie their different susceptibility to many or most common diseases, this map will play a critical role in identifying the genes responsible. The way SNP maps will be used is explored more fully in Chapter 6. Because of the importance of such maps, a targeted search was also made to discover as many as possible by a consortium called the SNP consortium (TSC) consisting of ten pharmaceutical companies and the Wellcome Trust. The genomes of a panel of 24 individuals from diverse ethnic backgrounds were sequenced to one-fold coverage, resulting in about 50% of the genome being resequenced in each individual. These sequences were then compared to the reference sequence. A combined total of 1.4 million non-redundant SNPs were identified by TSC and the original sequencing project.

Both Celera and the IHGSC estimated that the frequency of SNPs in each individual is 8×10^{-4} per nucleotide or about one SNP every 1.25 kb in the human genome, a figure that is in agreement with previous estimates. The actual SNPs identified in the combined TSC and IHGSC set occur one every 2–3 kb (the estimate varies according to how they were identified). It is estimated that the DNA region coding for each mRNA contains two of these SNPs. These are likely to be the most useful for studying phenotypic variation and disease susceptibility, especially if they are non-synonymous, i.e. the SNP results in a change in the amino acid sequence of the encoded protein. Of course, sequences upstream of coding regions may also be important. It is still difficult to decide how far upstream regulatory elements may extend. If 10 kb is taken as a rough estimate, then it is likely that 93% of all loci contain at least one SNP. The number of SNPs identified is likely to increase over the next few years

Interestingly, the level of SNP heterozygosity on X and Y chromosomes is only 75% and 21% respectively of the levels found on autosomes. This is to be expected if most of the SNPs are neutral, because the rate at which they will accumulate in a population is proportional to the effective population size (see Chapter 9). The effective population size of sex chromosomes is smaller than autosomes because on average in a population there are only three X chromosomes and one Y chromosome for every four autosomes. The spacing of SNPs also varies across the genome because of the different histories of each genomic region. The age of the most recent common ancestor or **coalescence** time of each region will vary by chance. There will be fewer SNPs where the coalescence time is shorter because there has been less elapsed time for mutations to occur.

4.6 Accessing the data

A page of this book contains approximately 3500 characters. If the sequence of the human genome was written in the same way it would require roughly 1 million pages or about 3000 books of the same size as this one! Moreover, to be useful the raw information would need to be **annotated**. This involves determining and recording where genes start and stop, the boundaries of exons and introns, the location of physical features such as chromosome bands, repetitive elements, CpG islands, mapping features such as STSs, genetic markers and SNPs and homologies to genes already studied in model organisms or duplicated copies of human genes already studied (paralogs, see below). Furthermore, this information needs to be available to a wide variety of laboratories worldwide which have access to the necessary tools to retrieve and analyse the data relevant to their particular research. Clearly, the printed page is not a suitable repository for this data, and it will probably never be printed out. However, these requirements make the World Wide Web an ideal way of accessing the data. Genome browsers have been developed to allow the sequence and the annotations to be accessed via a graphical interface. Figure 4.8 shows an example from the Ensembl site, constructed by the European Bioinformatics Institute at the Sanger Centre, showing the region around the *BRCA*1 locus on chromosome 13q12.3. The web page from which this was taken has facilities to zoom in or out, navigate in either direction and to view the DNA sequence itself. The information presented in Figure 4.8 illustrates the features of the human genome that emerged from analysis of its sequence described in the next section. A similar genome browser was constructed by the University of California at Santa Cruz. URLs to these and other websites relevant to this chapter are given in Table 4.2.

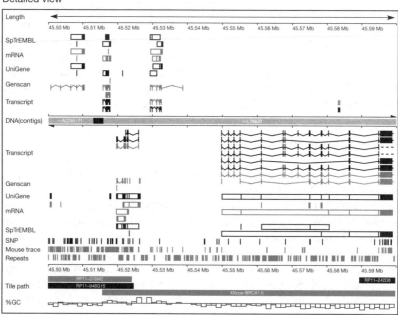

Figure 4.8 The entry for *BRCA1* in the Ensembl website. Illustrates how information is presented describing the position of known and unknown genes, expressed sequences, repetitive elements, SNPs, ESTs, chromosome location, homology with mouse DNA, contig coverage from the BAC clone library (labelled 'tile path'), the location of gaps and the %GC. Each DNA strand is annotated independently above and below the DNA bar. Transcripts refers to known or predicted transcripts and are colour coded according to status (e.g. experimentally verified, predicted, novel, etc). 'SpTrEMBL' refers to entries in the Swiss-Prot protein database, a non-redundant database with high quality annotation, or in TrEMBL, a subsection of the EMBL nucleotide database in which annotated coding sequences have been translated. TrEMBL entries will eventually be incorporated into SwissProt. 'Unigene' refers to a database curated by the National Center for Biotechnology Information (NCBI) that contains ESTs and full-length mRNA sequences of unique known or putative human genes. Genscan refers to a computer program that predicts genes from sequence *ab initio*. Mouse trace are matches to mouse DNA sequence derived from gel reads of the Mouse Genome Sequence Consortium. mRNA refers to homologies to mRNAs from other vertebrates. Marker refers to a marker in one of the genome map formats. Holding the mouse over each of these features produces a pop-up menu with links to further information. For example, holding the mouse over DNA (contigs) gives access to further contig information including sequence; holding the mouse over a SpTrEMBL hit provides a link to the database entry.

Table 4.2 Useful websites

URL	Website
www.ensembl.org/	The Ensembl genome browser constructed by the Sanger Centre and the European Bioinformatics Centre (EBI)
www.genome.ucsc.edu/	Contains draft sequence and genome browser constructed by University of California at Santa Cruz
www.ncbi.nlm.nih.gov/genome/guide/human/	National Center for Biotechnology (USA). Access to chromosome maps
www.ebi.ac.uk/genomes/mot/	EBI daily update on sequencing progress
www.ornl.gov/hgmis/	USA Department of Energy (DOE) Human Genome Project Information page
www.celera.com	Celera genomics homepage. Requires registration to view data
www.snp.cshl.org/data/	The SNP consortium website
www.hgrep.ims.u-tokyo.ac.jp/	Human genome reconstruction project. Provides an overview of genome structure.
www.ncbi.nlm.nih.gov/LocusLink/refseq.html	RefSeq database
www.geneontology.org	Database of protein and domain functions

4.7 Model organisms

As well as the human genome, the genome sequences of a variety of organisms have also been determined or are in the process of being determined. These are organisms that have been widely used as experimental models. Sequencing the genome of the following organisms has been completed: the budding yeast *Saccharomyces cerevisiae* (finished in 1996), the bacterium *Escherichia coli* (1997), the nematode worm *Caenorhabditis elegans* (1998), the fruit fly *Drosophila melanogaster* (2000), and the mustard weed *Arabidopsis thaliana* (2000). Sequencing of the mouse, rat, zebrafish and the pufferfishes *Fugu rubripes* and *Tetraodon nigroviridis* are currently in progress. Sequencing of **model organisms** has been a vital part of the Human Genome Project.

Initially, model organisms provided simpler genomes, that were used to test and improve sequencing strategies and technologies. For example the two-stage approach of draft and finished sequence was developed in the sequencing of the fly and worm genomes.

Secondly, they provide a context within which to analyse the human sequence. At the simplest level the sequence of functional proteins will be conserved across taxonomic categories. In contrast, with some limited exceptions such as regulatory sequences, the sequence of non-coding DNA is not under selective pressure and is not conserved. Thus conservation of sequence between model organisms and the human genome provides an essential tool to distinguish coding and non-coding sequences. The pufferfishes are remarkable because their genomes are much more compact than the human genome, with a much higher proportion of coding DNA. It will therefore be easier to identify these coding regions. As they are likely to have a similar complement of genes to humans, this will aid the identification of human coding regions.

Thirdly, comparison of genomes from different taxonomic categories illuminates evolution. As discussed below, comparison of the genomes of all eukaryotes allows the definition of a minimal set of proteins required for basic cellular function. Restricting the comparison to metazoan organisms defines the additional genes required to organise cells into animals. Comparing the fly and worm genomes to vertebrate genomes traces the evolutionary innovations that led to vertebrates. The zebrafish is used for the genetic analysis of developmental processes. The availability of the human and zebrafish genomes will provide instant access to the human counterparts of zebrafish genes as their roles emerge.

Fourthly, model organisms provide an opportunity for more sophisticated genetic analysis of gene function. The role of genes can be investigated by targeted deletions (gene knockouts), overexpression or ectopic expression, and *in vitro* mutagenesis. This has already played a vital role in the investigation of the role of genes affected by disease-causing mutations.

4.8 Analysis of the human genome sequence

With the determination of its sequence, the human genome becomes a finite entity that be can be completely and accurately described. In principle, we now know of the location and sequence of every gene and repetitive element. We can precisely enumerate the vital statistics of the genome: its size, the fraction that codes for proteins, the fraction that constitutes repetitive elements, the location of CpG islands, the correlation of physical and genetic maps, and so on. Before the sequence was determined all of these features could only be studied indirectly, albeit with increasing accuracy and confidence. This section summarises some of the main conclusions that emerge from an initial analysis of the sequence of the human genome. The reader should refer to Chapter 2 for an introduction to the main features of the genome.

Table 4.3 provides a summary of the vital statistics of the human genome. The values are mainly taken from the IHGSC analysis, because the Celera sequence contained a lower proportion (35%) of repeated DNA sequence, probably as a result of the whole genome shotgun strategy adopted. Other aspects of the Celera analysis gave a broadly similar picture, although the exact values varied slightly.

As always, real life is more complicated than principle and we shall see that there are still some areas of uncertainty, particularly identifying and defining the extent of all coding sequences. Both Celera and the IHGSC

Table 4.3 Vital statistics of the human genome. Data taken from the IHGSC analysis, except for values marked * which were taken from the Celera analysis. The repetitive elements *Alu*, LINE1, MaLR and MER1-Charlie are the most common in each respective class. The values for gene properties (size, exon number, exon size, etc.) were based on known genes in the RefSeq database.

Total genome size	3289 Mb
Chromosomes	
Largest (chromosome number)	279 Mb (1)
Smallest (chromosome number)	45 Mb (21)
X	163 Mb
Y	51 Mb
Fraction CpG	41%
Number of CpG islands	28 890
Proportion of genome that encodes proteins	1.5%
Proportion of genome transcribed	33%
Number of genes actually identified	26 500
Estimated total number of genes	31 000
Mean gene density*	9–14 genes/Mb
Mean of size of genes by genomic extent (median)	27 Kb (14 kb)
Largest gene by genomic extent (name)	2.4 Mb (DMD)
Mean gene transcript size (median)	1340 bp (1100 bp)
Largest gene transcript (name)	80.8 kb (Titin)
Mean number of exons per gene (median)	8.8 (7)
Mean size of internal exons (median)	145 bp (122bp)
Mean size of introns (median)	3365 bp (1023 bp)
Mean proportion of each gene that is coding	5%
Proportion of genome that consists of repetitive elements	44.8%
SINEs (number in 1000s)	13.4% (1558)
Alu elements	10.6 (1090)
LINEs	20.42 (868)
LINE1	16.89 (516)
LTRs	8.29 (443)
MaLR	3.65 (240)
DNA transposons	2.84 (294)
MER1-Charlie	1.39 (182)
Recombination per Mb (cM/Mb)*	
Male*	0.88
Female*	1.55
Sex averaged*	1.22

published their own independent analysis of the sequence, which the reader is referred to for a more complete description (see Further reading at the end of the chapter). Both groups broadly agreed over the main features, but inevitably where actual values are given there are some discrepancies. A further reason for regarding the figures as provisional is that the analysis was carried out on the draft sequence. Clearly actual values will be refined as the sequence is improved to finished standard. Nevertheless, the overall description of the genome provided here is likely to be generally accurate. This is confirmed by the use of chromosomes 21 and 22 as benchmarks, because the sequence of these chromosomes has already been determined to finished standard. Conclusions drawn from whole genome draft sequence were nearly always found to be consistent with the finished sequences of chromosomes 21 and 22.

4.8.1 Overall sequence architecture

4.8.1.1 GC content

The human genome has an average GC content of 41%. However, this is not uniformly distributed throughout the genome. Figure 4.9 shows that when examined in 50 kb windows, the GC content of each window can vary from 30% to over 60%. This is 15-fold greater than the variation that would be expected by chance in random sequence DNA. Using probes taken from regions of high and low GC content, FISH shows that G-bands of chromosomes correspond to regions of low GC content. GC content is strongly correlated with gene density, with GC-rich areas containing more genes than AT-rich regions (Figure 4.9).

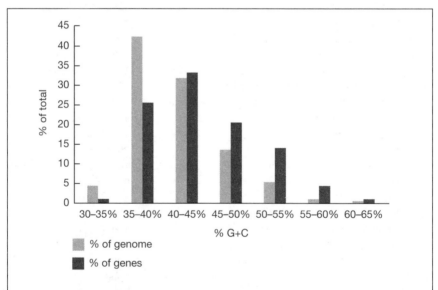

Figure 4.9 Distribution of GC content in the genome and the relationship between genes and GC content. Genome GC content was determined in 50 kb windows and divided into GC intervals of 5%. The histogram shows the fraction of total DNA (blue bars) and percentage of genes that is in each interval. There are proportionately more genes in the intervals above 50% GC and proportionately less in the intervals with less than 40% GC. (Reprinted with permission from Venter, J.C. et al. (2001) Science, **291**, 1304–1351, copyright 2001 American Association for the Advancement of Science.)

4.8.1.2 CpG islands

The dinucleotide CpG (i.e. C followed by G on the same strand) occurs less often than would be expected by chance in the human genome. As described in Chapter 2, the cytosine base in CpG is the target for DNA methylation, which inhibits gene transcription. Scattered throughout the genome are islands of sequence where the occurrence of CpG is not depressed from that expected and which are unmethylated. These **CpG islands** tend to occur at the start of genes, spanning the first exon and the promoter sequences immediately upstream. The genome sequence alone does not give information about methylation, but is informative of the number and distribution of CpG dinucleotides. There are about 28 000 CpG islands across the genome; the exact number is sensitive to the percentage GC used to define an island. This figure is close to the predicted number of genes. Most of the predicted islands are between 500 bp and 1500 bp long and there is a strong tendency to overlap the first exon and promoter sequence of known or predicted genes.

4.8.1.3 Repetitive DNA

About 45% of the human sequence is derived from one of four types of transposable element, described in Chapter 2. This figure is likely to be an underestimate, because many repetitive elements will have diverged so much that they are no longer recognisable. Furthermore, many of the gaps in the draft sequence will be repetitive DNA. The distribution of repetitive DNA among these types is shown in Table 4.3. The most common families are the *Alu* family, occupying 13% of the genome, and the LINE1 family, occupying 17%. Repeated sequences are interspersed continually throughout the genome, as illustrated in Figure 4.8.

Because the sequence of every copy in a family is known, it is possible to construct a phylogenetic tree and deduce the sequence of the original progenitor of each family. So for each individual element the number of nucleotide changes from the inferred progenitor can be determined, forming a molecular clock that allows the birth and subsequent fate of these elements to be determined. The clock can be calibrated by measuring the average difference in nucleotide sequence between humans and Old World monkeys, which are known from the fossil record to have diverged 25 million years ago. Figure 4.10 shows the range of substitutions found in each of the four types of transposable elements. Each interval of 3% substitution represents about 25 million years of evolution. This analysis not only provides the age of the elements, but also allows the periods during which the transposable elements were active to be calculated. In other words we can view the transposable elements as if they were independent organisms and chart the rise and fall of each type as if in a fossil record. A number of striking conclusions emerge from this study:

- In the lineage leading to humans transposition of all elements has been virtually absent in the last 25 million years. Nearly all the repetitive elements in the genome are no longer living, but are fossils. After transposition, they accumulated mutations that prevented further transposition events occurring. Most repetitive elements remaining in the genome are ancient and pre-date the eutherian radiation (mammals that aren't marsupials).

- There were two peaks of DNA transposon activity, one before and one after the eutherian radiation. These involved two families of DNA transposons called Charlie and Tigger, respectively. There has been very little activity of DNA transposons in the human lineage in last 50 million years. Only two classes of DNA transposons, MER75 and MER85, have been active in this period. Together they occupy less than 0.00041% of the genome.

- Only one transposition, an event involving an LTR retroposon, is known to have occurred since the divergence from chimpanzees 6 million years ago. This results in a single LTR element, found in the HLA locus, that is not common to all humans. LTR transposons were apparently most active more than 100 million years ago.

- The *Alu* and LINE1 families are 80 and 150 million years old, respectively. The last burst of *Alu* transposition was about 40 million years ago. Before that the most common elements were LINE2 and a SINE element called MIR. MIR transposition used the LINE2 reverse transcriptase and so became inactive alongside LINE2.

- Mechanisms for removing inactive elements are very inefficient. The half-life of a transposable element in the mammalian genome is 800 million years. This is in contrast to other types of animals. For example the half-life of a repeated sequence in Drosophila is only 12 million years.

- Transposable elements are still active in the mouse genome, and so most repetitive elements in the mouse are younger. This correlates with the observation that 10% of mouse mutations, but only 1 in 600 human mutations, are caused by repetitive elements. It is not clear why there should be such a difference in the evolution of hominids and rodents. It could be to do with the size of the breeding group, which is much smaller in hominids. This may have resulted in bottlenecks that would have reduced diversity (see Chapter 9).

The distribution of repetitive elements is not even across the genome. Some regions are very rich; others are almost totally devoid of elements. For example, a 525 kb region on the X chromosome, Xp11, consists of 89% repetitive DNA. A striking contrast is found in the four Hox regions that each contain a cluster of homeobox genes that play a crucial role in organising development. These four regions are thought to be the result of ancient duplications and are homologous to each other. All four regions

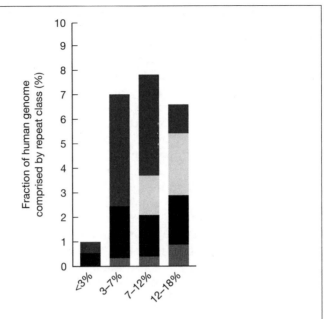

Figure 4.10 Sequence variation in repeated sequences. The percentage variation from the consensus sequence for *Alu* (blue), LINE1 (white), LTR elements (black) and DNA transposons (grey) is plotted in 3% intervals. Each interval represents about 25 million years of evolution. Few elements are in the <3% indicating that most are older than 25 million years. (Redrawn with permission from IHGSC (2001) *Nature*, **409**, 860–921).

contain less than 2% repetitive elements. The regions contain complex *cis*-acting regulatory elements. It is possible that insertion of transposable elements would disrupt the complex regulation shown by these genes. If so, it will be interesting to investigate the function of other areas of the genome that show a low density of repeat DNA. Another factor affecting the distribution of *Alu* elements is GC content. Young *Alu* elements are more frequently found in AT-rich regions, however progressively older *Alu* elements show a progressively increasing bias towards GC-rich DNA. Apparently, *Alu* elements preferentially target AT-rich DNA, but some form of selection operates to retain them in GC-rich DNA. The nature of this selective pressure is unclear, but its existence suggests that *Alu* elements are not selectively neutral.

The spread of repetitive elements throughout the genome is generally thought to have occurred not because they are phenotypically beneficial, but because they are essentially selfish elements that evolved only through selection within the genome that favoured their transposition. Nevertheless, repetitive elements have played a role in the evolution of genome function. The reverse transcriptase of LINE elements can act on mRNA to produce a DNA copy, which can insert into the genome to produce a copy of the original gene, called a **retrotransposon** (see Chapter 2). Because the mRNA has been spliced after transcription, the retrotransposon will lack introns. It will also have a polyA encoded in the DNA from the polyA tail of the mRNA. If it loses function as a result of the accumulation of inactivating mutation, it is known as a **processed pseudogene**. Retrotransposons can be identified because they will be single exon genes that show sequence homology to other multi-exon genes in the genome. Sequence analysis identified nearly 300 such examples, of which 97 are already known to be expressed and to have some function. In this set, genes involved in translation (ribosomal proteins and elongation factors) and nuclear regulation (non-histone chromatin proteins) are over-represented. Because of the mechanism by which they arose, retrotransposons do not carry upstream regulatory sequences from the gene that was copied. Thus at the new site where they insert they may acquire different regulatory regions, potentially changing the pattern of expression.

Repetitive elements may contribute to genome function in other ways. A total of 47 human genes are homologous to the proteins encoded by DNA transposons. There are also examples where repetitive elements are present in regulatory regions, suggesting that they have been co-opted to increase the repertoire of the patterns of gene expression.

4.8.1.4 Correlation of physical and genetic distance

Genetic maps are based on the proposition that the amount of crossing over between two polymorphic markers at meiosis is proportional to the distance separating them. The units of genetic maps are centimorgans (cM), corresponding to a recombinant fraction of 0.01. The location of polymorphic microsatellites used to construct genetic maps can be determined in the genomic sequence. This allows the correlation between physical and genetic distance to be examined. Across the whole genome a general correlation between genetic and physical distance is evident (Figure 4.4). It was already known that more recombination occurs in female than in male meioses. Averaging between sexes, 1 cM on the genetic map corresponds to 1.2 Mb of sequence. However, the actual correspondence varies at both small and large scales.

- At the local level when recombination is measured in windows of 3 Mb, there is an almost five-fold variation in the rate of recombination from the hottest to the coldest regions. This greatly exceeds the two-fold difference between male and females.

- Two types of variation are seen at the chromosomal level:

    ○ Recombination is higher near telomeres and lower near centromeres, especially in males.
    ○ Proportionately more recombination is seen on short chromosomal arms compared to long arms. This may be due to the requirement for at least one recombination per chromosome arm, which is necessary for disjunction of homologous chromosomes at meiosis.

Comparing the sequence of a segment with the rest of the genome makes it possible to identify examples of DNA duplication. However, segments that have duplicated recently will have almost identical sequences. This results in two problems:

1. During the assembly, sequences that are located in different parts of the genome may be combined into the same contig. This is much more of a problem for the whole genome shotgun approach, where there is no independent positional information. In fact it was one of the most difficult problems to solve, and required special programs that assessed whether the coverage of a contig was greater than would be expected, given the average coverage of the whole genome.
2. If the assembly of the final sequence was imperfect and gel reads that actually come from the same chromosomal location were assembled into different contigs, then the final sequence will appear to have a duplicated segment. The solution to this is to reject apparent duplications where the match is over 99.5% in finished sequence (98% in unfinished sequence), suggesting that they are really the same sequence that has been misassembled. The penalty is that some genuine recent duplications will be missed.

This analysis showed that about 5% of the genome is duplicated in blocks of between 10 and 50 kb. In addition the regions around the centromeres (**pericentric**) are very rich in interchromosomal duplications, giving rise to the suggestion that they may be involved in some repair mechanism for chromosome breakage.

An important practical reason for unravelling chromosomal changes during evolution is that the order of genes in chromosomal segments is likely to be conserved between humans and experimental animals such as the mouse. Genes whose chromosomal positions are conserved in this way are said to be **syntenic**. Experience with yeast and other organisms shows that the local order and orientation of genes can change, so the criteria for synteny is the presence of multiple homologous genes in the same chromosomal segment in both organisms or in duplicated blocks within the same organism. Knowledge of synteny between humans and model organisms will aid biomedical research in different ways. For example, intensive genetic analysis in mice may identify a genomic region involved in susceptibility to a mouse model of a complex disease. Synteny will allow the same region to be studied in humans to see if there is any evidence of involvement of a human counterpart.

Another reason to study chromosome evolution is to test a theory of vertebrate evolution, which holds that there were two rounds of whole genome duplication around 500 million years ago, at the time that jawed fishes appeared. This theory is based on the existence of the quadruplicated Hox gene clusters mentioned above. However, an alternative explanation is that they arose through localised interchromosomal duplications. Clearly this is an interesting question that may be resolved by study of the genome. A precedent for whole genome duplication is the yeast genome, where sequence analysis has revealed that about 100 million years ago it duplicated and then lost about 85% of the duplicated genes, leaving the remainder as duplicated blocks distributed around the genome.

Comparing DNA sequence allows only recent chromosomal events to be analysed. More ancient chromosomal changes cannot be followed because most non-coding DNA, which occupies >98% of the genome (see below), is not subject to sequence conservation. One way to approach the study of long-term chromosome and genome evolution is to look at the distribution of genes rather than compare DNA sequences. In this way larger blocks of duplicated chromosomal evolution emerge. For example the Celera analysis revealed the following:

- A total of 1077 blocks of duplicated sequence were identified in the human genome of which 781 contain five or more genes.

- Three very large blocks of duplication were detected, two of which involved chromosome 2.

  ○ Between 2p and chromosome 14. The duplication represents 70% of the sequence from the latter.
  ○ Between 2p and chromosome 12. This duplication includes two of the Hox gene clusters referred to above.
  ○ Between chromosomes 18 and 20 (Figure 4.11). This duplication includes 64 genes. After discounting a clear insertion between 'Krup rel' and 'collagen rel' on chromosome 18, the duplication covers about half of each chromosome. Only a fraction of each duplicated segment was found in the other chromosome. For example, the 64 genes occurring on both chromosomes 18 and 20 are found in a block of 217 genes on chromosome 18 and 322 genes on chromosome 20. The chances of this occurring by chance are very low. The most likely explanation is that a large segment was originally duplicated, then genes were lost from both segments, resulting in some genes still being present on both chromosomes, while other genes are present only on one or the other chromosome. Regions in the mouse genome can be identified that are syntenic to the duplicated blocks on chromosomes 18 and 20. Gene order and sequence conservation is much more similar in the pairs of sytenic mouse/human partners than in the pairs of duplicated blocks on the human chromosome. It follows that the duplication preceded the divergence of the lineages that led to rodents and primates. Although this picture is consistent with ancient whole genome duplications, it does not really distinguish between the competing hypotheses of piecemeal duplications of chromosomal segments. As more vertebrate genomes are sequenced, the evolution of chromosomes should become clearer.

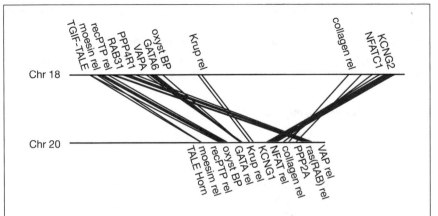

Figure 4.11 Duplicated blocks on chromosomes 18 and 20. The map compares a block of 36 Mb on chromosome 18 and 28 Mb on chromosome 20. Twelve of the 64 duplicated genes are shown. A large insert is present between Krup rel and collagen rel on chromosome 18. Apart from the insert it can be seen that there is a copy of each gene present on each chromosome. The order is not exactly conserved, but rearrangement of local gene order after duplication is commonly observed in chromosome evolution. (Reprinted with permission from Venter, J.C. *et al. Science*, **291**, 1304-1351. Copyright 2001 American Association for the Advancement of Science.)

Interspecies comparisons allow comprehensive maps of synteny to be constructed. Comparison with the mouse genome has led to the identification of 183 syntenic segments including some very large ones. For example the whole human chromosomes 17 and 20 are syntenic to segments on mouse chromosomes 13 and 2, respectively. Human and rodent lineages are thought to have diverged 100 million years ago, so there has been a combined total of 200 million years of evolution since then, giving an approximate overall rate of 1.0 rearrangement per million years of evolution. There is some evidence that the rodent genome shows an unusually high rate of rearrangement. Comparison of the human and zebrafish genomes identified 418 conserved segments. Human and zebrafish lineages diverged 400 million years ago, so the rate of rearrangement is about 0.5 rearrangements per million years of evolution, i.e. twice as high as in the rodent lineage. Although the genome sequences of other mammals are not yet available, much information can be gained from comparison of the cytogenetic, physical and genetic maps. Some startling conclusions emerge – for example, the organisation of the cat genome can be transformed into the human genome organisation by as few as 13 translocations. These and other similar observations suggest that the rate of rearrangements among some mammals may be as low as 0.2 rearrangements per million years of evolution.

4.8.2 Genes

The main interest of the human genome is clearly focused on genes which define our phenotype and whose malfunction leads to disease. The term 'gene' encompasses those regions which encode proteins and those regions that are transcribed into non-coding but functional RNA, such as rRNA, tRNA and other functions considered below. This section considers how these regions were identified, what they reveal about the parts list required to build the human organism and by comparison to the genomes of model organisms, what they reveal about evolution.

4.8.3 Identifying coding regions

Because coding sequences are interrupted by introns, it is not possible to recognise genes simply by searching for long ORFs. This makes the identification of human genes from the raw sequence data alone difficult and the results unreliable. It is necessary to supplement sequence analysis with experimental data and comparison with genes from other organisms. The size of the problem can be gauged using a database called RefSeq, that contains all the human genes for which the cDNA sequence has been determined in the piecemeal research that has been carried out over the years by a wide variety of different laboratories interested in different aspects of human biology and medicine. It contains about 10 000 sequences, which can be confidently annotated as genes. The sequence of RefSeq cDNAs can be mapped on genomic sequences to provide a description of a typical gene. An example of this is shown in Figure 4.8 for the *BRCA*2 gene. A typical gene will extend over 27 kb of the genome and contain a mean of 8.8 introns whose mean size is 145 bp. Some genes are much more extreme. The largest gene in terms of chromosomal coverage is the dystrophin gene covering 2.4 Mb on the X chromosome. The largest in terms of coding sequence is the titin gene, whose mRNA is 80 780 bp long, spliced together from 178 exons. Overall about 1.5% of the genome is coding sequence. The task of recognising genes is two-fold. Firstly, how can this 1.5% that constitutes the exons be identified? Secondly, once identified, how are the exons organised into genes? A further complication is differential splicing, which results in mRNAs containing different exons but derived from the same genomic region. Four types of evidence were used to identify exons and to determine which are part of the same gene:

1. Computer programs can be used that recognise exons on the basis of features such as length, ORFs, codon bias, splice sites (GT at the 5' splice site and AG at the 3' site). Such programs are said to work ***ab initio***, because they use only sequence and do not consider other types of experimental data. An example of such a program is GENSCAN, whose predictions can be viewed on the Ensembl website (see Figure 4.8). Because only a small proportion of the genome consists of exons, there is a very low signal-to-noise ratio and they are much less successful with human sequence than that of model organisms where exons are more easily recognised.
2. Exonic sequence will match the sequence of ESTs and cDNAs from human and other mammals. The major problem with this approach is that ESTs are contaminated with sequences derived from genomic

DNA or **hnRNA** that does not appear in spliced mRNA. It is usually necessary to guard against this by looking for evidence of splicing, i.e. the genomic sequence that matches the EST or cDNA is interrupted by sequences that correspond to introns. Matches to ESTs and cDNA are shown in the Ensembl database (see Figure 4.8).

3. Coding sequences tend to be evolutionarily conserved, whereas non-coding sequences are not (apart from regulatory regions). Thus the human sequence can be searched for segments that show conservation with other animals. The mouse and pufferfish genomes are used for this. The latter has a very compact genome, with exons constituting a much higher fraction. Areas of homology to the mouse genome are marked in the Ensembl genome browser (Figure 4.8); homologies to pufferfish DNA are marked on the UCSC genome browser (see Table 4.2 for URL).

4. The DNA sequence can be conceptually translated in all six reading frames (three on each strand). The resulting hypothetical proteins can be searched to see if they are similar to other known proteins from all other organisms or for amino acid motifs that are known to be characteristic of certain protein functions. Conceptual proteins that are recognisably similar to other proteins or contain known motifs are likely to be real.

Genes that have been reliably identified by sequencing of their cDNAs form a core set of known genes. In order to identify the remaining genes in the human genome the lines of evidence listed above were used in a structured search by both the IHGSC and Celera. The IHGSC used *ab initio* gene-finding programs to identify likely candidates, which they then examined for confirmation from ESTs, DNA or protein homology. The total of known genes and genes predicted in this way formed what they called the **Initial Gene Index** (IGI). Celera used a system which they called **Otto**, which marked the overall boundaries of likely genes using the experimental data and then constructed a model of gene structure with gene-finding programs. The degree of confidence in calling something a gene depends on the number of independent lines of supporting data. It should also be noted that genes will not be identified if they are novel, unique to humans, and not expressed in tissues used as the source of ESTs or cDNA libraries.

4.8.4 How many genes?

Because of the difficulty in gene identification it is not possible to completely answer the question of how many genes there are in the genome. There are 10 000 known genes in the RefSeq database of human cDNA sequences. According to Celera there are 26 000 genes that are either known or are predicted with two lines of supporting evidence and there are 39 100 genes with only one line of supporting evidence. The IHGSC has identified 24 500 genes and estimates that this represents about 60% of all genes, giving a likely total in the genome of 32 000. So both groups are in broad agreement, but there is still some way to go before there is a definitive list. The sequencing of chromosomes 21 and 22 has been completed to finished standard and extensive experimentation has been carried out to identify all coding sequences. A total of 539 confirmed genes have been identified on chromosome 22. Extrapolating to the whole genome, and adjusting for variations in gene density per chromosome (see below), this gives a global figure of 32–35 000. So overall it seems likely that the final list of human genes will be between 30 000 and 35 000. The number of different proteins is likely to be greater than this because of differential splicing. This results in the mRNA molecules with different combinations of exons arising from the same gene.

Celera calculated the overall average of gene density to be 9–14 genes/Mb or one gene every 71–111 kb, depending on whether gene number is based on predicted genes with either one or two lines of supporting evidence. Gene density is not uniform across the genome. We have already seen (Section 4.8.1.1) that gene density is higher in GC-rich regions. At the chromosomal level, chromosomes 17, 19 and 22 are gene rich, while chromosomes X, Y, 4, 13 and 18 are gene poor. The observed density of genes ranges from 5 to 11 genes/Mb on the Y chromosome, 6–10 on the least gene-dense autosomes to 23–29 per Mb on chromosome 19.

What is the significance of the estimated number of human genes? The yeast genome contains 6200 genes, the fly genome 13 000 and the worm genome 19 000. Thus it takes roughly 1.6, 2.6 and 5 times the number of genes to make a human compared to a worm, fly and yeast, respectively. Although an anthropocentric judgement, it is generally accepted that a human is far more complex than these organisms. Understanding how the encoded proteins combine to produce the final complex human form defines human research for the foreseeable future. We can make a start by examining the properties of these proteins and comparing them to the proteins encoded by these model organisms.

4.8.5 The human proteome

Just as the total aggregate of genetic material in an organism forms the genome, so the total sum of proteins encoded is known as the **proteome**. Much can be learnt about the likely functions of the proteome by comparing the predicted sequences of the proteins with those already studied, whether of human origin or from model organisms. The molecular function of a protein can be discerned if its sequence or parts of its sequence are similar to proteins whose function is known. On this basis, 58% of the proteins in the human proteome can be assigned a likely molecular function, a fraction that is similar for the yeast, fly and worm proteomes as well. Figure 4.12 summarises these functions.

The human proteome can be compared to the proteome of other organisms whose genomes have been sequenced. About 74% of human proteins are recognisably similar to proteins from other organisms. Conversely, 61% of the fly proteome, 43% of the worm proteome and 46% of the yeast proteome shows sequence similarity to proteins in the human proteome.

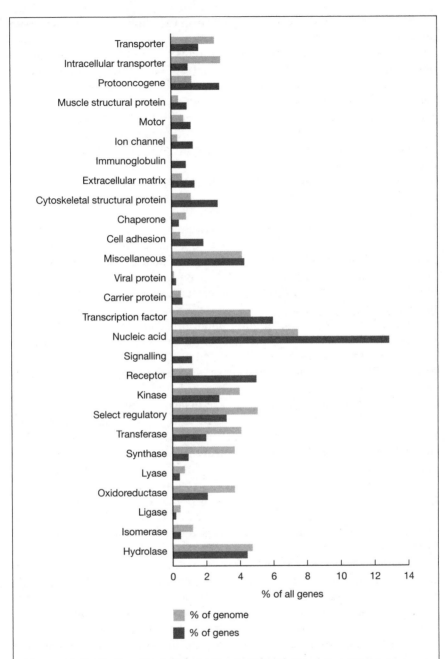

Figure 4.12 Distribution of molecular function in 26 000 known and predicted human genes and putative metazoan orthologs. Pale bars: the likely function of either known human genes or predicted genes where supporting evidence for the prediction was strong. Select regulatory refers to: i) Proteins involved in signal transduction such as heterotrimeric G proteins. ii) Cell cycle regulators. iii) proteins that modify kinases, phosphatases and G proteins. Dark bars: function of orthologs that are present in worm, fly and human genomes. (Data re-plotted from Venter J.C. *et al.* (2001) *Science*, **291**, 1304–1351.)

Similarity in the sequence of a protein in two different organisms arises because the encoding gene can be traced by descent to the common ancestor of the two organisms. Such proteins are said to be **orthologs**. An organism may have two or more genes that encode similar proteins because of gene duplication. Such proteins are said to be **paralogs**. The distinction between orthologs and paralogs is important because orthologs are likely to carry out similar functions in the two different organisms, whereas paralogs can diverge to acquire different functions. Often pairs of proteins showing sequence similarity are referred to as **homologs**. This term also implies descent from a common ancestor, but it is normally used in a way that does not attempt to distinguish between orthologs and paralogs. If the paralogs had already arisen in the common ancestor of two organisms, then the proteomes of both may contain the descendants of the paralogs. It can thus be difficult to distinguish between orthologs and paralogs. If the two proteomes each only contain one copy of a pair of homologous proteins then they are likely to be orthologs playing a similar role in each organism. Orthologs may also be recognised if a pair of proteins from the two different organisms are more similar to each other than to other paralogs in their respective proteomes.

When the yeast, fly, worm and human proteomes were compared, the IHGSC identified a set of 1308 proteins where there is at least one ortholog in each species. This set included cases where paralogs were present in one or more of the proteomes. When these were excluded, 564 proteins only contained one ortholog and no paralogs. This represents an evolutionarily conserved set of core proteins that are necessary for the operation of a eukaryote cell. A function could be assigned to 305 of these proteins. They mediate housekeeping functions in the cell such as metabolism, DNA replication, intracellular signalling, protein folding, membrane transport and the cytoskeleton. If the comparison is limited to multicellular animals – worm, fly and humans – a set of 1195 proteins can be identified where there is only one ortholog and no paralogs in each. The extra proteins are those that are required to organise cells into a multicellular organism and include proteins required for the complex intercellular signalling necessary for metazoans such as *src*-like tyrosine kinases.

A similar exercise by Celera identified a set of 1523 proteins that are each present in the human, fly and worm genomes and can be confidently defined as orthologs (this figure is higher than the one given above because it includes cases when paralogs are also present). The distribution of the functions of these conserved proteins is shown in Figure 4.12. When compared to the composition of the human proteome it can be seen that the following classes are over-represented in the conserved set:

- enzymes for nucleic synthesis and metabolism;

- enzymes involved in intermediary metabolism (transferases, oxidoreductases, ligases, lyases and isomerases);

- special regulatory molecules such as cell cycle regulators, heterotrimeric G-proteins, and protein kinases;

- protein transport and chaperones.

Analyses such as these do not identify all the conserved functions, because others will be mediated by members of protein families where the distinction between paralogs and orthologs can not be clearly defined. Nevertheless they offer us a glimpse of those functions conserved in evolution from the common ancestor of eukaryotes.

So far we have considered the functions of the human proteome that are conserved in the evolution of metazoans and eukaryotes in general. It is of equal interest to consider the nature of the innovations that marked the evolution of vertebrates, mammals and ultimately humans. About 7% of the human proteins and protein domains are only found in vertebrates. As might be expected, these are mainly made up of proteins involved in immune defence and the nervous system. So there has been relatively little innovation in the evolution of proteins with totally novel activities.

What has occurred is a great increase in the complexity of protein structure. Proteins have a modular structure in which domains with different activities are combined in different combinations to give a protein its final function. The human proteome contains more types of protein architecture, i.e. the linear order of domains in a protein, than the proteomes of the other eukaryotes. For example, a domain called the trypsin-like serine protease domain occurs with 18 other domain types in the human proteome to form proteins found in the complement fixation system, blood coagulation and fibrinolytic (clot dissolving) enzymes. This domain occurs with only eight other domains in the fly proteome, five in worm and one in yeast.

As well as an increase in protein complexity, there has been a selective expansion in protein families and domains associated with vertebrate specialisations, particularly with respect to immune function, intercellular signalling, olfactory receptors, hemostasis, apoptosis, neural function, splicing and translation. Some examples are given in Table 4.4. Expansion of protein families is not limited to human evolution; selective expansion can be seen in the proteomes of worm, fly and yeast genomes. Often there is evidence of independent expansion of the same family in the evolution of different groups of eukaryotes.

Table 4.4 Examples of expansions in protein families or domains in humans compared to other organisms. (Data from Venter, J.C. *et al.* (2001) *Science*, **291**, 1304-1351.)

Function	Example	Number of proteins			
		Human	Fly	Worm	Yeast
Intercellular signalling	Fibroblast growth factor	24	1	1	0
Immune defence	Immunoglobulin domain	765	140	64	0
Hemostasis	Matrix metalloprotease	19	2	7	0
Apoptosis	Calpain	22	4	11	1
Neural function	Voltage gated K⁺ alpha	33	5	11	0
Transcription factor	Zinc finger containing	607	232	78	28
	Homeotic	168	104	74	4

Therefore, although there are only 2–3 times more genes in the human genome compared to the fly and worm genomes, the biological complexity that can be programmed by this genome is greatly increased by novel combinations of protein domains and selective expansion of protein families. Moreover, about 40% of human genes show alternative splicing, which gives these genes the capability of producing a number of different proteins. Finally, complexity can arise from patterns of gene regulation both at the transcriptional and post-transcriptional level, chemical modifications of proteins after translation and regulation of protein stability. The expansion of proteins involved in transcription, splicing and translation will all contribute to an increase in the overall capacity for regulating gene expression.

4.9 Exploiting the human genome sequence

The determination of the human genome sequence will have a profound effect on the future direction of both fundamental and biomedical research. Knowing the complete sequence of the human genome, and the genome of other representative organisms, has led to the emergence of a new science called **genomics**. Genomics allows the properties of the genome to be considered holistically instead of one gene at a time. We have already seen how we can define the human proteome, which represents a complete list of proteins present in the human organism. The human proteome can be compared with that of other eukaryotes to define the minimum protein set needed to build a eukaryote cell or a metazoan animal and to investigate the innovations that arose in the evolution of vertebrates, mammals and humans.

As well as the proteome, we can define the human **transcriptome**, i.e. the sum total of all the RNA transcripts. We can investigate the range of transcripts produced in each tissue and how that varies during development or in response to environmental changes. We can also investigate the range of differently spliced transcripts to define the different proteins that can be expressed by each gene.

The function of each gene is investigated by **functional genomics**. Although many types of investigation are precluded using humans, 74% of human genes have a homologue in a model organism, where sophisticated molecular genetics can be used to investigate their function. Another way to investigate function is **structural genomics**, which aims to elucidate the three-dimensional structure of each protein. Proteins with no discernible sequence similarity may nevertheless have a similar structure and therefore function. So structure determination may help reveal the function of the 40% of genes whose sequence gives no clue as to their function. Determination of structure is carried out by X-ray crystallography, or for small proteins (<20 kDa), by NMR. Both methods require milligram quantities of purified protein. Expression of human genes in heterologous hosts such as bacteria, yeast or insect tissue will provide recombinant proteins for these studies.

Hopefully, it may not be necessary to determine the structure of every protein. As the database of determined structures expands, algorithms to convert primary sequence into secondary and tertiary structure may become efficient enough to allow reliable predictions of protein structure to be made from the primary sequence. Although long sought after by protein chemists, such algorithms are presently, at best, imperfect.

Thus we can anticipate listing each gene, its pattern of expression and the structure and function of the encoded protein. Some of this knowledge, such as expression profiles, may be available quite soon, as we shall see below. A complete description of structure and function may be some way off. Even when such a description can be made, it is wrong to assume that we will understand all aspects of human biology. The biological systems formed by these proteins will form complex non-linear networks and interactions, and we will need new tools to investigate them. Nevertheless, our understanding of human biology will clearly be greatly advanced.

Although we may look forward to long-term advancement of our understanding of human biology, advances in biomedicine may be more immediate. Some important information can be immediately gleaned from the sequence. Some genes already known to play a role in human disease have hitherto unsuspected paralogs that are also involved in disease. For example, a mutation in the *CNGA3* gene causes complete colour blindness (achromatopsia). Searches of the draft sequence revealed a homolog, *CNGB*, in which mutations caused achromatopsia in families that did not have mutations in *CNGA3*. More generally, as we shall see in Chapter 6, SNP maps are expected to play a crucial role in identifying alleles that contribute to the risk of complex disease. SNP may also be used in pharmacogenetics, which is concerned with understanding and predicting individual variation in the effectiveness of drugs and occurrence of adverse reactions.

In this section we shall investigate some of these issues in more detail. We shall start with oligonucleotide microarrays or DNA chips, an enabling technology that underpins many of the foreseeable applications of the genome sequence.

4.9.1 Oligonucleotide arrays

The photolithography techniques used in the construction of computer chips can be used in conjunction with solid phase oligonucleotide synthesis to build very precise, regular arrays of oligonucleotides in a very small surface area (Figure 4.13). As many as 10^6 different oligonucleotides can be arrayed on a 1.28×1.28 cm glass surface in regular 50 µm square cells. These figures represent the present state-of-the-art, but just as the power of computer chips continually increases, so the capacity of these arrays will also increase. The array is hybridised to a fluorescently labelled DNA probe and automated laser confocal microscopy is then used to determine in each cell of the array, whether hybridisation has occurred and to record the result (Figure 4.14). Different coloured fluorophores can be used to distinguish between control and experimental samples. This technique allows a very large number of hybridisations to be carried out in parallel. In the following sections we shall see how this technique can be applied to experimentally test *in silico* predictions of coding regions discussed in Section 4.8.3, investigate patterns of gene expression in different tissues and allow large-scale population surveys of sequence variation in SNPs. In Chapter 8 we shall see how DNA chips could be used in mutation detection.

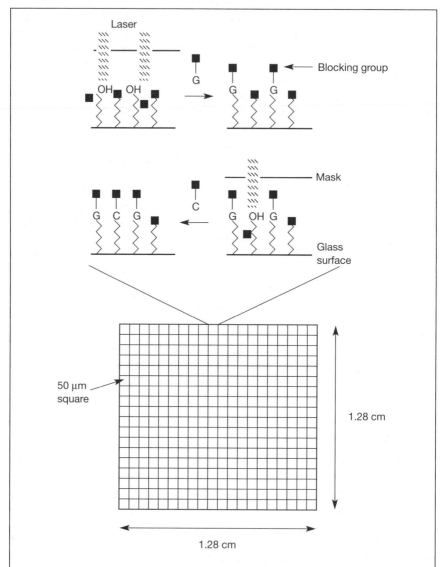

Figure 4.13 Construction of an oligonucleotide microarray (DNA chip). Solid-state oligonucleotide synthesis is carried out in cycles. In each cycle a light is used to remove a photolabile blocking group from a linker molecule attached to the glass substrate. A mask is used to direct the light to the selected cells and a protected nucleotide is coupled to the unprotected linker. The mask is then moved to initiate the next cycle where a different oligonucleotide will be coupled.

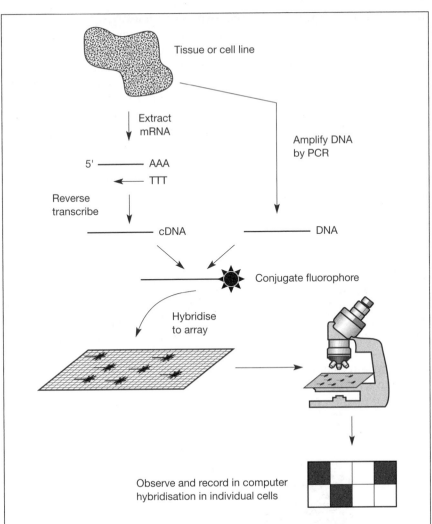

Figure 4.14 Hybridisation of fluorescently labelled probes to an oligonucleotide array.
Genomic DNA probes are amplified by PCR, RNA probes are prepared by RT-PCR from mRNA.
They are conjugated to fluorophores that are excited at particular wavelengths by lasers and
emit or fluoresce at a different wavelength, resulting in a characteristic colour. Commonly
used fluorophores are cy3 (fluoresces red), fluorescein and cy5 (both fluoresce green).
Because of the limitations of colour availability, blue is used in this diagram as a generic
representation of fluorescence. The labelled probes are allowed to hybridise to the array and
are then examined with a laser scanning confocal microscope. This illuminates the array with
laser light filtered to the excitation wavelength of the fluorophore. The emitted light in each
cell is then recorded and can be quantified.

The sequence of the human genome

4.9.1.1 Identifying DNA sequences expressed in mRNA

As discussed in Section 4.7.3, various lines of evidence can be used to predict the locations of exons. The existence of some of these putative genes is supported by more than one line of evidence, other predictions are more tentative, and there is quite a large number of predictions made by computer programs analysing sequence *ab initio* that have no supporting experimental evidence. It is clearly a priority to test the veracity of each prediction and to investigate whether there are any other expressed regions that the gene-calling programs have missed. To do this, arrays consisting of oligonucleotides from each predicted exon are constructed and hybridised to fluorescently labelled cDNA prepared from cell lines. Because of the high density of oligonucleotides, it is possible to survey the whole genome with a relatively small number of arrays. Figure 4.15 shows how this works in more detail to experimentally test gene predictions made by the IHGSC.

This experiment successfully identified about two-thirds of the experimentally verified exons and about one-third of exons predicted only on the basis of *ab initio* analysis. The experiment described in Figure 4.15 only used mRNA from two tissue culture cell lines. The limitation of this analysis are the mRNAs that are only expressed in particular tissues, perhaps at limited times during development. This may be one of the main reasons why only two-thirds of experimentally verified exons were recognised. Nevertheless it demonstrates that in principle the complement of human genes can be efficiently investigated using oligonucleotide arrays. In addition, it suggests that a substantial proportion of exons, predicted solely on the basis of sequence analysis, do in fact exist. This was actually a surprising result, because there had been considerable scepticism about the performance of these programs.

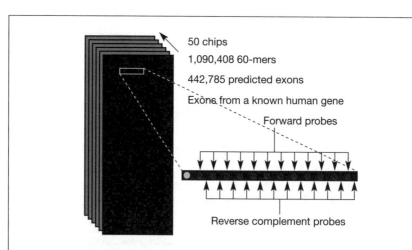

Figure 4.15 Testing whether predicted exons are expressed as mRNA. 60bp oligonucleotides are synthesised from each exon. As control for non-specific hybridisation, a reverse complement for each oligonucleotide is also included in the array. mRNA was prepared from two cell lines, reverse transcribed to cDNA and labelled with either a red (cy3) or green fluorophore (cy5). (Redrawn with permission from Shoemaker, D.D. *et al.* (2001) *Nature*, **409**, 922-927.)

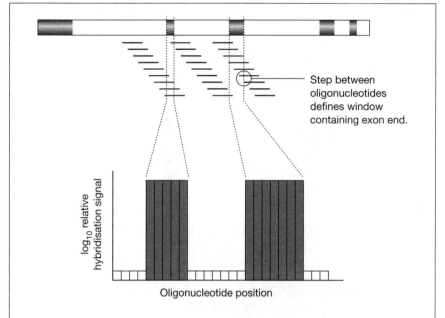

Figure 4.16 Use of a tiling array to identify and define exons. 60-mer oligonucleotides are synthesised so that the start of each oligonucleotide is staggered by 10 bp forming a tiling array that spans a genomic region. Labelled cDNA is hybridised to the array. Only oligonucleotides that partially or completely overlap an exon (shown in blue) will hybridise. The beginning and end of the exons will be defined to within the 10 bp step within each nucleotide. The exact borders of the exon will be evident from the location of the GT/AG rule.

As the experimental verification of exons of the human proceeds, the results can be used to train the computer programs to become more accurate. As well as genome-wide surveys of expressed sequences, oligonucleotide arrays can be used to investigate the detailed structure of individual genes. A series of overlapping oligonucleotides, called a **tiling array**, is synthesised. These scan a sequence to detect the start and stop of exons (Figure 4.16). Oligoncueotide arrays can also be constructed to investigate whether there are any additional genes that have been missed in the initial analysis. This would involve the synthesis of oligonucleotides from unique or single copy sequence that have not so far been predicted to be expressed.

4.9.2 Expression profiling

Once the identity of genes and their boundaries has been established, oligonucleotide arrays can be used to investigate the pattern of gene expression. Oligonucleotides from each exon are synthesised in an array and hybridised to fluorescently labelled cDNA prepared from different issues. The pattern of hybridisation reveals which genes were represented in the mRNA pool of each cell type. Figure 4.17 shows an example in which the expression of all the genes on chromosome 22 was investigated in 59 separate cell types (heart, trachea, lymph node, cerebellum, foetal brain, foetal

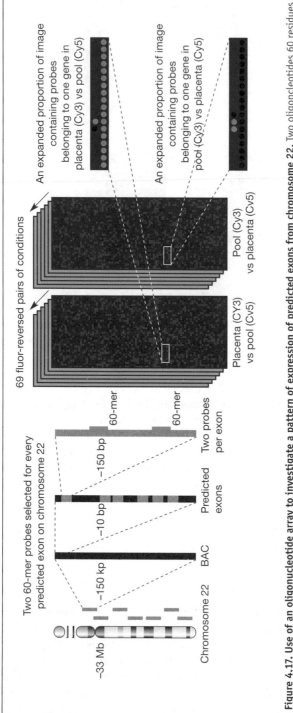

Figure 4.17. Use of an oligonucleotide array to investigate a pattern of expression of predicted exons from chromosome 22. Two oligoncleotides 60 residues (60-mers) long are synthesised from each experimentally verified exon. The arrays are hybridised to mRNAs extracted from 59 cell types, reverse transcribed into cDNA and labelled with either Cy3 or Cy5 fluorophores. As a positive control a pool of cDNA is prepared from a mixture of de-differentiated cell lines that are expected to express most genes. In the experiment shown the placenta cDNA is labelled green with Cy3 and the pool is labelled red with Cy5. The experiment was repeated with the labelled fluorophores reversed to control for any effect on hybridisation that may be caused by the labelling of cDNA with the fluorophore. (Redrawn with permission form Shoemaker, D.D. *et al.* (2001) *Nature,* **409**, 922–927.)

heart etc.). The experiment illustrates the identification of an exon that is preferentially expressed in placenta cells. Experiments of this type can also be very useful to decide which exons belong to which genes. Adjacent exons that belong to the same gene will appear co-ordinately regulated in different tissues, since they are part of the same mRNA molecule. Conversely, lack of coordinate regulation experimentally assigns two adjacent exons to separate genes. This picture may be complicated by differential splicing, which results in different exons from a gene remaining in the mRNA in different tissues.

4.9.3 Detecting sequence variation

As discussed in Section 4.5, a large number of SNPs were identified during the course of genome sequencing. These will be used in population surveys to map alleles contributing to the risk of complex disease (see Chapter 6). SNPs are also very important in the newly emerging science of pharmaco-genetics (see below). Both applications of SNPs will require technology that can efficiently and cheaply detect genetic variation in a large number of experimental subjects at a large number of loci. Oligonucleotide arrays are ideally suited to this purpose. A related application is the detection of mutation in genes responsible for single-gene disorders where there are no common disease-causing alleles; we shall see an example of detecting mutations using oligonucleotide arrays in Chapter 8.

Oligonucleotide arrays can be used to detect sequence variation, because a mismatch between the bound oligonucleotide and the fluores-cently labelled probe will reduce or abolish hybridisation. This can be exploited in two separate ways to detect a nucleotide that differs from the reference genome sequence:

- A gain of hybridisation in one cell of a subarray of oligonucleotides that contains all possible sequence variants at the position being investigated.

- A loss of hybridisation of the experimental probe hybridised in competition to a reference probe that matches the reference sequence.

In both cases a series of overlapping oligonucleotides, called (a **tiling array**) can be used to scan a particular sequence for variants. The way the two different approaches operate is illustrated in Figure 4.18. Alternatively, the array can be constructed to scan for variants at a large number of separate sites that might show variation. For example the set of 1.4 million mapped SNPs could be arrayed and used in a single experiment to determine the genotype of each polymorphism in an experimental subject.

An exciting new development is to use arrays to literally re-sequence DNA on the chip (Figure 4.19). A tiling array of oligonucleotides is synthe-sised to leave a free 3'OH. This then serves as a primer for *in vitro* DNA synthesis using the DNA from the subject as a template and four dideoxynu-cleotides, each labelled with a different colour. The dideoxynucleotides

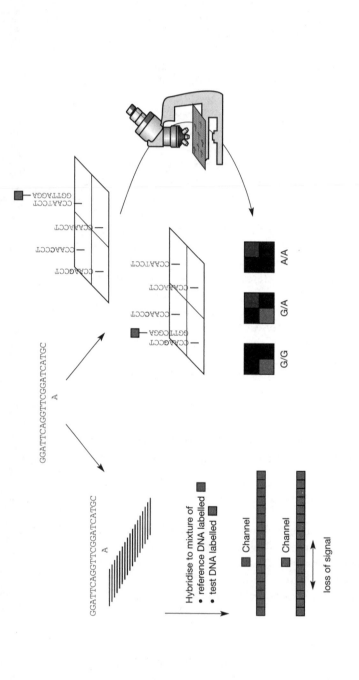

Figure 4.18 Detection of sequence variation by oligonucleotide arrays. At the top of the diagram a polymorphic sequence is depicted where C in the reference sequence may be replaced by A. On the left the variant is detected by loss of the hybridisation signal when the test DNA (fluorescently labelled blue) is hybridised in the presence of competitor reference DNA (labelled grey). Oligonucleotides in the tiling array that overlap the variant site (black lines) will all hybridise less effectively than the reference DNA, whereas oligonucleotides that do not overlap the variant site (blue lines) will hybridise effectively. This will be visible as a lack of fluorescence in the cells containing the relevant oligonucleotides. When the microscope filters are set to detect the grey fluorescence from the reference DNA (grey channel) all cells will show fluorescence. On the right the variant is detected by a change in hybridisation signal. For each nucleotide in the reference sequence a set of oligonucleotides is synthesised, which contains each of the four possible nucleotides at the position being interrogated (shown as a blue letter). The reference DNA will hybridise to the cell containing G at the relevant position, whereas the variant hybridises to the cell containing T. DNA from a heterozygote will hybridise to both cells. In practice the array interrogating each position can include oligonucleotides that will detect deletion and insertion of a nucleotide. This is illustrated in Chapter 8 for the detection of a mutation in the *BRCA1* gene.

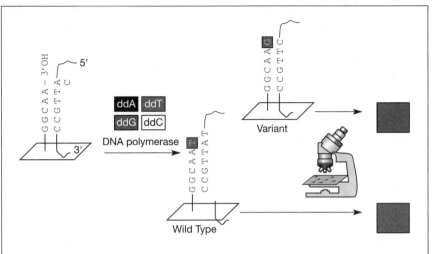

Figure 4.19 Determining DNA sequence on a chip using a single nucleotide extension reaction. The primer is attached to the chip surface at its 5′ end. This leaves the 3′OH end free to be extended after annealing to the template being interrogated. Four dideoxynucleotide terminators are available for the extension, each tagged with a different coloured fluorescent dye. The base incorporated is determined by examining the cell on the chip with a fluorescence microscope. In this example the results from two different DNA sequences are shown. The wild type or reference sequence incorporates a T residue opposite the A in the template, and is coloured grey. A variant with C at the interrogated position incorporates a blue coloured G residue.

prevent the synthesis proceeding beyond the addition of the next base in the sequence, whose colour will reveal its identity. There are serious plans to develop this technology so that individual human genomes can be re-sequenced in a couple of days.

4.9.4 Pharmacogenetics and pharmacogenomics

Much of the impetus for sequencing the human genome has come from pharmaceutical companies in the expectation that the information will aid the discovery and development of new drugs and allow drug treatment regimes to be personalised.

4.9.4.1 Identification of new drug targets

Traditionally, drug discovery depends upon the identification of targets that a small chemical drug either activates (agonist) or inhibits (antagonist). Large combinatorial libraries of chemicals are examined in high throughput screens. **Hits** in the initial screen are then examined more closely to investigate their effect in detail and to modify the chemical to improve its effectiveness. Such optimised molecules are described as **leads**. There are about 500 targets for existing drugs and drug discovery programs. For the most part these consist of either enzymes or G-coupled protein receptors that bind extracellular ligands and induce a specific cellular response. The targets are chosen either because they are already the targets for existing successful drugs or on the basis of the molecular pathology of the disease. Availability of drug targets is a major limitation in the development of new drugs. It has been estimated that there are about 5000 targets for common diseases for which new drugs are sought. It is hoped that genomics will aid the discovery of new targets in three different ways.

- As described in Chapter 6, it will aid the discovery of susceptibility alleles for complex diseases. The protein encoded by the susceptibility allele may itself be a target for pharmaceutical intervention. If it is not directly suitable as a target, investigation of its role may reveal other targets which are part of the same biochemical pathway or interact in some other way.

- Examination of the proteome may reveal targets that are paralogs of existing targets. For example, a receptor for the neurotransmitter serotonin is known to be encoded by the 5-HT_{3A} gene. However, although the 5-HT_{3A} protein is functional *in vitro*, the conductance of the ligand-gated channel it controls is lower than that measured *in vivo*. A homolog, called 5-HT_{3B}, was detected in the draft sequence. A heterodimer of 5-HT_{3A} and 5-HT_{3B} functions *in vitro* with identical properties as the *in vivo* receptor. Serotonin plays a major role in mood disorders and schizophrenia, so clearly the heterodimer will be a much improved target to use in screens for new drugs. The IHGSC identified 18 possible novel paralogs of existing drug targets in the human proteome.

- Expression profiling using microarrays can be used to investigate the pattern of gene expression in diseased and healthy tissues, including cancers. Genes whose expression is different in diseased tissues could be involved in the pathogenesis and thus could be potential targets. Clearly, expert biological knowledge will need to be to applied to evaluate the likely relevance of such potential targets. A major problem is to decide whether the change in gene expression is a cause or a consequence of the disease. Deciding that a potential target is in fact useful is known as **validation**.

There is considerable debate as to which of these approaches is more useful. Proponents of susceptibility alleles argue that targets revealed in this way are automatically validated and so this approach avoids the expensive and time-consuming mistake of undertaking a high throughput screen with an invalid target. However, as we shall see in Chapter 6, the identification of susceptibility alleles is still by no means straightforward.

4.9.4.2 Individual variation to drug responses

It is a commonplace observation that many, or indeed most drugs, show great variation in effectiveness, toxicity and adverse side-effects when administered to different individuals. Much of this variation is genetically determined. Pharmacogenetics is the study of this variation at the level of the single gene, while pharmacogenomics is concerned with the analysis of the variation at a genome-wide level.

Individual variation in the outcome of drug treatment arises from the interplay of a number of separate factors:

- inactivation by oxidation by cytochrome P450s;
- in some cases oxidation by cytochrome P450s results in drug activation;
- conjugation to water-soluble moieties by enzymes such as glutathione-S-transferase leading to their clearance from the bloodstream through the kidney;
- target sensitivity;
- toxicity;
- heterogeneity of disease mechanisms.

All of these factors are subject to individual variation that is under genetic control, and often the alleles can be associated with SNPs. Some examples are given in Box 4.1. The complexity of the interaction between these factors can be seen by considering the following possible outcomes when a drug is administered to a group of patients:

- In the first patient the drug is inactivated and cleared from the blood-stream slowly and the target is sensitive to the drug – the drug is therefore effective at low doses.

- In a second patient the target is sensitive but the drug is rapidly inactivated and/or cleared from the bloodstream. Any favourable effect is transient at best. However, if the frequency or size of the dose is increased the drug may work.

- A third patient clears the drug slowly and is very sensitive to toxic side-effects. There is a very adverse outcome to the drug treatment.

- The molecular cause of the disease in the fourth patient is different, the target of the drug is not relevant to the disease and there is no benefit from the drug treatment. A drug with a different target would have been effective. The patient is suffering because of the wrong choice of treatment.

- In a fifth patient, the drug is relevant to the particular form of the disease, but the target is insensitive. There is no benefit from the treatment.

BOX 4.1: EXAMPLES OF INDIVIDUAL VARIATION IN DRUG TREATMENT RESPONSE

Inactivation

Drugs are subject to enzymic modification, particularly by the cytochrome P450 superfamily found in the liver and small intestine. These enzymes catalyse the oxidation, peroxidation or reduction of a wide variety of drugs, as well as protoxins, procarcinogens and natural endogenous molecules. Large variations in P450s levels between individuals affect the activity and half-life of a wide variety of drugs as well as contributing to cancer susceptibility through the activation of carcinogens in cigarette smoke or the modification of oestrogen to promote oestrogen-mediated cancers. Cytochrome P450s are encoded by 55 different CYP genes. These are divided into subfamilies on the basis of sequence homology. The most abundant P450s are encoded by members of the CYP3A family, which consists of CYP3A itself, as well as CYP3A4 and CYP3A5. This family is responsible for inactivation of HIV protease inhibitors, calcium channel blockers, cholesterol lowering drugs, cancer chemotherapeutics and immunosuppressants such as cyclosporin and FK506. CYP3A5 accounts for 50% of the activity of this family. Recently it has been shown that individual variation in CYP3A5 levels is due to SNPs that result in differential splicing. *CYP3A5*3* and *CYP3A5*6* alleles cause the inclusion in the mRNA of an exon carrying a nonsense mutation. Only individuals with the *CYP3A5*1* allele produce a full-length, active protein. This is just one example of many, where individual variation in drug clearance can be traced to single allelic differences.

Activation

Although oxidation by cytochrome P450s is normally inactivating, sometimes it is required for activation. A well-known example of this variation is the effectiveness of opiates such as codeine, which is activated by a cytochrome P450 called CYP2D6. Between 3 and 10% of Caucasians are so-called poor metabolisers due to recessive alleles in this gene and thus fail to gain pain relief from this analgesic. Only 1–2% of African–Americans are poor metabolisers, a common observation that genetic variation to drug response is strongly affected by ethnicity. Interestingly, poor metabolisers have low pain thresholds because of a defect in synthesising endogenous morphine (endorphins).

Target sensitivity

Opiates bind to the μ opioid receptor which is the primary site of action for most pain-relieving drugs. An SNP is located at position 118 in the polypepetide chain affecting a glycosylation site. The less common allele is present at a frequency of about 10% in the population, although it varies in different ethnic groups. The less common allele has the same response to exogenous opiate drugs, but shows a three-fold elevated response to β-endorphin. This variation could be significant in the causation of drug addiction and is an example of genetically determined variation in drug targets.

Continued

Another example of variation of drug targets is the β_2-adrenergic receptor. Agonists of this receptor are widely used in asthma treatment. Several SNPs have been described in the β_2-adrenergic receptor gene. One is responsible for a five-fold difference in the likelihood of response to the anti-asthmatic drug albuterol.

Toxicity

During the Second World War, American troops serving in tropical regions were given anti-malarial drugs such as primaquine. African–American soldiers were found to be much more liable to develop haemolytic anaemia. The cause was shown to be X-linked mutations leading to a deficiency of glucose-6-phosphate dehydrogenase, which affect about 10% of African–Americans. Ironically female carriers of G6PD deficiency are protected against malaria, which is why such mutations are prevalent in the African–American gene pool.

Heterogeneity of disease mechanisms

In Chapter 6 we shall see that the allele ApoE*ε4 increases the risk of Alzheimer's disease. However, many Alzheimer's patients do not carry the ApoE*ε4 allele. As well as conferring susceptibility, this allele affects the response to tacrine, a drug that is used to treat Alzheimer's disease: 80% of patients who did not carry the ApoE*ε4 allele improved after tacrine treatment, whereas 60% of patients who did carry the allele actually deteriorated. As susceptibility alleles are documented to complex diseases this type of observation may become more common.

It is not difficult to see why finding the right drug and dose regime is often more a matter of trial and error than rational evaluation. SNP profiling with oligonucleotide arrays provides the promise of personalised drug treatment regimes. Figure 4.20 provides a hypothetical example of how this might work. A patient is suffering from lymphoblastic leukaemia. In order to design a rational drug treatment the pharmacogeneticist would need information on the following:

- What changes have occurred in genes involved in the cancer? The genotype of the lymphoblastic cells is already known to influence an important prognostic indicator and is therefore a guide as to the intensity and aggression required of the chemotherapy.

- What is the drug sensitivity status of likely targets such as ion channels and cell surface receptors?

- What is the drug metabolism profile of deactivating and clearing enzymes such as cytochrome P450s and GST?

- What is the likelihood of toxicity to the cardiovascular and endocrine systems?

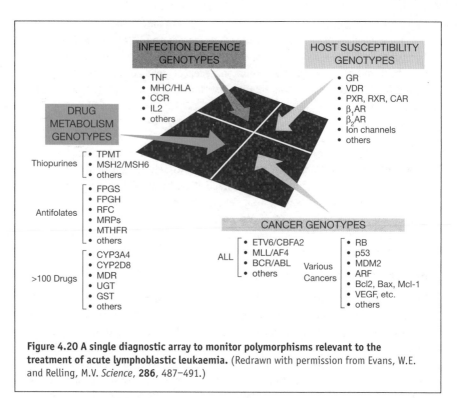

Figure 4.20 A single diagnostic array to monitor polymorphisms relevant to the treatment of acute lymphoblastic leukaemia. (Redrawn with permission from Evans, W.E. and Relling, M.V. *Science*, **286**, 487–491.)

● What is the status of the host immune system in providing defence against invading pathogens to which the patient will be vulnerable once the treatment starts?

A single chip will provide information on the status of SNPs and gene expression profiles relevant to these questions (Figure 4.20). Armed with this information, an individual drug regime can be designed that is most effective against the properties of the particular cancer, minimises the risk of adverse side-effects, optimises dose and avoids the use of drugs that will be rapidly inactivated or to which the cancer cells will be resistant. As shown in Box 4.1, many of the genetic polymorphisms required to carry out this exercise are already characterised. Many more will be discovered by empirical screens of SNP haplotypes correlated to observation of the outcome of the treatment. Personalised medicine may be one of the first practical benefits of the human genome sequence and associated technology.

A related application may be in the development of new drugs. Often the initial trials of drug effectiveness and safety (phase II trials) will reveal that a drug is effective, but only in a minority of cases. Alternatively, they may reveal serious adverse side-effects, but again only in some cases. If SNP profiles are collected at the same time as the phase II trials and correlated with the outcome of the drug treatment, then it may be possible to build up a profile of SNP haplotypes of the likely non-responders or those prone to the adverse side-effects. This information may then be used to select patients for the large-scale phase III trials. In this way the experimental treatment will be targeted at those most likely to benefit and least likely to suffer adverse side effects. This will result in the eventual definition of a set of patients for which the drug should be used. Although extensive profiling will be necessary at the trial stage, only a few SNP sites will turn out to be predictive, so the test for drug suitability could be quite simple. Health managers are often reluctant to allow the use of drugs where a limited proportion of patients benefit. Personalised medicine will select the patients who will benefit, so the cost/benefit equation will be transformed. Furthermore, there are many medicines that are generally effective but cannot be used because of severe adverse side-effects in a few patients. Again if these can be accurately and easily predicted in advance it may extend the repertoire of drugs that are available for use. There is some ethical concern about the use of drugs that may be ineffective or even dangerous to a group of patients, because, in the event of inadequate prior testing, they could be used on the wrong patients. This is clearly an area that will need stringent regulation.

4.10 Summary

- The human genome has now been sequenced to draft standard, i.e. about 90% of the sequence has been determined with an accuracy of less than 1 error in 10 000 bp and 95% of the sequence with an error rate of less than 1 error in 1000 bp. There are still gaps, mostly consisting of repeated DNA sequence. By 2003 the quality will be improved to finished standard, where all the sequence is determined to an accuracy of less than 1 error in 10 000 bp and gaps are kept to a minimum.

- The sequencing was a race between an international consortium of publicly funded laboratories called the IHGSC and a private company called Celera Genomics. The IHGSC took a clone-by-clone approach. Celera shotgun sequenced the whole genome in one operation.

- The complete sequence allows an overall view of genome structure and evolution. This allows:
  - The long-range variation of GC content to be determined. Genes are more dense in GC-rich regions
  - The total number of GC islands to be obtained
  - A complete census of transposable repetitive elements. Most transposable repetitive elements are no longer active in the human genome.
  - Identification of duplicated blocks of chromosomes
  - Comparison with the mouse and other vertebrate genomes to identify syntenic regions.

- The total number of genes is not yet completely known, but both IHGSC and Celera conclude that there are between 30 000 and 35 000. This is much lower than previous estimates.

- Conceptual translation allows the amino acid sequence of the encoded proteins to be determined. The total sum of encoded proteins is called the proteome and it constitutes a parts list to make a human organism. Comparison with other proteins allows the function of about 60% of the proteins to be guessed.

- Comparison of the yeast, worm, fly and human proteomes allows the evolution of proteins to be studied and identifies a core group of proteins that are present in all eukaryotes and a second group that is common to all metazoans.

- Determination of the nucleotide sequence of entire genomes has established a new science of genomics, which analyses genomes holistically. Functional genomics studies the functions of genes and structural genomics aims to elucidate the structure of the encoded proteins.

- Exploitation of the genome will be based on oligonucleotide arrays. These can be used to:

  ○ Define the extent of DNA sequences transcribed into mRNA, i.e. define the position and boundaries of genes.
  ○ Study the expression of all genes in the genome in one operation.
  ○ Carry out mass screening of sequence variants such as SNPs or the status of alleles that determine the response to drug treatments.

Further reading

Sequencing the human genome

THE INTERNATIONAL HUMAN GENOME CONSORTIUM (2001) Initial sequencing and analysis of the human genome. *Nature*, **409**, 860–921.

VENTER, J.C., ADAMS, M.D., MYERS, E.W., *et al*. (2001) The sequence of the human genome. *Science*, **291**, 1304–1351.

These two papers describe the sequencing strategy and analysis of the draft sequence of the human genome. Other papers in the same issues of *Science* and *Nature* provide further commentary and analysis.

GREEN, P. (1997) Against a whole-genome shotgun. *Genome Research*, **7**, 410–417.

WEBER, J.L. and MYERS, E.W. (1997) Human whole-genome shotgun sequencing. *Genome Research*, **7**, 401–409.

The Weber and Myers paper proposes the strategy for shotgun sequencing used by Celera and presents computer simulations to show that it would work. The Green paper marshals the arguments against this strategy.

SHOEMAKER, D.D., SCHADT, E.E., ARMOUR, C.D. *et al*. (2001) Experimental annotation of the human genome using microarray technology. *Nature*, **409**, 922–927.

Describes the use of oligonucleotide microarrays to experimentally verify gene predictions.

SIMPSON, A.J.G., DE SOUZA, S.J., CAMARGO, A.A. and BRENTANI, R.R. (2001) Definition of gene content of the human genome: the need for deep experimental verification. *Comparative and Functional Genomics*, **2**, 169–175.

Reviews the difficulties in identifying genes in the draft sequence.

Construction and use of oligonucleotide arrays

HACIA, J.G. (1999) Resequencing and mutational analysis using oligonucleotide microarrays. *Nature Genetics*, **21**, 42–47.

LIPSHUTZ, R.J., FODOR, S.P.A., GINGERAS, T.R. and LOCKHART, D.J. (1999) High density synthetic oligonucleotide arrays. *Nature Genetics*, **21**, 20–24.

These two papers review the basic technology for the construction and use of oligonucleotide arrays.

CHEN, J.W., IANNONE, M.A., LI, M.S. *et al.* (2000) A microsphere-based assay for multiplexed single nucleotide polymorphism analysis using single base chain extension. *Genome Research*, **10**, 549–557.

HACIA, J.G., FAN, J.B., RYDER, O. *et al.* (1999) Determination of ancestral alleles for human single-nucleotide polymorphisms using high-density oligonucleotide arrays. *Nature Genetics*, **22**, 164–167.

Pharmacogenetics and pharmacogenomics

EVANS, W.E. and RELLING, M.V. (1999) Pharmacogenomics: Translating functional genomics into rational therapeutics. *Science*, **286**, 487–491.

MARCH, R. (2000) Pharmacogenomics: the genomics of drug response. *Yeast*, **17**, 16–21.

ROSES, A.D. (2000) Pharmacogenetics and the practice of medicine. *Nature*, **405**, 857–865.

These are excellent reviews of pharmacogenetics and pharmacogenomics. They discuss the use of oligonucleotide arrays in mapping complex diseases and predicting individual outcomes of drug treatment. They present many examples of polymorphisms that affect the stability, effectiveness and toxicity of drugs.

DREWS, J. (2000) Drug discovery: A historical perspective. *Science*, **287**, 1960–1964.

Describes the drug development process and explains how the figure of 5000 possible drugs targets is derived.

BOND, C., LaFORGE, K.S., TIAN, M.T. *et al.* (1998) Single-nucleotide polymorphism in the human mu opioid receptor gene alters beta-endorphin binding and activity: Possible implications for opiate addiction. *Proceedings of the National Academy of Sciences USA*, **95**, 9608–9613.

KUEHL, P., ZHANG, J., LIN, Y. *et al.* (2001) Sequence diversity in CYP3A promoters and characterization of the genetic basis of polymorphic CYP3A5 expression. *Nature Genetics*, **27**, 383–391.

These two papers are references to the primary literature for two cases cited in the chapter.

CHAPTER FIVE

Single-gene disorders

Key topics

- Cloning disease genes
- *CFTR* gene
- Dystrophin gene
- Trinucleotide repeat expansions
- Haemoglobinopathies
- Inherited predisposition to cancer
  - Retinoblastoma and the tumour suppressor hypothesis
  - Li–Fraumeni syndrome
  - Hereditary breast cancer: *BRCA*1 and *BRCA*2
  - Colorectal cancer: FAP and HNPCC
  - Neurofibromatosis
  - Ataxia telangiectasia

5.1 Introduction

Cloning human genes can be immensely difficult. In some cases it has required large-scale collaborations between many hundreds of scientists and has taken years to accomplish. Nevertheless, the potential rewards are equally high and it has transformed research into many inherited diseases. Often, cloning a gene makes it possible to know for the first time the nature of the protein whose malfunction causes a disease. The type of mutation in affected individuals may be determined, leading to diagnostic tests for the presence of the mutation in families at risk. Finally, it may become possible to cure the disease by **gene therapy**, in which a functioning copy of the gene is introduced to complement the defective gene. This chapter considers how genes are cloned and examines what we learn about the nature of some of the major monogenic diseases by cloning the genes responsible.

and progress towards gene therapy in Chapter 7.

The amino acid sequence of the protein encoded by the gene can be deduced from the nucleotide sequence of the gene. The amino acid sequence may show similarities to other proteins that have been studied either directly in humans or in model organisms such as bacteria, yeast, roundworm, *Drosophila* and mice. This will provide important clues to its cellular function and illuminate the molecular pathology arising from its malfunction. Knowing about the molecular defect helps research to develop better and more rational therapies for the disease.

The nature of mutations can also provide important clues about the protein's function. Frameshift, nonsense and deletion mutations are likely to prevent any active protein being made. We may therefore deduce that the defect arises through a lack of function. Lack of function is not the only way in which a gene may be damaged. Sometimes mutations occur that lead to an increase in activity, perhaps through the loss of regulation, so that the protein is active in inappropriate circumstances. We shall also see that cloning genes led to the discovery of a totally unexpected and previously unimagined type of mutation called a **trinucleotide repeat expansion.**

Identification of common mutations can be used to develop diagnostic tests for the presence of a damaged gene in families at risk. One mutation may be very common. Sometimes these are **founder mutations,** which can be shown to have occurred in a common ancestor. Diagnostic tests for these mutations are usually straightforward. In other cases, many different mutations may occur and diagnostic tests are more difficult.

5.2 Cloning disease genes

If the protein encoded by a gene has been characterised, a number of methods are available to identify the correct clone in a gene library. Firstly, a partial amino acid sequence may be used to predict the nucleotide sequence of part of the gene. A synthetic oligonucleotide may then be synthesised to use as a hybridisation probe to recognise a clone with cDNA sequences. Alternatively, the protein can be used to raise an antibody. This can be used to screen a **cDNA library** constructed in an expression vector to ensure the encoded protein is expressed in the host. Such methods have been used to clone important genes, such as that encoding human blood clotting factor XIII where previous research had managed to purify some of the protein.

In most cases, however, the protein has not been characterised prior to cloning the gene that encodes it. So how can the gene be cloned in the first place? The only information that may be used to find the gene is its position in the genetic map. This is exploited in an approach known as positional cloning, which has been the method used to clone many genes in the last few years.

In positional cloning, families in which the disease is segregating are studied to detect linkage with polymorphic genetic markers. This involves examining a large number of markers to find one where a particular allele and the occurrence of disease are co-inherited in the family. Such linkage suggests that the marker and the disease gene are in close proximity so that the frequency of meiotic recombination has been reduced. The lower the recombination frequency the closer together they are.

Another important strategy for locating the chromosome region of a gene is to look for cytogenetic rearrangements associated with the disease, such as deletions, inversions and translocations. These events require that the chromosomes are physically broken. A breakpoint within a gene will often result in its inactivation. Using Giemsa and Q-bands (see Chapter 2) it is possible to map the breakpoints and therefore the gene affected. A translocation event was very helpful in locating the DMD gene to Xp21 (Figure 5.1).

Linkage analysis or cytogenetic rearrangements narrow down the location of a gene to a particular chromosomal region. But how is the gene itself to be recognised and isolated? A meiotic recombination fraction of 0.01

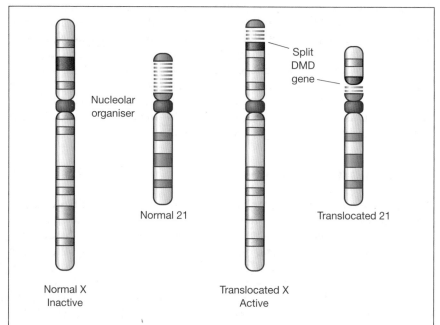

Nucleolar organiser

Normal 21

Split DMD gene

Translocated 21

Normal X
Inactive

Translocated X
Active

Figure 5.1 Localisation of the DMD gene to Xp21 using a translocation. A female patient suffering from DMD was found to have a translocation between the X chromosome and chromosome 21. The breakpoint on chromosome 21 was within the 28S ribosomal locus. The breakpoint on the X chromosome was at the cytogenetic band Xp21. Although there was still a normal X chromosome, DMD resulted because of unfavourable Lyonisation, i.e. in most of her cells the normal X chromosome was inactivated. The reason for this is not clear, but it could be connected with some form of dosage compensation: if the translocated X chromosome was inactivated there would be two active copies of the X chromosome between Xp21 and the telomere of the p arm (one on the X chromosome and one on the translocated 21 chromosome). Probing a gene library made from this individual with a 28S rDNA probe yielded a junction clone called XJ. This clone contained sequences from the Xp21 region, which proved to be closely linked to DMD mutations.

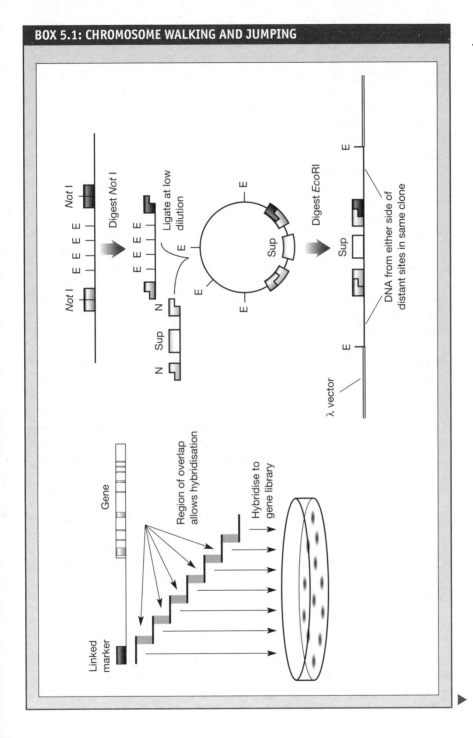

Chromosome walking (shown left) is a way of moving along the chromosome from a linked genetic marker to a target gene. The DNA forming the genetic marker, e.g. a probe detecting an RFLP, is used as a probe in a gene library constructed so that the clones overlap. Unique-sequence DNA isolated from a clone that hybridises with the original clone is used to probe the library again to detect further overlapping clones. The process is repeated many times, each step resulting in a clone that extends towards the target gene. Overall direction of the walk may be monitored using long-range restriction maps made with rare cutting enzymes (see Section 3.4.4). Alternatively, new genetic markers may be found in the newly isolated clones. These may be used to measure recombination with mutations in the target gene. Before the use of YAC and BAC vectors, genomic libraries were constructed in λ or cosmid vectors where the insert size was 20 or 40 kb respectively. It required many steps to walk 1 Mb along a chromosome. Chromosome crawling might be a more appropriate term! Sooner or later the walk would be halted because a section of DNA was missing from the library. Such gaps could be bridged by a special technique called chromosome jumping (shown right), which connected adjacent target sites for rare cutting enzymes such as *Not* I. In this technique, chromosomes are digested with *Not* I, which has target sites separated by up to 500 kb of DNA. The fragments are ligated at low concentrations in the presence of higher concentrations of a linker DNA containing a nonsense suppressor and *Not* I compatible ends. The low concentration of fragments favours the formation of very large circles in which the sticky ends from adjacent *Not* I sites are joined together via the linker. The circles are digested with an enzyme such as *Eco*RI, for which there will be many sites between adjacent *Not* I sites, and ligated to a λ vector containing a nonsense mutation in an essential gene. Plaques can only form upon transfection into an *E. coli* host if the linker is present, because it contains the nonsense suppressor. The linker and suppressor are necessary to select the *Eco*RI fragments that contain the *Not* I site and not the other fragments generated.

(1 cM) corresponds to an average physical separation of 1 Mb, so the linked marker could still be a considerable distance from the gene. The next stage is to isolate overlapping DNA clones from the region. During the cloning of many important disease genes, a process known as chromosome walking was used (Box 5.1). However, the availability of detailed clone maps (discussed in Chapter 3) has made this step increasingly unnecessary.

5.2.1 Identifying the gene

Perhaps the most difficult step in cloning disease genes is identifying the gene responsible in the ordered library collection of clones that result from the chromosome walking. Some genes, such as those for dystrophin, CFTR and factor VIII, are scattered across several hundred kilobases of the chromosome on which they reside. The coding exons sometimes amount to only a few per cent, or even less, of the locus. A number of strategies can be employed to find exons in such genes.

1. Exons encode mRNA, so unique-sequence DNA can be isolated from genomic clones and used to probe **Northern blots** or cDNA libraries to identify regions that are expressed. Any genomic sequences identified in the near vicinity of a gene will be a candidate to be part of the coding sequence. If the **pathophysiology** of the disease indicates that the gene is likely to show tissue-specific transcription, then mRNA from different tissues can be probed to see if the mRNA encoded by the genomic fragment matches the expected pattern.

2. Special techniques such as **exon trapping** can be employed to search for exons that have splice acceptor sites. A number of ways have been developed to do this. Figure 5.2 shows the method used in the search for the HD gene, which played a critical role in its discovery.

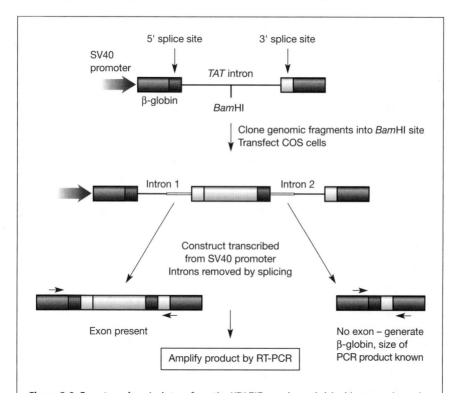

Figure 5.2 Exon trapping. An intron from the HIV *TAT* gene is sandwiched between the end of one exon and the start of the next from the rabbit β-globin gene. Fragments from a genomic region being scanned for exons, approximately 4 kb in size, are shotgun cloned into a *Bam*HI site in the *TAT* intron and transfected into COS cells. The construct is transiently expressed from the SV40 early promoter and the resulting mRNA transcripts amplified by **RT-PCR**, a form of PCR that uses an RNA template, using primers that anneal in the exons. The 5' and 3' splice sites (see Section 2.2.1) from the β-globin sequences ensure that the *TAT* intron is removed by splicing; the size of the RT-PCR product will therefore be known. If the genomic insert contains an exon with flanking introns and attendant 5' and 3' splice sites, it will in effect generate two novel introns labelled intron 1 and intron 2. These will be removed by splicing to leave an mRNA transcript containing the exon. RT-PCR now results in a novel product containing the exon.

3. Unique-sequence DNA isolated from genomic clones can be used to probe DNA from other mammals in **Southern blots.** This is known as **zoo blotting.** The rationale is that only protein-coding sequences are likely to be conserved, whereas non-coding sequences are not. DNA from exons will thus cross-hybridise to DNA from other mammals.

4. Some types of mutation, such as deletions and trinucleotide repeat expansions, are relatively easy to identify. If such a mutation co-segregates with the disease, this will map the location of the gene affected.

5. CpG islands (see Chapter 2), which often mark the beginning of genes, can be hunted using enzymes such as *Hpa* II that only cut DNA containing an unmethylated CCGG target site.

6. DNA sequences can be analysed with computer algorithms that detect systematic differences between coding and non-coding DNA. This is more than a question of looking for ORFs, because exons are often not much larger than the length of ORFs that would be expected to occur by chance.

7. Genomic maps are increasingly plotting the nature and position of all expressed sequences (see Chapter 3), thus identifying the location of candidate genes. Sometimes a gene in the region will encode a biologically relevant product. Such a gene would be a candidate for the disease gene; this is known as the **positional candidate** approach. A good example of the positional candidate approach is the cloning of four mismatch DNA repair genes responsible for hereditary non-polyposis colon cancer (see below). This approach is becoming increasingly predominant because of the availability of high-resolution maps.

The application of some of these techniques is illustrated by the cloning of the *CFTR* gene, described in Box 5.2.

Once possible genes have been identified, it is necessary to demonstrate which one is actually responsible for the disease. There may be several ways of making a connection. One of the best is to demonstrate co-segregation of mutations in a gene with the occurrence of the disease itself. One problem here is that we can expect variation in base sequence as a result of neutral polymorphisms. Mutations that prevent the protein from being formed, such as nonsense, frameshift and deletion mutations, are unlikely to be neutral to protein function. The co-segregation of these with the disease is a powerful indication that the right gene has been identified. The presence of an expanded trinucleotide repeat is another very useful indicator. Finally, if cells from the affected individual show some phenotype in tissue culture, the ultimate proof is to transfect such cells with the wild-type gene and demonstrate that the defect is corrected.

The first gene to be cloned by positional cloning was that for chronic granulomatous disease, in 1986. This was followed by DMD in 1987 and CF in 1989. The early gene hunts were extremely difficult, largely because of the low resolution of the genetic maps and the absence of physical and expression maps. The hunt for the CF gene was estimated to have cost US$200 million! HD was the first of the major monogenic disorders to be linked to a genetic marker in 1983. It took a further 10 years and required a large international collaboration to finally isolate the gene. The construction of detailed physical and expression maps of the genome has made this process much easier and arguably most of the important monogenic disorders have now been cloned. The focus is shifting to the more difficult task of cloning polygenic loci that may contribute to the risk of multifactorial disease (see Chapter 6).

5.3 Cystic fibrosis

CF is a common monogenic disorder among northern Europeans. It affects approximately 1 in 2000 people and thus has a carrier frequency of 1 in 22. Its primary symptoms are chronic bacterial infection, inflammation of the lungs and an elevated electrolyte level in sweat. It may also result in pancreatic exocrine insufficiency, obstruction of the bowel in the newborn (**meconium ileus**), diabetes mellitus, liver cirrhosis and male infertility due to CBAVD. The primary defect is decreased chloride ion export and increased sodium ion absorbance leading to insufficient hydration of epithelial surfaces. In the lungs this results in a sticky mucus that cannot be cleared by the action of the cilia that line the epithelia. The failure of this mucociliary clearance mechanism means that foreign particles such as bacteria cannot be removed. For reasons that are not understood at present, infections with *Pseudomonas* species predominate.

5.3.1 CFTR gene and protein

The gene responsible for CF was cloned exclusively from its genetic map position without the aid of cytogenetic rearrangements or knowledge of the protein involved. Box 5.2 describes the process in detail as an example of the difficulties encountered in such an undertaking. The protein encoded by the gene is now known as CFTR.

In Chapter 2 we saw that human genes can be very large and the gene encoding human blood clotting factor VIII was cited as an example. The *CFTR* gene is another example of an extremely large gene. It occupies approximately 230 kb of chromosome 7 and consists of 27 exons ranging in size from 38 to 724 bp. Like other large genes, only a small fraction is actually coding sequence. It produces a 6.5 kb transcript that is found in those tissues affected by CF, such as the lungs, pancreas, sweat glands, liver, nasal polyps, salivary glands and colon (Figure 5.3).

Although existing research had suggested that the primary defect in CF was chloride ion transport, prior to cloning the gene there was no direct knowledge of the protein affected. Conceptual translation of the nucleotide sequence showed that the CFTR protein consists of 1480 amino acid residues. It is similar to the ABC (ATP-binding cassette) superfamily of membrane transporters. These proteins are found in all living organisms, from bacteria to humans. As their name implies, they are

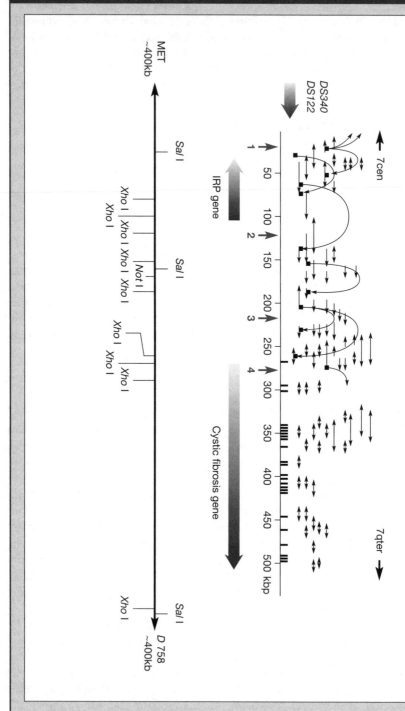

The gene responsible for CF proved exceptionally difficult to clone. This was partly because of the lack of any cytogenetic rearrangements that could be used to map its location accurately and partly because of technical difficulties encountered during cloning. The first step was to map the gene to the chromosomal region 7q21. This was done by observing linkage between CF and two RFLP markers (MET and D7S8). An intensive search for new markers yielded two more, D7S122 and D7S340, that were more closely linked to the *CFTR* gene. Unfortunately, they were only 10 kb apart so effectively they marked the same point on the chromosome. Bidirectional chromosome walks were initiated from the landing places of chromosome jumps from these two markers. In total this involved cloning 280 kb in 49 λ clones and making nine chromosome jumps. In the figure each clone is shown by an arrow, indicating the direction of the step that was made with it. Double-headed arrows indicate clones that were used to walk in both directions. Each jump is shown as an arc. A long-range restriction map was also generated using rare cutting enzymes and pulsed field gel electrophoresis (see Section 3.4.4). The map showed that the *CFTR* gene (the name now given to the gene responsible for CF) was about 250 kb in size and was entirely contained within a 380 kb *Sal* I fragment. Progress of the walk was monitored as sites were passed in this long-range restriction map. As well as these clones recovered by walking and jumping (shown in the left-hand part of the diagram), further clones were recovered by probing genomic libraries with cDNA clones (those clones on the right above the *CFTR* gene), so that in total 500 kb of DNA was cloned.

As the walk proceeded, most of the techniques described earlier were applied to identify the *CFTR* gene. Zoo blots proved to be effective and identified four regions whose sequence was conserved in other mammals. Any transcripts encoded by these regions were sought by using the cross-hybridising clones to probe Northern blots and cDNA libraries prepared from tissues affected by CF. Four clones cross-hybridised to other species, shown by the numbered vertical arrows in the figure. Region 1 could be eliminated on genetic grounds. Region 2 corresponded to a gene called *IRP* (INT-related protein) that was known not to be responsible for CF. However, because *IRP* was known to map to the D7S8 side of CF this indicated the direction in which walking and jumping should continue. Although region 3 contained a CpG island that often marks the start of a gene (see Chapter 2), no transcripts could be detected.

Region 4 eventually turned out to be the 5' end of the *CFTR* gene, but this only became apparent after a frustrating series of experiments. The region was identified by two clones called H1.6 and E4.3. The sequences of these clones were rich in undermethylated CpG dinucleotides, suggesting that they were CpG islands. However, they did not contain any ORFs nor did they detect any transcripts in Northern blotting. Eventually, after screening seven different libraries, clone H1.6 hybridised to a single cDNA clone, called 10.1, in a cDNA library made from sweat glands. Clone 10.1 hybridised to a 6.5-kb transcript in a Northern blot and the signal was stronger in cells expected to express the *CFTR* gene. Sequencing showed that the 920 bp insert in clone 10.1 contained a long ORF. Only 113 bp of this sequence overlapped with H1.6. The small overlap explained the difficulty experienced in identifying a transcript. The *CFTR* gene had been found, but only just! Less perseverance or a slightly smaller insert in clone H1.6 and it would have been missed.

Obtaining a full-length cDNA clone again proved more difficult than expected. Clone 10.1 was used to re-probe cDNA libraries; 18 clones were recovered but none contained the full 6.5 kb

Continued ▶

gene identified in Northern blots. The number of clones recovered was less than expected, and they grew slowly and contained rearrangements. This suggested that clones containing the full-length gene were selected against in the formation or propagation of the cDNA libraries. The consensus sequence of the gene had to be pieced together from the partial sequences of the inserts of these 18 clones. A clone containing the full sequence was then assembled by subcloning from these fragments.

This gene was re-isolated from cDNA libraries prepared from sweat glands of affected and unaffected individuals. A common 3-bp deletion (ΔF508) was exclusively associated with the occurrence of the disease: 145/214 CF chromosomes carried this mutation (68%); in contrast 0/198 normal chromosomes had the mutation ($P<10^{-57.5}$!). The protein encoded by this gene showed strong similarities to the protein superfamily known as the ABC (ATP-binding cassette) membrane transporters. These two observations strongly suggested that the gene responsible for cystic fibrosis had indeed been isolated. The final proof came later with the demonstration that when transfected into epithelial cell cultures derived from CF patients it corrected the chloride ion conductance defect.

The work was carried out as a collaboration between the groups of Lap-Chee Tsui and John Riordan in the Hospital for Sick Children, Toronto, and Francis Collins in Michigan. Lap-Chee Tsui's group identified the closely linked D7S122 and D7S340 markers and carried out the chromosome walking and physical mapping in collaboration with the group of Francis Collins, who had developed the chromosome jumping technique. John Riordan's group provided the cDNA libraries and identified the crucial cDNA clones. The outcome was victory in a race with many other groups around the world who were seeking the gene.

responsible for pumping small molecules in and out of cells. Members of this family can be responsible for multidrug resistance in human cells undergoing cancer chemotherapy.

The CFTR protein consists of five domains (Figure 5.3). There are two membrane-spanning domains each consisting of six transmembrane segments. These domains are thought to form a membrane pore through which the chloride ions are transported out of the cell. There are two domains, called nucleotide-binding folds (NBFs), that bind and hydrolyse ATP, essential for the transport process. A domain that has a high proportion of charged amino acid residues separates the two halves of the molecule. This domain is unique to the CFTR protein and has a regulatory function. It contains several sites that are phosphorylated by cAMP-dependent protein kinase (protein kinase A). This phosphorylation is essential for the protein to function and accounts for the prior observation that chloride ion transport is dependent on cAMP-dependent protein kinase.

5.3.2 Mutations affecting the *CFTR* gene

Over 550 different mutations of the *CFTR* gene have been catalogued (Figure 5.3). It is therefore quite astonishing that over two-thirds of CF mutant alleles

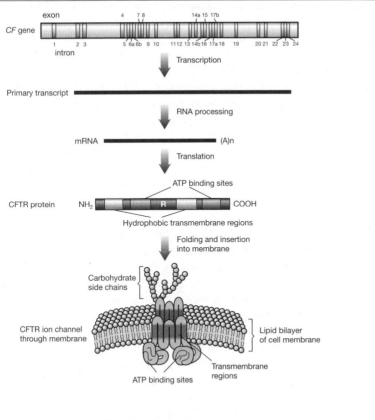

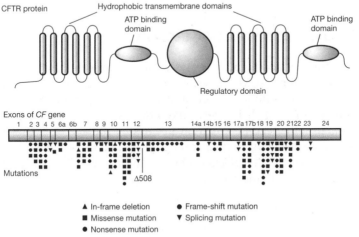

Figure 5.3 Structure of the CFTR gene, the transmembrane protein it encodes and the spectrum of mutations that affect its function.

are exactly the same mutation, known as ΔF508. As discussed in Chapter 9, there is evidence that this mutation is extremely ancient, possibly occurring in one of the very first anatomically modern humans that lived in Europe 50 000 years ago and passed on to succeeding generations from this founder. However, this claim is controversial (see Chapter 9).

It has been suggested that carriers of a CF mutant allele enjoy some selective advantage. One possibility is that as CF carriers have a reduced amount of the CFTR protein, there may be a decreased flow of water across the intestinal mucosa. This may make heterozygotes better able to survive intestinal infections that result in infantile diarrhoea. If this is so, then it tells us something about the living conditions of early human populations. Early Europeans must have been exposed to different diseases than other humans as they were the only populations where CF alleles became frequent.

The F508 mutation is a 3-bp deletion that results in the loss of a phenylalanine residue at position 508 in the polypeptide chain in one of the two NBFs. (Δ is a symbol commonly used to signify a deletion and F is the standard single-letter abbreviation for phenylalanine). The CFTR protein undergoes a complex maturation and folding process before the active form is finally located in the cell membrane. ΔF508 polypeptides are defective in this process as they do not undergo an ATP-dependent folding step in the **Golgi apparatus** and are rapidly degraded. However, if the protein does fold, the mutation does not affect its activity. When cells carrying the ΔF508 mutation are cultured at subphysiological temperatures, folding takes place and the protein is active. This provides a target for therapeutic intervention as small drugs may be found that aid the folding of ΔF508 CFTR polypeptides *in vivo*.

5.4 Duchenne muscular dystrophy

DMD is characterised by progressive failure of muscle growth and wasting (dystrophy), leading to weakness, paralysis and respiratory difficulties. In addition the heart, smooth muscle and central and peripheral nervous systems may be affected. It is a sex-linked disease affecting 1 in 3300 boys. Symptoms usually become apparent in boys by the age of 3, they become confined to a wheelchair in their teenage years and usually die by the age of 30. A milder form, called Becker muscular dystrophy (BMD), is caused by mutations that map to the same locus.

5.4.1 DMD gene

The gene was cloned by positional cloning in 1987. Cytogenetic rearrangements resulting in DMD always have one breakpoint in the region Xp21. One rearrangement resulted in a fusion of Xp21 with the 28S ribosomal RNA locus on chromosome 21, which allowed a marker called XJ1.1, closely linked to DMD, to be isolated (see above). Another important strategy was the use of DNA from a boy identified as 'B.B' who suffered from three

X-linked disorders: DMD, chronic granulomatous disease and retinitis pigmentosa. Because the genes responsible for these disorders are closely linked, it seemed likely that B.B.'s X chromosome carried a deletion that affected at least part of all three genes. DNA from the region spanned by this deletion was isolated from an XXXXY cell line using a technique called subtractive hybridisation, which preferentially enriched for the region missing in B.B.'s DNA. One of these clones, called pERT87, was shown to be closely linked to the site of DMD mutations. Clones such as pERT87 and XJ1.1 were used to map the genomic locus, including the construction of a long-range restriction map. pERT was also used to probe cDNA libraries, resulting in the isolation of a number of cDNA clones that together spanned the 14-kb mRNA of the locus. These cDNA clones were then used to make a detailed map of the genomic locus including the positions of exons.

The DMD gene is notable for its huge size. It is composed of 79 exons that, together with promoter regions, are scattered across 2.4 Mb, about 2% of the whole X chromosome! Transcription of this region is complex, with different cell types using at least seven distinct promoters between them. A full-length 14-kb mRNA is produced from different promoters in muscle cells, cortical cells and Purkinje cells. A shorter transcript is produced by internal promoters in Schwann cells and glial cells, encoding a protein that is correspondingly smaller.

5.4.2 Dystrophin

The protein encoded by the DMD gene, called **dystrophin**, is composed of 3685 amino acid residues. Its role is shown in Figure 5.4.

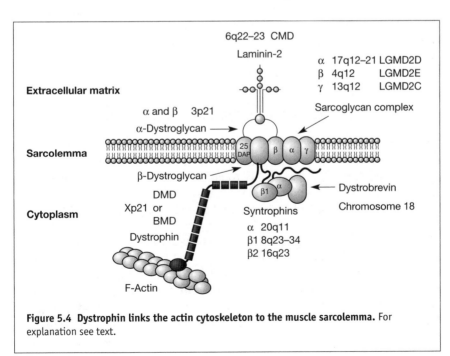

Figure 5.4 Dystrophin links the actin cytoskeleton to the muscle sarcolemma. For explanation see text.

Dystrophin is a long rod-like molecule that links actin fibres in the cortex of muscle cells to the extracellular basal lamina (connective tissue). It does this by forming a bridge between actin and a transmembrane glycoprotein complex called the **dystrophin-associated sarcoglycan complex** (DASC) located in the **sarcolemma**, a membranous sheath that surrounds muscle fibres. DAGC is required for the function and maintenance of muscle cells. It provides a signalling pathway between the connective tissue and the cytoskeleton of muscle cells. In the absence of dystrophin this complex does not form properly. Mutations that affect proteins of the DASC complex also result in a form of muscular dystrophy, called limb girdle muscular dystrophy. In addition, some people suffering from congenital muscular dystrophy (CMD) harbour mutations affecting α-laminin-2, to which DASC is attached outside the cell.

5.4.3 Mutations affecting the DMD gene

DMD mutations arise at a rate of 1×10^{-4} per generation, two orders of magnitude higher than for other X-linked loci. This high rate of mutation is probably due to the target presented by its extreme size. Deletions account for 60% of these mutations. The end-points of the deletions are clustered in two particular introns. The reason for this is not clear, but it could be related to the presence of a mobile genetic element in both introns. Where deletions disrupt the reading frame of the protein, the severe Duchenne form of the disease results. Where the reading frame is preserved, the milder Becker form results. Surprisingly, one mutation has been identified that results in the Becker form, even though the protein is less than half its original size. A similar pattern is seen in small deletions and point mutations: where the mutation allows a full-length protein to form the milder Becker form generally results, but nonsense and frameshift mutations producing truncated proteins cause the severe Duchenne form.

5.5 Trinucleotide repeat expansion mutations

We saw in Chapter 2 that microsatellites consist of tandemly repeated dinucleotide, trinucleotide or tetranucleotide sequences and that the number of repeats is polymorphic. The ability to clone genes and to analyse the mutations responsible for genetic disease led to the unexpected discovery that polymorphic trinucleotide repeats can dramatically increase in length to form a novel type of mutation called a **trinucleotide repeat expansion** (TRE). A total of 11 such mutations have been identified (Table 5.1). In TRE mutations the number of repeats is unstable in meiosis, in some cases even in mitosis, usually causing a further increase in repeat length. For this reason they are sometimes called **dynamic mutations.** Dynamic mutations are responsible for a phenomenon called **anticipation**, where either the severity or the degree of penetrance of a disease apparently increases with each succeeding generation.

Table 5.1 Disorders caused by trinucleotide repeat expansions. **155**

Fragile sites caused by CCG expansion in 5' UTR
FRAXA (fragile-X syndrome)
FRAXE (mental retardation)
FRAXF
FRA16A
FRA11B (Jacobsen syndrome)

CTG expansion in 3' UTR
Myotonic dystrophy

Neurodegenerative disorders caused by CAG expansions in coding regions
Huntington's disease
Spinal and bulbar muscular atrophy
Spinocerebellar ataxia
Dentatorubral–pallidoluysian atrophy
Machado–Joseph disease

TREs can broadly be divided into two classes. In one class, exemplified by fragile-X syndrome, the trinucleotide concerned is CCG, which expands in the **untranslated regions** (UTRs) of genes. This typically results in chromosome fragile sites: cytogenetically visible, non-staining gaps in metaphase chromosome spreads. In the other class, the trinucleotide concerned is CAG, which occurs in a number of neurodegenerative disorders such as HD. In this class, the number of expanded repeats is much lower and occurs within the coding region of the gene affected. One TRE mutation that does not easily fit in this classification is responsible for MD. In this case the trinucleotide concerned is CTG and the expansion site is located in the 3' UTR of the gene that is probably affected.

5.5.1 Fragile sites caused by CCG expansions

In TRE mutations involving CCG the normal repeat number of ~30 increases to several hundred or even thousands. It results in **fragile sites** in chromosomes. These are non-staining gaps in chromosomes visible in metaphase spreads of cells cultured under conditions such as folate deprivation or chemical inhibition of DNA synthesis. Five loci involved in fragile sites have been cloned (Table 5.1). All have been shown to involve expansions of a CCG repeat. Some of these fragile sites have no apparent phenotypic consequences. However, one of them, FRAXA, results in fragile-X syndrome, characterised by moderate to severe mental retardation. Two other fragile sites, FRAXE and FRA11B, are also associated with mental retardation.

Fragile-X syndrome is the commonest cause of congenital mental retardation after Down's syndrome, affecting 1 in 2000 children. There are a number of unusual features in its pattern of inheritance.

1. It shows incomplete penetrance as, on average, 20% of boys who are known to carry the allele, because they transmit the disorder to subsequent generations, are not affected. These boys are classified as normal transmitting males. The degree of penetrance increases with each succeeding generation in a pedigree. This phenomenon is called anticipation and, in one form or another, is characteristic of TRE mutations.
2. It shows at least partial dominance because 30% of female heterozygotes are also affected. These affected females always inherit the fragile site from their mothers, never from their fathers.
3. In males the disorder is never caused by a new mutation during gametogenesis in their mothers.

FRAXA, the fragile site associated with fragile-X syndrome, is located at Xq27.3. The gene affected is called *FMR1* (fragile-X mental retardation 1). The expansion affects a polymorphic site in which the number of CCG repeats normally ranges from 6 to 52. Individuals affected by the disease are said to carry a full mutation when the number of repeats increases to 230–1000 copies. The expansion site is located upstream of the coding region near to a CpG island. The full mutation induces **DNA methylation,** which extends into this CpG island and consequently inactivates the *FMR1* gene. Some boys, who suffer from a severe form of the disorder, have been found to have deletions of the *FMR1* gene. This confirms that the expansion does indeed cause loss of gene function and is not acting in some other way to change the activity of the *FMR1* gene product. The role of the FMR1 protein has not yet been fully elucidated. It contains an amino acid sequence that is found within proteins known to bind RNA. The FMR1 protein binds specifically to approximately 4% of mRNA molecules found within brain cells, so it may have a role in controlling translation.

The number of repeats at the expansion site can show intergenerational instability, which accounts for the unusual pattern of inheritance. Normal transmitting males harbour **premutations** in which the number of repeats ranges from 60 to 230. These can expand into full mutations during meiosis in a carrier female but not in a normal transmitting male. The frequency with which the premutation expands into a full mutation is dependent upon its length, which can only increase in a female carrier. With each passage of a premutation through a carrier female its length may increase, consequently raising the chance of expansion into a full mutation in the next female carrier in the pedigree. This explains the phenomenon of anticipation. The requirement for a premutation to be expanded by passage through a carrier female explains why the disorder is never caused in boys by a new mutation during gametogenesis in their mothers. The observation that expansion can only occur in females explains why affected females always receive the mutant allele from their mothers. Both males and females bearing the full mutation show somatic instability where the repeat length and degree of methylation can show wide variation in different cells.

A total of five neurodegenerative disorders have been shown to be caused by expansion of polymorphic CAG repeats located within the coding region of the gene affected (Table 5.1). All of the disorders are late onset, autosomal dominant and progressive in nature. They are all characterised by similar symptoms: ataxia, chorea, dementia and sometimes psychosis. The effect of each disease is limited to a different group of cells in the brain, yet each gene is expressed in many different classes of brain cells; there is currently no explanation for this apparently paradoxical observation. CAG encodes glutamine, so the expansion mutation will cause an increase in a polyglutamine tract in the encoded proteins. Since these are all dominant mutations this must in some way alter the activity of the protein rather than reduce its function.

The best-known and most common of this group of diseases is HD, which affects 1 in 10 000 Caucasians. The symptoms are generally manifested in mid life, but it can show juvenile onset in which case the course of the disease is more rapid and the symptoms more severe. A particularly distressing aspect of HD is that, as a late-onset, autosomal dominant disease, sufferers will normally have had children before the onset of the disease. Each of these children will have a 50% chance of being affected.

In 1983, HD was one of the first human genes to be mapped by linkage to polymorphic DNA markers. However, the gene responsible was only cloned in 1993 after an exceptionally prolonged and difficult gene hunt. Two strategies finally led to the gene.

1. Linkage disequilibrium showed that although multiple mutations have occurred to cause HD, one-third of HD chromosomes probably descend from the same ancestral chromosome. The haplotype analysis narrowed the location of the gene to a 500-kb region.
2. Exon trapping (see Figure 5.2) was used to identify every transcript that arose from the region identified by the linkage disequilibrium analysis. In HD patients the gene encoding one of these was affected by an expansion of a trinucleotide repeat $(CAG)_{\sim 15}$.

The gene affected, known as *IT15*, is located near the telomere on chromosome 4. It encodes a protein of unknown function called **huntingtin**. The coding region contains a polymorphic CAG repeat that expands in those affected by the disease (Figure 5.5). The normal range of repeat lengths is 15–35; in affected individuals this increases to 36–121. It is remarkable that the lowest number of repeats that causes the disease is only one more than the highest found in normal individuals, although it should be said that repeats in the 30–40 range are extremely rare. Apparently increasing the polyglutamine tract from 35 to 36 residues is enough to trigger the onset of the disorder.

HD does not show anticipation in the formal sense, but there are some related phenomena. Longer repeat lengths are associated with juvenile onset and more severe symptoms. These are generally transmitted by affected fathers, suggesting that expansion takes place during male gameto-

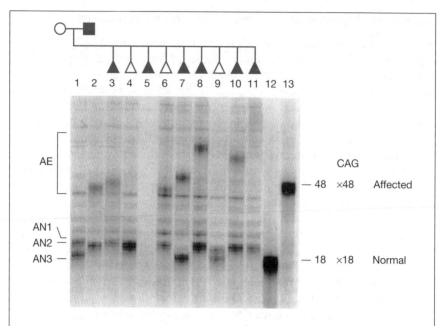

Figure 5.5 Trinucleotide repeat expansion in HD. The figure shows a large family in which the disease is segregating. The region surrounding the CAG repeat site has been amplified using PCR. All individuals in the pedigree have a normal allele with the CAG trinucleotide repeated about 18 times. Affected individuals also have a mutant allele where the repeat number has increased. Lanes 12 and 13 show reference cDNA clones where the repeat length is 18 (wild type) or 48 (mutant). The PCR reaction in lane 5 failed. Individual number 11 apparently only has a normal allele. Subsequent analysis showed that this individual carried an expansion so large that the PCR reaction failed. As a consequence of the very large repeat, the disease was particularly severe and showed juvenile onset.

genesis. This is reminiscent of fragile-X syndrome, where the expansion can only take place in the female germline. Age of onset is also inversely correlated with repeat length.

5.5.3 Myotonic dystrophy

MD (also known as **dystrophia myotonica**) is an autosomal dominant neuromuscular disorder. It affects 1 in 8500 individuals, making it the most common neuromuscular disorder showing adult onset. It also shows a range of symptoms from cataracts and frontal baldness in mildly affected individuals to multisystem failure in severely affected individuals with symptoms including muscular dystrophy, myotonia (an inability to relax muscles after voluntary contraction), endocrine dysfunction, mental retardation, cardiac abnormalities and testicular atrophy in males. It also shows a variable pattern of onset, ranging from a congenital condition with severe symptoms to a classic adult-onset form and a mild, late-onset form. The disorder shows anticipation, with increasing severity and decreasing age of onset with each succeeding generation. Often the symptoms are so mild in the grandparents or parents that the disease had not been diagnosed prior to its occurrence in a severe form in the grandchildren.

The disorder is caused by the expansion of a CTG repeat located in the 3' UTR of a gene known as the **dystrophia myotonica protein kinase** (*DMPK*). The repeat is polymorphic in normal populations with repeat numbers in the range 5–35. This expands in affected individuals to repeat numbers in the range 50 to >3000. There is an approximate correlation between repeat length and severity and age of onset. The number of repeats increases with each succeeding generation, explaining the observed anticipation. Severely affected congenital cases show the highest number of repeats. Mildly affected or asymptomatic forebears of more severely affected individuals will have between 50 and 80 repeats. The expansion is not sex-limited as it has been observed in parents of both sexes.

There have been conflicting reports as to whether the expansion increases or decreases *DMPK* expression, so it is not clear whether the disorder results from a gain or loss of gene function. Moreover, neither *DMPK* knockout mice nor mice that overexpress a human *DMPK* transgene show full symptoms of the disease; thus there is still some doubt as to whether the *DMPK* gene is solely responsible for the disorder. The expansion occurs near the CpG island for a second gene called **DM locus associated homeo-domain protein** (*DMAHP*), which may also play a role in the disease as its transcript level is reduced by the expansion allele and its pattern overlaps substantially with that of the *DMPK* gene in mice.

5.6 Haemoglobinopathies

Haemoglobin is the oxygen-carrying molecule in the blood. It forms 90% of the protein content of red blood cells and is by far the most abundant blood protein. It is a tetramer consisting of two α-globin and two β-globin polypeptide chains (α_2 or β_2) each complexed with a molecule of iron-containing haem. It binds oxygen cooperatively, which is important to its physiological function of binding oxygen at the high levels of dissolved oxygen found in the lungs and releasing it at the lower levels found in body tissues. Homotetramers of the same globin type (α_4 or β_4) show hyperbolic oxygen binding and are unable to release oxygen under physiological conditions.

The haemoglobinopathies are a group of diseases that affect haemoglobin production or function leading to anaemia. Genetic defects directly affecting the globin genes can be divided into two classes of almost equal prevalence: sickle cell disease and the thalassaemias. As well as sickle cell disease and the thalassaemias, glucose 6-phosphate dehydrogenase deficiency is another common hereditary disease that can affect haemoglobin. This results in favism (haemolysis after eating beans), anaemia in response to certain drugs and neonatal jaundice.

In sickle cell disease, β-globin is produced in normal quantities but an amino acid substitution causes it to precipitate into inactive aggregates within red blood cells, resulting in the characteristic sickle morphology. In the thalassaemias, one of the globin chains is reduced or absent. This results in an excess of the other type, which is responsible for the clinical manifestations. β Thalassaemias describe disorders in which β-globin is reduced or absent; α thalassaemias describe corresponding disorders in which α-globin is reduced or absent. A complete absence of α-globin (α^0 thalassaemia) causes **hydrops fetalis,** which is fatal at birth due to oxygen starvation. Sickle cell anaemia, milder forms of α thalassaemia and the β thalassaemias show a wide range of clinical severity that may be treatable by regular blood transfusions.

Collectively, the haemoglobinopathies are the most common form of disease traceable to a monogenic defect. Approximately 7% of the world's population are thought to be a carrier for one of the disorders. Sickle cell anaemia is most common in Africa where it affects 100 000 infants each year. There are also significant numbers of cases in other regions where there are populations of African descent, such as the USA (1500 new cases annually), the Caribbean (700 new cases annually) and the UK (140 cases annually). The thalassaemias are especially prevalent in South East Asia. In Thailand, for example, 500 000 suffer from some form of hereditary anaemia. β Thalassaemia is also common in Mediterranean peoples. Some Mediterranean islands have a particularly high rate: the carrier frequency is 20% in parts of Sardinia and Cyprus, resulting in a population incidence of 1%.

High frequencies of the haemoglobinopathies correlate closely with the worldwide incidence of malaria, indicating how they have arisen. Heterozygotes are more resistant to infection by *Plasmodium falciparum,* the causative agent of the most severe form of malaria. This leads to a polymorphism, where selection for mutant alleles when carried by heterozygotes is balanced by selection against the same alleles when carried by homozygotes.

5.6.1 Globin genes

Both α-globin and β-globin are encoded by a family of related genes that are expressed at different times during development. The α-globin gene cluster occupies 30 kb of chromosome 16 (Figure 5.6). There are two α-globin genes, α_1 and α_2, encoding identical polypeptides and differing only in small changes in non-coding parts of the gene, such as the 3' UTR and introns. The two genes are thought to have arisen by duplication and this is reflected by blocks of homology in the surrounding regions. As well as the α-globin genes, there is a related α-globin-like gene called ζ_2-globin which is expressed in early embryonic development (see below). In addition there are pseudogenes (see Chapter 2) of both α-globin and ζ-globin called $\psi\alpha_1$ and $\psi\zeta_2$ respectively. Finally, there is a gene that resembles α-globin called θ-globin, the status of which is uncertain. It is transcribed, but the resulting protein is apparently not incorporated into haemoglobin. It may be a pseudogene in the early stages of evolution.

The β-globin gene cluster is 70 kb in size and located on chromosome 11 (Figure 5.7). It consists of five genes and one pseudogene. There are two similar adult forms, β-globin itself and δ-globin, which constitutes about 6% of adult β-globin-like chains. During the foetal stage of development Aγ and Gγ are expressed; they encode β-globin-like proteins that differ only at posi-

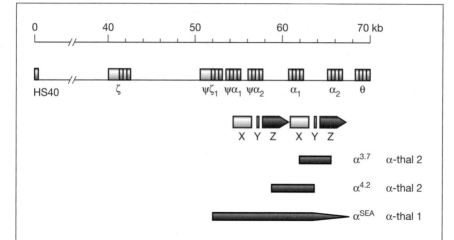

Figure 5.6 Structure of the α-globin cluster and deletions causing α thalassaemia. X, Y and Z are blocks of sequence homology repeated in the regions surrounding α_1 and α_2 genes. The blue bars show the extent of deletions resulting in α thalassaemia. The figures in superscripts show the size of the deleted regions. SEA, South East Asia.

tion 136, which is alanine in Aγ and glycine in Gγ. Finally, ε-globin is expressed in early embryonic development.

As a result of the changing pattern of expression during development, a variety of haemoglobin molecules appear that differ in the composition of the haemoglobin subunits (Table 5.2). These different haemoglobins meet the changing oxygen transport requirements at different developmental stages.

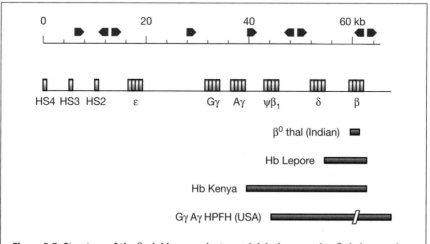

Figure 5.7 Structure of the β-globin gene cluster and deletions causing β thalassaemia. The solid bars indicate the extent of deletions found in particular deletion mutations. HS4, HS3 and HS2 are locus control elements. The arrows above the scale indicate the location and orientation of *Alu* elements.

Table 5.2 Haemoglobins made at different times during development.

| | β-globins | | | | |
| | Embryonic | | Foetal | Adult | |
α-globins	ε	γ	γ	δ	β
ζ	$\zeta_2\varepsilon_2$ Hb Gower1	$\zeta_2\gamma_2$ Hb Portland			
α			$\alpha_2\gamma_2$ Hb F	$\alpha_2\delta_2$ Hb A$_2$	$\alpha_2\beta_2$ Hb A

During development the pattern of expression of both the α-globin and β-globin clusters is under the control of locus control regions (LCRs). HS2–HS4 control the expression of β-globin and HS40 controls the expression of α-globin (see Figures 5.6 and 5.7). LCRs are enhancers that bind erythroid-specifc transcription factors. Without the LCR there is very little expression of the genes, but exactly how they act is still unclear. One striking fact about both clusters is that the physical arrangements of the genes reflect the order with which they are expressed during development, but the significance of this is still unknown. The LCRs map a considerable distance upstream of the clusters. For example, HS40 is located over 60 kb upstream of the α_1 and α_2 genes. Thus they are an extreme example of the way that enhancers can act at distance to activate transcription of the gene they control.

5.6.2 α Thalassaemia

α Thalassaemias are caused by deletions of one or both α-globin genes. An α-thalassaemia 1 (α-thal 1) chromosome has one remaining gene while an α-thalassaemia 2 (α-thal 2)chromosome has both genes deleted. These chromosomes can combine in different ways to result in the presence of zero, one, two or three α-globin genes (Figure 5.8). The severity of the resulting thalassaemia is determined by the number of remaining copies.

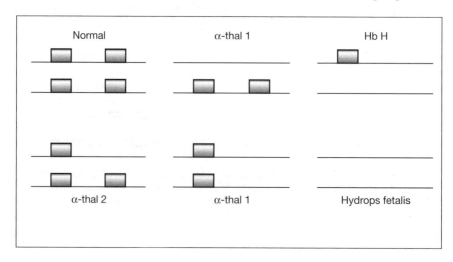

- Individuals with three α-globin genes (α-thal 2) are asymptomatic carriers.

- Two functioning α-globin genes (α-thal 1) results in mild anaemia.

- One functioning α-globin gene results in moderate anaemia with inclusions in red cells consisting of β_4 tetramers known as Hb H.

- Individuals with no α-globin genes die shortly after birth from a condition known as hydrops fetalis. The primary haemoglobin is Hb Barts (γ_4), which has such a high oxygen affinity that it fails to release any oxygen in the tissues.

The deletions apparently result from unequal crossing-over between the duplicated segments which originally gave rise to the two α-globin genes (Figure 5.9). Reciprocal chromosomes with three α–globin genes have been observed.

5.6.3 β Thalassaemia

The severest form of β thalassaemia is β^0 thalassaemia where there is a complete lack of β-globin chains in adult life. The resulting excess of α-globin chains precipitate, leading to red blood cell damage and also inhibiting the production of new red blood cells (**erythropoiesis**). In β^+ thalassaemia there is a large reduction but not complete absence of β-globin synthesis. A relatively benign form of β thalassaemia is hereditary persistence of foetal haemoglobin (HPFH), which occurs when the absence of β-globin is compensated by the continued synthesis of the foetal form. Sometimes, synthesis of foetal haemoglobin persists, but at insufficient levels to compensate for the loss of β-globin; this condition is known as δβ thalassaemia because there is also an absence of δ-globin.

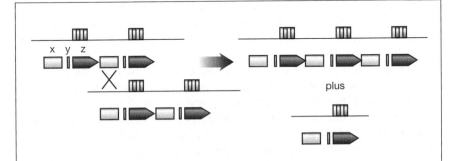

Figure 5.9 Unequal crossing-over results in a deletion of an α-globin gene. Unequal crossing-over between X regions produces one chromosome with a triplicated α-globin locus and a chromosome containing only one. In this example, the resulting deletion would be 4.2 kb in size, as observed in some α-thal 2 chromosomes (Figure 5.6). Chromosomes carrying three α-globin loci have been observed in populations where α-thal 1 is common. (Redrawn with permission from Roses, A.D. (2000) *Nature*, **405**, 857–865.)

Virtually every type of point mutation that could lead to a reduction or absence of β-globin synthesis has actually been observed in cases of β^0 and β^+ thalassaemia. These include:

- a deletion of the whole β-globin gene;

- promoter mutations preventing or reducing transcription;

- mutations of splice sites preventing the removal of introns;

- mutation of the poly-A acceptor site affecting RNA processing;

- nonsense and frameshift mutations causing termination of translation.

Many deletions affecting the β-globin cluster have been analysed; these are represented in Figure 5.7. One of them, Hb Indian, is the deletion affecting the β-globin gene that results in β^0 thalassaemia. Some of the others are apparently the result of unequal crossing-over between two β-globin genes, resulting in their fusion and the deletion of intervening DNA. For example, Hb Lepore results from a deletion that fuses the 5' portion of δ-globin with the 3' portion of β-globin. The reciprocal product of this recombination event, known as Hb anti-Lepore, has been observed.

5.7 Inherited predisposition to cancer

All cancers can be said to be genetic diseases, in that they arise through damage to 100 or so cellular **oncogenes** which, in their unmutated state, normally cooperate to maintain control over cell proliferation (see Box 5.3). The controls are complex so that, for the most part, failure of one component is compensated by the action of a different part of the control network. Before a cell is completely liberated from proliferation controls, multiple parts of the control system must be damaged. The origin of a tumour, and its progression to an invasive malignant state, is thus a multi-stage process: successive mutations to different oncogenes are necessary. For this reason the risk of cancer increases progressively with age as the mutations accumulate.

The mutations may be caused by endogenous cellular processes or they may be inflicted by environmental mutagens. Rarely, individuals are born with germline mutations affecting key components of the regulatory network. These are responsible for hereditary predispositions to cancer. Many of the products of these genes have also been found to be mutated in sporadic cancers and the study of the genes affected by these mutations has provided critical insights in understanding the mechanisms that normally control cell division as well as allowing diagnostic tests for families at risk.

Products of oncogenes operate at different levels. Some function in signal transduction pathways, ensuring that cells only divide when positively stimulated to do so by external **growth factors** collectively known as **mitogens**. Others may encode cellular receptors of the external signal, components of the intracellular signalling apparatus or nuclear transcription factors that change the pattern of gene expression. Mutations in these pathways are often dominant at a cellular level: they change the pattern of activity or result in hyperactivity of the gene product so that the signalling pathway is activated inappropriately. One of the most commonly occurring types of mutation in this category are mutations to members of the **ras** family, which act at the interface of growth factor receptors and intracellular signalling pathways.

Other types of cell division regulators act to inhibit cell cycle progression. They provide mechanisms, called **checkpoints**, that prevent cell division when it would be inappropriate, for example they may halt the cell division cycle to allow damage to DNA to be repaired. In presence of severe damage to DNA they even direct the cell to commit suicide, a process known as **apoptosis**. The way in which they operate is described in Box 5.3. Finally, other types of gene product are involved in the maintenance of genome

BOX 5.3: THE CONTROL OF CELL DIVISION

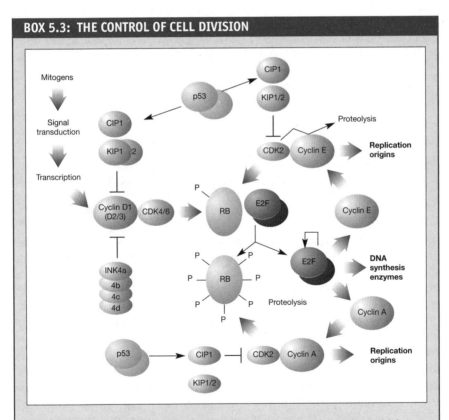

Progress through the cell cycle is controlled by the **cell cycle engine**, which consists of a family of **protein kinases** called cyclin-dependent protein kinases (CDKs). As their name implies, these kinases must be associated with a cyclin partner to be active. The periodic appearance and disappearance of different

Continued ▶

cyclin families controls the action of the different CDKs, which in turn regulates the passage of cells to division. Cells are normally only able to divide in the presence of mitogens; in the absence of mitogens cells remain in a quiescent state with unreplicated DNA (a state sometimes called G_0). Stimulation of a quiescent cell results in the rise of the cyclin D family of proteins, of which there are three members, cyclin D1, D2 and D3, which differentially respond to various signals in different tissues. Cyclin D molecules associate with CDK4 or CDK6. This complex phosphorylates the retinoblastoma protein (RB), causing it to release a heterodynamic transcription factor called E2F. Free E2F activates transcription of genes required for DNA synthesis. E2F may also activate its own transcription and that of cyclin E and cyclin A. Cyclin E expression is periodic, normally peaking in late G_1. The catalytic partner of cyclin E is CDK2 which also phosphorylates RB, making this process independent of cyclin D/CDK4/6 and hence the presence of mitogens. Cyclin A normally appears in **S phase** and levels remain high throughout the rest of the cycle. It associates with CDK2 in early S phase and with CDC2 in G_2. Cyclin E/CDK2 and cyclin A/CDK2 carry out key phosphorylation steps that activate replication origins and so trigger DNA replication. Once this occurs, E2F is inactivated through phosphorylation by cyclin A/CDK2, and cyclin E is destroyed by proteolysis to prevent further rounds of DNA synthesis. Cyclin E may be targeted for proteolysis by phosphorylation by its own CDK2 partner.

A key part in the regulation of this process are two families of CDK inhibitors called the INK4 family (four members) and the CIP1/KIP family. The INK4 proteins specifically inhibit cyclin D/CDK4. Mutations that inactivate INK4 or cause amplification of the cyclin D gene have been found in a wide variety of commonly occurring cancers. CIP1 is induced by p53, which mediates the DNA damage checkpoint pathway, pausing the cell cycle to allow DNA repair. In the presence of severe damage p53 provokes apoptosis. DNA damage is sensed by p53 through a pathway that includes the AT protein, which is inactivated in ataxia telangiectasia.

integrity and DNA repair. Failure of checkpoint and DNA repair functions allows other genetic changes to accumulate, which may further incapacitate the mechanisms that control cell proliferation. Oncogenes therefore occupy a key position in the regulatory network and their failure has been implicated in virtually every type of cancer.

Mutation to genes encoding checkpoint and repair functions are generally recessive at a cellular level because inactivation of one copy of the gene can be compensated by the function of the other copy. Such genes are known as **tumour suppressors** or **anti-oncogenes** because they act to prevent tumours from occurring. Many of the germline mutations that give rise to hereditary predisposition to cancer are tumour suppressors. Although they are recessive at the cellular level, germline mutations to these genes result in a predisposition to cancer that is inherited in an autosomal dominant pattern. This comes about because a cell in which one copy of the gene is inactivated by a germline mutation requires one further mutation event to inactivate the other. The chances that this will occur during somatic development are high and tumours arise at a high frequency. In contrast two independent mutation events are required in somatic cells not carrying a germline mutation; such events will occur and result in sporadic cancers, but at a lower frequency.

Germline mutations to tumour suppressor genes commonly result in predisposition to tumours in diverse types of tissue. Sometimes, as in the case of retinoblastoma, one tissue predominates, presumably because of additional controls that protect cells in other tissues. However, in such cases tumours do occur in other tissues. In other cases, such as Li–Fraumeni syndrome, the predisposition is to multiple tumour types.

5.7.1 Retinoblastoma

The study of **retinoblastoma** allowed Alfred Knudson to formulate the tumour suppressor paradigm described above, which has been central to the study of hereditary cancers. Retinoblastoma is a rare childhood cancer of the retina, affecting about 5 in 100 000. About 40% of the cases show an autosomal dominant pattern of inheritance while the remaining 60% are sporadic. Several features of the inherited form distinguish it from the sporadic form. The penetrance of the hereditary form is 95%, the mean number of tumours is three and tumours in both eyes are common. Survivors are susceptible to other tumours, particularly osteosarcoma and soft-tissue sarcomas, at a rate of about 15% by age 30. In contrast the children with the sporadic form will normally suffer from a single tumour in one eye and are not susceptible to other types of cancer.

Knudson proposed that the hereditary form occurs when one copy of the gene involved, called *RB1*, is inactivated in the germline and the other is inactivated by mutation during somatic development. He showed that the known rate of mutation at the *RB* locus, the number of target embryonic **retinoblasts** (the embryonic precursor cells), the penetrance of the hereditary form and population frequency of the sporadic form were all consistent with such a hypothesis. Multiple tumours occur in the hereditary form because of the high probability that more than one retinoblast will be affected by a second mutation hit. The increased occurrence of other tumour types reflects a common role for the *RB1* gene product in proliferation control in other tissues. In contrast to the hereditary form, multiple tumours are rare in sporadic cases because it is extremely unlikely that more than one retinoblast will be independently affected by two hits.

This model predicts that when the second copy of the *RB1* gene is examined it will be found to have been inactivated. This can happen in a variety of ways: (i) point mutation, (ii) deletion, (iii) chromosome non-disjunction or (iv) mitotic crossing-over. Three of these mechanisms will result in tumour cells becoming homozygous for markers near the *RB1* locus on the chromosome bearing the germline mutation (Figure 5.10). Such a situation is known as **loss of heterozygosity** (LOH).

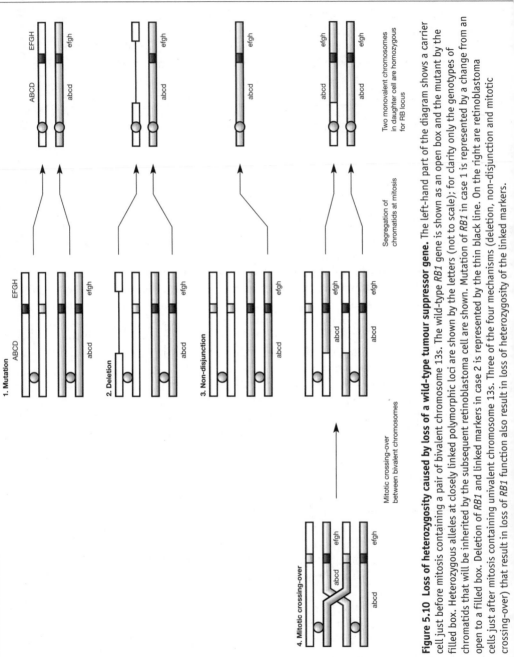

Figure 5.10 Loss of heterozygosity caused by loss of a wild-type tumour suppressor gene. The left-hand part of the diagram shows a carrier cell just before mitosis containing a pair of bivalent chromosome 13s. The wild-type *RB1* gene is shown as an open box and the mutant by the filled box. Heterozygous alleles at closely linked polymorphic loci are shown by the letters (not to scale); for clarity only the genotypes of chromatids that will be inherited by the subsequent retinoblastoma cell are shown. Mutation of *RB1* in case 1 is represented by a change from an open to a filled box. Deletion of *RB1* and linked markers in case 2 is represented by the thin black line. On the right are retinoblastoma cells just after mitosis containing univalent chromosome 13s. Three of the four mechanisms (deletion, non-disjunction and mitotic crossing-over) that result in loss of *RB1* function also result in loss of heterozygosity of the linked markers.

all of chromosome band 13q14. This observation identified the location of *RB1*.
The isolation of *RB1* allowed these predictions to be tested. The use of RFLP
markers showed LOH was common in tumour cells from hereditary cases.
Moreover, the introduction of the wild-type *RB1* gene into such tumour cells
suppressed their tumour characteristics, providing further experimental verification of the tumour suppressor hypothesis. LOH is now used as a key
diagnostic test of whether a gene is acting as a tumour suppressor.

Although the *RB1* gene was first identified through a rare form of
cancer, subsequent research has shown that its product (RB) plays a key
role in regulating the cell cycle (see Box 5.3). It acts to block cells from ini-
tiating a round of mitotic cell division unless stimulated by mitogens. Other
mutations that affect components in the same pathway as RB have also
been found in hereditary cancers. For example, INK4a, which encodes one
of the p16 family of CDK inhibitors (see Box 5.3), is mutated in familial
melanoma. Many sporadic cancers have a high frequency of mutations in
components of the RB pathway. For example, homozygous loss of *RB1* is
found in nearly all small cell lung carcinomas.

5.7.2 Li–Fraumeni syndrome

Li–Fraumeni syndrome is a rare autosomal dominant disease characterised by
predisposition to multiple forms of cancer. In women, breast cancer is the most
common of these. The gene affected by Li–Fraumeni syndrome is *TP53*, which
encodes p53, a key cell cycle regulator (see Box 5.3). p53 provides a checkpoint
function in cells with damaged DNA. If the damage is repairable, p53 stalls the
cell cycle by stimulating inhibitors of CDK4/cyclin D and CDK2/cyclin E,
which are protein kinase/cyclin complexes responsible for triggering the start of
the cell cycle (see Box 5.3). If the damage is severe, p53 triggers cell death by
apoptosis. The removal of this function allows cells with damaged DNA to
divide, increasing genome instability and increasing the chances of damage to
another component of the control machinery. Inactivation of p53 is probably
the most frequent mutational event in sporadic cancers.

5.7.3 Hereditary breast cancer

Breast cancer is the most common form of malignant cancer among
women. Each year in the USA an estimated 182 000 women will develop
breast cancer and 46 000 women will die from the disease. Estimates of the
cumulative lifetime risk vary from 1 in 8 to 1 in 12. The risk of breast
cancer is two- to three-fold higher in women with first-degree relatives
(mother or sister) with the disease, with the higher risks associated with an
earlier age of diagnosis in the relative (<40 years of age). Despite this, no
specific genetic factors have been identified in the majority of these cases.
However, a subset of between 5 and 10% of cases, known as **hereditary
breast cancer** (HBC), are due to germline mutations. HBC shows a number
of characteristics that distinguish it from the sporadic form.

- HBC clusters in families, showing a pattern of inheritance that is characteristic of an autosomal dominant trait.

- As well as breast cancer, there is an increase in other forms of cancer, particularly ovarian cancer and male breast cancer, otherwise a rare disease.

- Age of onset in HBC is much lower: the mean age of diagnosis in HBC is 47 compared to 62 in the sporadic form.

- In HBC there is a much higher incidence of bilateral cancers, i.e. tumours in both breasts.

5.7.3.1 *BRCA1* and *BRCA2*

Mutations in two genes, *BRCA1* and *BRCA2*, have been shown to be responsible for 60–80% of all HBC cases. A woman carrying either of these mutations has about a 40% chance of suffering from breast cancer by the age of 40, a 66% chance by age 55 and has an overall lifetime risk of 82%. *BRCA1* mutations also confer a similarly high risk of ovarian cancer while *BRCA2* mutations are associated with an increased risk of male breast cancer, although the risk of ovarian cancer is also elevated.

Both *BRCA1* and *BRCA2* were mapped by searching for co-inheritance of polymorphic genetic markers and the disease in families in which the incidence of breast cancer is high. *BRCA1* is located at chromosome 17q21 and *BRCA2* at chromosome 13q12–13 (this is described in more detail in Chapter 6). Both genes were cloned by positional methods, concentrating on transcripts produced in the region to which the genes were mapped. *BRCA1* is composed of 22 exons distributed over more than 100 kb of genomic DNA. It encodes a protein of 1863 amino acids whose biochemical role is not yet clear, although evidence is accumulating that it is involved in maintaining genome integrity. *BRCA2* is composed of 27 exons spread over 70 kb of genomic DNA; it encodes an extremely large protein of 3418 amino acids.

Cloning the *BRCA1* and *BRCA2* allowed mutations that predispose to breast cancer to be identified. A very large number of different mutations have now been characterised which, together with the large size of both genes, will make testing for mutations very difficult. For the most part, the mutations are frameshift and nonsense mutations that would be expected to destroy gene function and thus be recessive at a cellular level. Since the mutations in both genes are inherited as autosomal dominants, it was expected that both genes would be tumour suppressors following the paradigm established for retinoblastoma (see above). This was confirmed when it was found that LOH occurred in nearly all cases examined. Thus HBC comes about when one copy of the gene is inactivated by a germline mutation and the other is lost by a deletion in somatic cells. Furthermore, *BRCA1* has been shown to function as a tumour suppressor *in vitro*. The proliferation of breast and ovarian cancer cell lines is inhibited by transfection with wild-type, but not mutant, *BRCA1*. This action is tumour specific: the proliferation of lung or colon cell lines is not inhibited by *BRCA1*.

It was hoped that identifying tumour suppressors specific to breast and ovarian cancers would help elucidate the molecular pathology of sporadic breast cancers. It is still not clear whether this hope will be realised because the evidence is conflicting. Expression of *BRCA*1 and *BRCA*2 is reduced in sporadic breast cancers, which might be expected of a tumour suppressor. Moreover, a significant proportion of sporadic cases show LOH at either 17q21 or 13q12–13, where *BRCA*1 and *BRCA*2 are located. However, there is a conspicuous absence of mutations in the remaining *BRCA*1 or *BRCA*2 genes, which would be predicted by the tumour suppressor model. It is possible that the mutations in the remaining copy are occurring in non-coding regions and are acting to reduce expression.

5.7.3.2 Population distributions of *BRCA*1 and *BRCA*2 mutations

The size of *BRCA*1 and *BRCA*2 and the large number of different mutations that can occur make it difficult to undertake population screens to measure the mutation frequencies of these genes. An indirect estimate of the frequency of *BRCA*1 alleles may be obtained from the incidence of ovarian cancer. First-degree relatives of breast cancer patients are significantly more likely to suffer from ovarian cancer compared with the general population. Since *BRCA*1 mutations are the main cause of hereditary ovarian cancer, the frequency of such mutations may be calculated from the excess risk of ovarian cancer in first degree relatives of breast cancer patients. The allele frequency arrived at by these means is 0.0006. This corresponds to a carrier frequency of 1 in 800. The carrier frequency for all HBC mutations may be as high as 1 in 300. Although such mutations account for only 5–10% of all breast cancer cases, the frequencies are very high for germline mutations and breast cancers due to *BRCA*1/2 mutations may be said to be one of the most common monogenic disorders.

Two particular HBC mutations, *BRCA*1 185delT and *BRCA*2 6174delT, have each been found to present at a frequency of about 1% in the Ashkenazi Jewish population in the USA. *BRCA*1 185delT was found to be responsible for about 28% of early-onset breast cancers in this population, but the *BRCA*2 mutation was only found to be responsible for 8% of such cancers. The proportion of cases caused by *BRCA*1 is consistent with estimates of its penetrance based on its segregation in HBC families. However, the penetrance of the *BRCA*2 mutation based on these figures is about four-fold lower than previous estimates. Another study with the same population showed that the risk of cancer in all carriers of both *BRCA*1 185delT and *BRCA*2 6174delT mutations was 56% by age 70, significantly lower than the 85% risk estimated from high-risk breast cancer families.

This discrepancy highlights a problem with the estimates of penetrance derived from high-risk families, i.e. there may be other factors contributing to the risk in these families that are not present in normal families. For example, the penetrance of a *BRCA*1 or *BRCA*2 mutation may be modified by alleles at other loci. In the breast cancer families there may be a higher frequency of unfavourable modifying alleles, leading to a higher penetrance of *BRCA*1 and *BRCA*2 mutations. In other words, there has been a bias of ascertainment that led to unusual families being selected because in these families the *BRCA1/2* alleles behaved as simple autosomal dominants. *BRCA1* and *BRCA2* may have a lower penetrance in genetic backgrounds that are more common. If this is the case, counselling a woman who carries a *BRCA1* or *BRCA2* mutation may need to take account of her family history.

5.7.4 Colorectal cancer

Colorectal cancer is the second or third most common cancer of the western world with a lifetime risk of between 5 and 6%. It is rarer in the developing world and Japan. About 90% of cases are sporadic. However, there are two well-characterised hereditary forms: **familial adenomatous polyposis** and **hereditary non-polyposis colorectal cancer**. Both forms fulfil the classic criteria of tumour suppressors.

5.7.4.1 Familial adenomatous polyposis

Familial adenomatous polyposis (FAP) affects between 1 in 8000 and 1 in 15 000 of the population. It is characterised by the presence of hundreds to thousands of benign adenomas or polyps covering the colon and rectum. These polyps are clonal in origin. Some of these polyps are larger than others and from these malignant carcinomas arise. FAP is inherited as an autosomal dominant disease. The gene involved is called **APC** (adenomatous polyposis coli). It was mapped by linkage studies to 5q21–22 and positionally cloned in 1991. *APC* is composed of 15 exons, producing a 9 kb mRNA. The cellular role of the gene is currently unknown but it is thought to be involved in cell-to-cell signalling or adhesion. Most cells in polyps are homozygous for mutations or deletions of the *APC* gene. Occasional polyps are found in normal individuals and these also bear mutations in the *APC* gene. However, loss of the *APC* gene is not sufficient for carcinomas to develop. Mutations in other genes, such as *TP53* encoding p53, *KRAS* (a member of the *ras* family) and another tumour suppressor gene known as *DCC* (deleted in colon cancer) are found in full-blown carcinomas.

5.7.4.2 Hereditary non-polyposis colorectal cancer

Hereditary non-polyposis colorectal cancer (HNPCC), also known as Lynch syndrome, is a hereditary cancer that is not associated with the formation of polyps. It is characterised by an earlier age of onset than sporadic colorectal cancer and frequently involves cancer in other organs. Estimates of its incidence vary widely because of difficulties in distinguishing it from the sporadic form. The classical form of the disease is defined by its occurrence in three members of a family in successive generations, with one affected individual diagnosed before age 50. Some estimates are as low as 1 in 10 000, others as high as 1 in 200. Estimates of what proportion of all colorectal cancers are caused by HNPCC range from 0.7 to 13%. A mean of these figures would indicate a population incidence of 1 in 400, making it one of the most common of monogenic disorders.

HNPCC is characterised by instability of multiple microsatellite loci located throughout the genome. This suggests a defect in mismatch repair of

DNA. Mismatch repair operates by excising one strand of a stretch of DNA that contains a base-pair mismatch. The missing strand is then replaced by the action of DNA polymerase, using the remaining strand as a template. Mutations in genes mediating this process characteristically result in a decrease in the stability of repeated DNA sequences.

After an HNPCC susceptibility locus was located to chromosome 2p by linkage analysis, a search was undertaken for genes that might provide this function. A candidate gene, human *MSH2*, was cloned that is a homologue of the yeast *MSH2* gene and the bacterial *mutS*, both of which are known to mediate mismatch repair. It was shown that mutations affecting *MSH2* co-segregated with the disease in several HNPCC kindreds. Subsequent to the identification of *MSH2*, three other genes have been cloned that are also involved in mismatch repair and in which germline mutations co-segregate with the disease in different HNPCC kindreds. These genes are known as *MLH1*, *PMS1* and *PMS2*. Most cases (~80%) of HNPCC are caused by mutations in *MLH1* or *MSH2*. Most mutations affecting these genes result in truncated proteins or substitutions at highly conserved residues. In both cases an inactive protein would result. Thus it seems likely that the two-hit tumour suppressor model applies to HNPCC. Individuals with germline mutations inactivating one copy of the gene develop tumours when the other copy is lost due to somatic mutation.

5.7.5 Neurofibromatosis

Neurofibromatosis (NF) is characterised by so-called *café-au-lait* pigmented skin patches and disfiguring benign tumours called neurofibroma, which are outgrowths of **neural crests**. It exists in two forms, the more common, NF type 1 (NF1), being one of the most widespread autosomal dominant diseases, affecting 1 in 3000 children. Mutations in the *NF1* gene predispose to a number of tumours that originate in cells of the nervous system.

The *NF1* gene is located at chromosome 17q11 and encodes a protein known as neurofibromin. Neurofibromin is an inhibitor of the *ras* oncogene, acting in the same manner as another protein, GTPase-activating protein (GAP; Figure 5.11). The tissue specificity of *NF1* mutations may reflect an absence or lowering of GAP function in neural crest cells, making them more dependent on neurofibromin. Tumours arising in patients with *NF1* frequently have inactivating mutations in the remaining copy of the *NF1* gene. Therefore the *NF1* gene behaves as a classic tumour suppressor gene, in terms of both its cellular function as an inhibitor of an oncogene and its genetic properties.

5.7.6 Ataxia telangiectasia

Ataxia telangiectasia (AT) is another disease, like Li–Fraumeni syndrome and retinoblastoma, where rare germline mutations help identify an important part of the normal machinery that protects against cancer. AT is a rare autosomal recessive disease, affecting between 1 in 40 000 and 1 in 100 000 individuals worldwide. It is a complex disease, characterised by a variety of

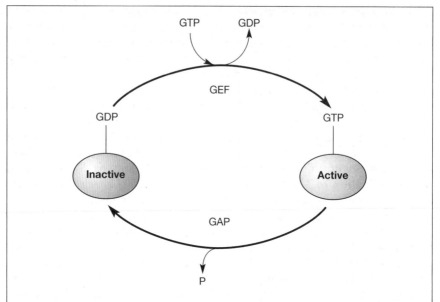

Figure 5.11 The *ras* cycle. When bound to a molecule of GTP, *ras* is inactive. Hydrolysis of GTP mediated by GTPase-activating protein (GAP) converts it to the inactive state. Exchange of GDP for GTP, mediated by guanosine exchange factor (GEF), converts it back into the active state. Oncogenic mutations lock *ras* into the active form.

apparently unconnected symptoms that first become apparent at the age of 3 and lead to death in the second or third decade:

- **ataxia** caused by cerebellar degeneration, which leads to progressive neuromotor deterioration;

- dilated blood vessels in the eye and face (telangiectasia);

- severe immune deficiency and absence of the thymus;

- acute sensitivity to ionising radiation;

- predisposition to multiple cancers;

- heterozygotes are mildly affected but exhibit predisposition to cancer and radiation sensitivity.

Genetic studies were confused by there being apparently four complementation groups, each manifesting distinct phenotypic characteristics. A single gene was located by linkage mapping, so the complementation observed is apparently an example of intragenic complementation. At a cellular level, AT cells behave as if they have a defect in the p53-mediated checkpoint pathway. The gene encodes a large protein with similarities to phosphoinositol 3-kinases, which have already been shown to be part of intracellular signalling pathways and to be involved in the control of DNA synthesis and

5.8 Summary

Benefits of cloning genes
- It tells us about the role of the affected protein within the cell, allowing research into new drugs to treat the disease.

- Characterisation of mutations that result in the disease allows diagnostic tests to be developed.

- Availability of the wild-type gene may lead to gene therapy.

Positional cloning
- The gene is mapped by looking for linkage between polymorphic genetic markers and the occurrence of the disease.

- Clones for the area are located in existing clone maps or isolated by chromosome walking and jumping.

- Possible genes are located by examining transcripts arising from the region.

- Co-segregation of mutations with the disease or features such as trinucleotide repeat expansions are used to decide which of the possible genes is responsible for the disease.

- The positional candidate approach examines genes known to map in the same area and which have a role consistent with the biological defect.

Cystic fibrosis
- CF is the most common autosomal recessive disease that affects Europeans.

- The gene affected is called *CFTR*. It encodes a protein that transports chloride ions across the cell membrane.

- Two-thirds of mutant alleles are the same mutation, called ΔF508.

Duchenne muscular dystrophy
- DMD is a sex-linked disease.

- The chromosomal locus is exceptionally large (2.5 Mb).

- The gene affected encodes a protein called dystrophin that links the actin cytoskeleton to the extracellular cortex via a complex called DASC.

Trinucleotide repeat expansions
- Expansions involving the trinucleotide CCG are responsible for fragile sites in chromosomes, including fragile-X syndrome. These occur in the 5' non-coding area of genes and result in their inactivation.

- Expansions involving CAG in the coding region are responsible for neurodegenerative disorders, including HD.

- MD is caused by an expansion of a CTG in the 3' UTR of a protein kinase.

- Trinucleotide repeats are unstable between generations, leading to anticipation.

Haemoglobinopathies

- These are diseases affecting haemoglobin that result in severe anaemias.

- Sickle cell disease results from a single amino acid substitution.

- Thalassaemia results in a reduction of α-globin or β-globin levels

- β^0 Thalassaemia results from mutations that prevent any β-globin expression. Milder forms of α thalassaemia can arise when deletions remove the β-globin gene but there is a compensatory persistence of foetal haemoglobin.

- There are two copies of the α-globin gene in the cluster so there are four copies in each diploid individual. The severity of α thalassaemias correlates with the number of remaining α-globin genes. The most severe is hydrops fetalis in which there are no remaining copies.

Hereditary cancers

- A complex network controls cell proliferation. Cancer results from the accumulation of somatic mutations that disrupt this network.

- Tumour suppressors are genes that normally suppress cancers; cancers result when both copies are knocked out by mutation.

- Many hereditary cancers result when one copy of a tumour suppressor gene is inactivated by a germline mutation and the other copy is lost by somatic mutation.

- Retinoblastoma is a rare form of cancer, the study of which led to the identification of a key tumour suppressor gene that is mutated in many sporadic cancers.

- About 5–10% of breast cancer is caused by mutations in two genes, *BRCA*1 and *BRCA*2, that act as tumour suppressor genes. Whether they are affected in sporadic cases is still uncertain.

- Hereditary colon cancers can also be caused by mutations in tumour suppressor genes:
  - FAP is caused by mutations in the *APC* gene.
  - HNPCC is caused by mutations in genes responsible for mismatch repair.

- NF is caused by mutations in a gene that is a negative regulator of the *ras* oncogene that is commonly mutated in sporadic cancers.

Further reading

Positional cloning

COLLINS, F.S. (1995) Positional cloning moves from perditional to traditional. *Nature Genetics*, **9**, 347–350.

POUSTKA, A. and LEHRACH, H. (1986) Jumping and linking libraries: the next generation of molecular tools in mammalian genetics. *Trends in Genetics*, **2**, 174–178.

Cystic fibrosis

COLLINS, F.S. (1992) Cystic fibrosis: molecular biology and therapeutic implications. *Science*, **256**, 774–779.

KEREM, B.S., ROMMENS, J.M., BUCHANAN, J.A. *et al.* (1989) Identification of the cystic fibrosis gene: genetic analysis. *Science*, **245**, 1073–1080.

MARX, J.L. (1989) The cystic fibrosis gene is found. *Science*, **256**, 923–925.

RICH, D.P., ANDERSON, M.P., GREGORY, R.J. *et al.* (1990) Expression of cystic fibrosis transmembrane conductance regulator corrects defective chloride channel regulation in cystic fibrosis airway epithelial cells. *Nature*, **347**, 358–363.

RIORDAN, J.R., ROMMENS, J.M., KEREM, B.S. *et al.* (1989) Identification of the cystic fibrosis gene: characterisation of complementary DNA. *Science*, **245**, 1066–1073.

ROMMENS, J.M., IANUZZI, M.C., KEREM, B.S. *et al.* (1989) Identification of the cystic fibrosis gene: chromosome walking and jumping. *Science*, **245**, 1059–1065.

ZIELENSKI, J. and CHEE, L.C. (1995) Cystic fibrosis: genotypic and phenotypic variations. *Annual Review of Genetics*, **29**, 777–807.

Duchenne muscular dystrophy

AHN, A. and KUNKEL, L.M. (1993) The structural and functional diversity of dystrophin. *Nature Genetics*, **3**, 283–291.

BROWN, R.H. (1997) Dystrophin-related proteins and the muscular dystrophies. *Annual Review of Medicine*, **48**, 457–466.

KOENIG, M., HOFFMAN, E.P., BERTELSON, C.J., *et al.* (1987) Complete cloning of the Duchenne muscular dystrophy (DMD) cDNA and preliminary genomic organisation of the DMD gene in normal and affected individuals. *Cell*, **50**, 509–517.

MONACO, A.P. and KUNKEL, L.M. (1987) A giant locus for the Duchenne and Becker muscular dystrophy gene. *Trends in Genetics*, **3**, 33–37.

ROBERTS, R.G. (1995) Dystrophin, its gene; and the dystrophinopathies. *Advances in Genetics*, **33**, 177–231.

Trinucleotide repeat expansions

ASHLEY, C.T. and WARREN, S.T. (1995) Trinucleotide repeat expansion and human disease. *Annual Review of Genetics*, **29**, 703–728.

Huntington's Disease Collaborative Research Group (1993) A novel gene containing a trinucleotide repeat that is expanded and unstable on Huntington's disease chromosomes. *Cell*, **72**, 971–983.

OOSTRA, B.A. and WILLEMS, P.J. (1995) A fragile gene. *Bioessays*, **17**, 941–947.

WILLEMS, P.J. (1994) Dynamic mutations hit double figures. *Nature Genetics*, **8**, 213–215.

Globin gene clusters

COLLINS, F.S. and WEISSMAN, S.M. (1984) The molecular genetics of human haemoglobin. *Progress in Nucleic Acid Research and Molecular Biology*, **31**, 314–462.

GROSVELD, F., ANTONIOU, M., BERRY, M. *et al.* (1993) The regulation of human globin gene switching. *Transactions of the Royal Society of London Series B*, **339**, 183–191.

Inherited cancers: general

BROWN, M.A. and SOLOMON, E. (1997) Studies on inherited cancers: outcomes and challenges of 25 years. *Trends in Genetics*, **13**, 202–207.

KNUDSON, A.G. (1993) Anti-oncogenes and human cancer. *Proceedings of the National Academy of Sciences USA*, **90**, 10914–10921.

SHER, C.J. (1996) Cancer cell cycles. *Science*, **274**, 1672–1677.

BOYD, J. (1995) BRCA1: more than a hereditary breast cancer gene? *Nature Genetics*, **9**, 335–336.

EASTON, D. (1997) Breast cancer genes: what are the real risks? *Nature Genetics*, **16**, 210–211.

MIKI, Y., SWENSEN, J., SHATTUCKEIDENS, D. *et al.* (1994) A strong candidate for the breast and ovarian cancer susceptibility gene *BRCA1*. *Science,* **266**, 66–71.

ODDOUX, C., STRUEWING, J.P., CLAYTON, C.M. *et al.* (1996) The carrier frequency of the BRCA2 617delT mutation among Ashkenazi Jewish individuals is approximately 1%. *Nature Genetics*, **14**, 188–190.

ROA, B.B., BOYD, A.A., VOLCIK, K. and RICHARDS, C.S. (1996) Ashkenazi Jewish population frequencies for common mutations in *BRCA1* and *BRCA2*. *Nature Genetics*, **14**, 185–187.

TAVTIGIAN, S.V., SIMARD, J., ROMMENS, J. *et al.* (1996) The complete *BRCA2* gene and mutation in chromosome 13-linked kindreds. *Nature Genetics*, **12**, 333–337.

WOOSTER, R., BIGNELL, G., LANCASTER, J. *et al.* (1995) Identification of the breast cancer susceptibility gene *BRCA2*. *Nature*, **378**, 789–792.

WU, L.C., WANG, Z.W., TSAN, J.T. *et al.* (1996) Identification of a RING protein that can interact *in vivo* with the *BRCA1* gene product. *Nature Genetics*, **14**, 430–439.

Hereditary non-polyposis colorectal cancer

DE LA CHAPPELLE, A. and PELTOMÄKI, P. (1995) Genetics of hereditary colon cancer. *Annual Review of Genetics*, **29**, 329–348.

Neurofibromatosis type 1

SHEN, M.H., HARPER, P.S. and UPADHYAYA, M. (1996) Molecular genetics of Neurofibromatosis type-1 (NF1). *Journal of Medical Genetics*, **33**, 2–17.

Ataxia telangiectasia

SAVITSKY, K., BARSHIRA, A., GILAD, S. *et al.* (1995) A single ataxia gene with a product similar to PI-3 kinase. *Science*, **268**, 1749–1753.

CHAPTER SIX

The genetic components of complex diseases

Key topics

- Evidence for genetic components in complex disease
- The genetic architecture of complex diseases
- Identifying genes involved in complex diseases
- Candidate genes
  - Parametric studies
  - Non-parametric studies
  - Population association studies
- The genetic components of Alzheimer's disease.

6.1 Introduction

Most of the genes involved in the major monogenic disorders have now been isolated by positional cloning. Monogenic disorders are often devastating in their consequences, but their population incidence is low so their overall contribution to public health is relatively minor. Diseases such as cancer, asthma, migraine, diabetes, rheumatoid arthritis, multiple sclerosis, hypertension, cardiac disease, obesity and psychiatric disorders such as schizophrenia, manic depression and autism are much more common. Such diseases do not show any clear pattern of Mendelian inheritance. They do, however, cluster in families, indicating that a genetic component is likely to be operating. The aetiology of such diseases is complex, being a mix of environmental and genetic causes. For this reason they are often referred to as **complex** or **multifactorial diseases**. Research in human molecular genetics is now focused on identifying the genetic factors involved in complex diseases. As we shall see this is not an easy task. However, success in this enterprise promises to prevent, or lead to early diagnosis and better treatment of, diseases that may affect us all at some stage in our lives. These benefits may be listed as follows.

- Identification of individuals with an increased susceptibility to a disease will allow lifestyle changes to lower the risk and more intensive medical surveillance to detect the first signs of its onset should it occur. Early diagnosis is nearly always a strong factor in successful treatment.

- Establishing the underlying genetic risk factors removes one variable from the study of the genotype–environment interaction. Identification of environmental risk factors thus becomes easier and more precise.

- Many diseases with apparently similar clinical features may be heterogeneous in their molecular **aetiology** (causes). This may affect the prognosis and be the reason why some patients respond well to one particular treatment and others do not. If the different types of defect that can lead to the same disease can be elucidated, and suitable diagnostic tests developed, then treatments can be better fitted to the actual disease.

- Cloning the genes contributing to the risk will lead to the characterisation of the encoded proteins. These proteins will be components of the cellular processes that are malfunctioning in the disease. This will clearly lead to a better understanding of the normal state and what is going awry in the pathogenic state. Such knowledge will be the starting point for the development of more rational therapies. In particular the proteins affected may be used to establish screens for small-chemical drugs that might modify their action. For example, suppose a particular disease is shown to occur through the hyperactivity of a cell surface receptor. The receptor may be expressed as a **recombinant protein**, that is where it is expressed from the cloned gene in another host such as *E. coli*, yeast or mammalian tissue culture cells. This will provide a plentiful supply of the protein, which can be used in an *in vitro* screen for small drugs that reduce its activity.

In this chapter we shall first of all review the evidence that genetic factors do contribute to the aetiology of common complex diseases. We shall then consider the possible ways in which such factors will be operating, drawing upon models for continuous variation elaborated by quantitative geneticists working with experimental organisms. We shall then be in a position to consider experimental strategies used in human research to identify the genetic loci involved and to compare the strengths and weaknesses of each. However, right from the outset, it is important to realise there is no one certain method that can be relied upon to identify all the factors involved and that in reality a combination of approaches are used. Each of the individual methods will be illustrated with case studies. Finally at the end Alzheimer's disease will be used to illustrate how the approaches were combined to make real progress in the analysis of an important and common disease.

6.2 Evidence for a genetic component in complex disorders and phenotypic traits

Because family members are likely to share the same environment as well as the same genes, it is not easy to disentangle the relative contributions of genetic and environmental factors which may determine what we are like and what diseases we are likely to suffer from. For the most part, susceptibility to common diseases and phenotypic characteristics do not follow simple patterns of inheritance, so neither is likely to be determined monogenically. Nevertheless, family, adoption and twin studies clearly show a strong genetic component.

6.2.1 Family studies

Members of a family share a proportion of genes in common that may be simply predicted from their degree of relatedness. For example, siblings or parent/child pairs share 50% of their genes. Pairs of individuals in an extended family may be classified as first, second or third degree relatives according to the proportion of genes they share (Table 6.1). If genetic factors play a role in the occurrence of the disease then the average risk that a relative of an affected individual will also suffer from the same disease should be correlated to the degree of relatedness to the affected person. Table 6.2 shows two examples of congenital disorders where this has been shown to be true.

Table 6.1 Genes shared in common by different classes of relatives

Class of relative	Proportion of genes shared	Examples
First degree	50%	Parent/child, siblings
Second degree	25%	Grandparent/grandchild, aunt or uncle/niece or nephew
Third degree	12.5%	Cousins, great-grandparent/great-grand child

Table 6.2 Observed risks to relatives of probands suffering from congenital malformations. Pyloric stenosis affects new born babies and infants. It results in projectile vomiting caused by a malfomation of the pyloric sphincter at the junction between the oesophagus and stomach.

Category affected	Cleft lip	Pyloric stenosis
General population	0.001	0.001
First degree relatives	x40	x10
Second degree relatives	x7	x5
Third degree relatives	x3	x1.5

One way to dissect the different contributions of nature and nurture is to measure the recurrence of a disease in biological relatives who have been reared apart through adoption. This may then be compared with the recurrence risk in a control group of non-adoptive families. For example the risk that the offspring of a schizophrenic parent will be affected is 13%, compared to a population incidence of 1%. In one investigation, the frequency of schizophrenia in adopted offspring of schizophrenic parents was compared to the frequency of schizophrenia in a control group of adopted individuals of unaffected parents. In the 47 adopted children of schizophrenic mothers, 5 also suffered from the disorder. In contrast none of the 50 individuals in the control group were affected. Apparently there is little change in the rate of recurrence, regardless of whether a child is reared by its biological or adopted parent. Thus the reason why schizophrenia is familial is that family members share the same genes not that they share the same environment.

6.2.3 Twin studies

Twin studies provide a natural experiment with which to examine the relative contributions of nature and nurture. Monozygotic or identical twins are genetically identical, dizygotic twins share only 50% of their genes. Since both types are born at the same time, in the same family, the environment in which they grow up is likely to be similar. Of course in practice there will be differences in the actual environment they experience. There may be differences in the position of the foetus, difficulties during birth, they may contract different diseases, their interaction with other family members, friends, teachers, etc. may be different, they will experience different life events such as accidents and so on. It is also important to remember that some differences may come about simply through the action of chance and some perceived differences may be the result of error in experimental measurement. All these diverse effects are collected together under the umbrella term non-shared environment.

The parameter that is usually measured in twin studies is concordance. This is defined as the percentage proportion of identical twins that both suffer from a disease or exhibit a phenotypic trait when that disease or trait occurs in one member of the pair. Both monozygotic and dizygotic twins share the same environment. However, dizygotic twins share only 50% of their genes, whereas monozygotic twins share all their genes. So the difference in concordance between monozygotic and dizygotic twins is a measure of the extent to which genetic differences contribute towards the population variability of a trait, a parameter known as **heritability**. We will consider heritability in more detail in the next section. In certain cases, monozygotic twins are separated at birth and reared apart. While the number of such cases studied is much smaller than the bulk of twin studies, they provide an additional type of experiment where genetically identical individuals experience different environments.

Table 6.3 shows some examples of twin concordance studies. The difference in concordance between monozygotic and dizygotic twins shows that there is a strong genetic component to many common diseases. Clearly, lifetime health is strongly influenced by genetic constitution. It is interesting to note that a disease such as tuberculosis which may have been assumed to be wholly environmental, because it is caused by a bacterial infection, shows a high concordance between identical twins. This illustrates how resistance to infection is influenced by genetic factors. For most of the human race, for most of its history, infectious diseases were a common cause of death. It is hardly surprising that natural selection should operate to favour genes that increase resistance. Another interesting feature is that breast cancer shows relatively low concordance, despite the identification of two genes, *BRCA*1 and *BRCA*2 that effectively result in simple autosomal dominant inheritance for the disease. The reason for is that *BRCA*1 and *BRCA*2 only account for a small fraction of the total number of breast cancer cases.

Table 6.3 Examples of twin concordance values for some common diseases

Disorder	Concordance %	
	Monozygotic	Dizygotic
Breast cancer	6.5	5.5
Cleft lip	35	5
Insulin-dependent diabetes mellitus	50	5
Non-insulin-dependent diabetes mellitus	100	10
Multiple sclerosis	20	6
Peptic ulcer	64	44
Rheumatoid arthritis	50	8
Tuberculosis	51	22

6.2.4 Behavioural disorders and personality traits

Twin studies also show that some behavioural disorders such as schizophrenia have at least in part a genetic basis (Table 6.4). One disorder that seems to be strongly influenced by genetic factors is autism. This disorder was pre-

Table 6.4 Examples for twin concordance values for common psychiatric disorders

Disorder	Concordance %	
	Monozygotic	Dizygotic
Alcoholism	40	20
Autism	60	7
Schizophrenia	44	16
Alzheimer's disease	58	26
Major affective disorder	60	20
Reading disability (dyslexia)	64	40

viously considered to be entirely environmental, said to be caused by cold and aloof parents. Sparing such parents the additional trauma of such a label is a powerful benefit of such studies.

The application of twin and adoption studies to the study of normal phenotypic variation has revealed strong genetic components to the observed phenotypic variation of many fundamental human attributes. Table 6.5 shows that twin concordance is high for general intelligence, personality (extraversion/neuroticism) and even general happiness. These studies on the contribution of genetic factors to innate personality are controversial. In part the controversy is based on technical factors such as the design of the experiments and what is actually being measured. However, the controversy also arises out of a fundamental misconception as to how these studies can be interpreted. They do not show that health, intelligence or happiness are unalterably predetermined by genetics; rather they allow the observed phenotypic variation in a particular population in a particular environment to be partitioned into genetic and environmental components. The difference may seem semantic, but we shall see in the next section that it is actually fundamental.

Table 6.5 Some examples for twin concordance values for some fundamental human attributes

Trait	Concordance %	
	Monozygotic	Dizygotic
General intelligence	80	56
Perceived happiness	50	8
Neuroticism	44	18
Extraversion	50	18

6.3 The genetic architecture of complex diseases

As we saw in the preceding section, genetic factors are likely to contribute to the risk of complex disease. However, these diseases do not segregate in a Mendelian fashion. In fact many, or even most, phenotypic characteristics such as height or weight show continuous variation, which is apparently inconsistent with Mendel's laws, which predict discrete phenotypic classes. This inconsistency caused a great controversy at the beginning of the 20th century between those such as Francis Galton, who analysed traits showing continuous variation, and others such as William Bateson who had rediscovered Mendel's laws of inheritance. The two schools were reconciled by Ronald Fisher in 1918 who showed that continuous variation could arise if a trait were controlled by a number of genes called **polygenes**. Each allele of the individual polygenes segregates in a Mendelian fashion but only makes a small contribution to the overall phenotype. The combined effect of many polygenes, modified by environmental factors, would produce the continuous variation observed. Mendel's experiments were carried out on traits that were special cases, in that they were controlled by pairs of alleles at single genes. In human genetics these correspond to monogenic diseases, considered in the previous chapter. To understand the genetics of complex diseases we have to consider what research in quantitative genetics has revealed about the nature of polygenes. Some medically important conditions, such as hypertension, do show continuous variation, however most are discontinuous – a person either does or does not have diabetes or Alzheimer's disease. We shall see how the polygenic model can be developed to take account of this.

6.3.1 Genetic and environmental components of continuous variation

When plotted in a frequency diagram, measurements of a phenotypic characteristic in a population, such as height or weight, will show a continuous distribution that conforms to a normal or Gaussian distribution. Figure 6.1 shows a hypothetical example of the height of wheat plants in a field. There are two important parameters of this distribution that can be defined: the **mean**, *m*, and the **variance**, V_p. As its name implies, variance is a measure of the amount of phenotypic variation observed. As shown in Figure 6.1, variance is related to the standard deviation, the limits within which two-thirds of the samples will be located. The phenotypic variation could come about through variation in the local environments within which each plant grows or because of variation of the genotypes in the plant population. In a plant such as wheat, a conceptually simple experiment will allow the variation

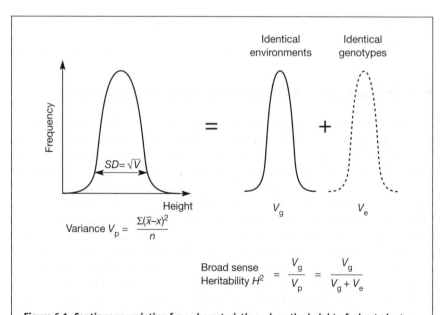

Figure 6.1 Continuous variation for a characteristic such as the height of wheat plants.
SD = standard deviation, V_p = phenotypic variance, V_e = environmental variance, V_g = genetic variance, x = height of each plant, $\bar{x}$ = mean height , n = number of samples.

due to the environmental and genetic factors to be measured. In a pure breeding population of wheat all the plants will have the same genotype; any residual variation will thus be due to environmental variation V_e (Figure 6.1). Conversely, if the original population were grown in a uniform environment then any residual variation would be due to variation in the genotypes of the wheat population, that is the genetic variation, V_g. Thus the total phenotypic variation is the sum of the environmental and genetic variation. The proportion of total variation that is due to genetic variation is known as **heritability**.

Biometrical geneticists distinguish between broad and narrow-sense heritability. Broad-sense heritability is the sum of genetic variation due to all causes, that is variation due to additive, dominant and epistatic components (we shall discuss these further in the next section). Of these, only variation due to additive causes responds to selection and is known as narrow-sense heritability.

In the hypothetical experiment described above, a visitor from Mars observing the field of genetically identical wheat may conclude that variation in height is entirely environmental. However, observation of the wheat growing in a constant environment may lead to the converse conclusion that the variation is entirely genetic. This leads to an important conclusion: heritability is not a fixed or intrinsic property of a character – it depends on the particular population being observed and the particular environments in which the organism exists. To place this in a human context, consider another partly hypothetical example. Cigarette smoking is a major environmental risk in the development of lung cancer. However, not all smokers develop lung cancer. In part this is a reflection of the nature of risk, but also it reflects genetic factors that modify the environmental risk. For example a polymorphism in *CYP2D6*, a gene encoding a cytochrome P450 that metabolises carcinogens, increases the risk of cancer from cigarette smoke. Consider two populations – in one cigarette smoking is universal, in the other health education and restrictions on tobacco advertising mean that only 30% of the population smoke. For simplicity in this example, let us assume that cigarette smoking is the only environmental variable. In the first population the environmental variation is zero and the observed variation in the incidence of lung cancer is entirely genetic. The observed heritability of lung cancer would be 100%, but this does not mean that lung cancer is predetermined by each individual's genetic constitution; it would be almost entirely prevented by not smoking. In the second population, the heritability would be relatively low and the environmental variation high. This example emphasises a critical point – a high heritability does not mean that a trait is genetically predetermined. In a medical context identifying genetically at-risk individuals may allow interventions that prevent the disease from developing. In terms of general phenotypic traits, such as height, weight, etc., it may indicate that there is little variation in exposure to an environmental factor that may nevertheless still have an important influence on the trait.

6.3.1.1 Relative risk λ

It is not possible to carry out controlled experiments in a human population similar to those for an experimental organism such as wheat. As discussed above, heritability can be estimated from twin concordance studies. Alternatively, a parameter that is often used to measure the influence of genetic variation in human populations is called **relative risk**, λ. This is defined as:

$$\lambda_r = \frac{\text{Frequency in relatives of affected person}}{\text{Population frequency}}$$

where r denotes the degree of the relationship: i.e. first degree (sibs), second degree (uncles/nephews, grandparents/grandchildren), or third degree (cousins).

For example the frequency of insulin-dependent diabetes mellitus (IDDM) in sibs of affected individuals is 6% and the population incidence is 0.4%. The λ_s for IDDM is therefore 15 (where the subscript s indicates sibs). In contrast, λ_s for the monogenic disease cystic fibrosis is 500.

6.3.2 Polygenic variation and the threshold hypothesis

The genetic component of the variation shown in Figure 6.1 can be explained by proposing that there is more than one gene that influences the phenotypic trait. Figure 6.2 shows the possible genotypes and their relative frequency that arise from the simplest case where just two genes (each with two alleles) segregate in a population. It is assumed that one allele of each gene shifts the phenotype by an equal amount towards one end of the phenotypic spectrum and that there are no dominance or additive effects. As we shall see in the next section, these are very simplistic assumptions, highly unlikely to apply in reality. Environmental factors also influence the phenotype, so individuals with the same genotype will not have identical phenotypes, rather they will form a subpopulation that also shows a continuous distribution around a mean determined by the genotype. The observed overall population distribution of phenotypes will be the sum of these separate subpopulations.

Even with two genes the distribution starts to resemble a continuous curve after environmental effects are considered. With more genes contributing, the match to a continuous curve becomes better. When two individuals with phenotypes towards one end of the phenotypic distribution mate, then the phenotypes of their progeny will form a distribution where the mean is closer to the population mean than the mean phenotype of the two parents, but still shifted relative to the population mean (Figure 6.3). Because the mean phenotypes of the progeny are shifted from the population mean, the chances that the phenotype of some of the progeny will be extreme are higher than for a member of the general population. So phenotypes will tend to run in families.

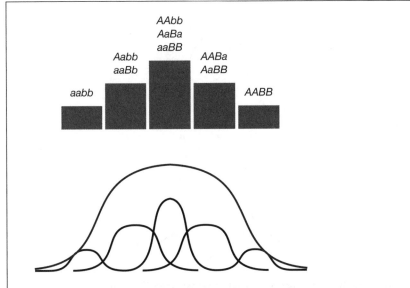

Figure 6.2. Two loci controlling a trait with environmental variation. A quantitative phenotype is controlled by two loci each with two alleles, *A* or *a* and *B* or *b*. Each copy of *A* or *B* shifts the phenotype by an equal amount. There are thus five possible genotypic classes depending on the number of these alleles. The mean phenotype of each class is determined by the number of *A* and/or *B* alleles. However, environmental variation results a continuous distribution about the mean of each class, shown by the black curves. The observed population variation will be the sum of these subgroups and approaches a continuous distribution, shown by the blue line.

For medically important traits that show a continuous distribution, such as hypertension or obesity, this model is sufficient to account for the observed behaviour of such traits to run in families without showing Mendelian inheritance. In such a model we assume that one of the alleles at each locus increases the phenotypic value of the quantitative character being studied. If this increase is medically unfavourable it is called a **risk allele** or **susceptibility allele**. The overall phenotype is thus determined by the sum of the effects of all the risk alleles, modified by environmental variation. However, most diseases are discontinuous in that an individual either does or does not suffer from the disease. To account for this, it is supposed that the influence of polygenes and the environment combine to produce an underlying distribution of liability to the disease in the population. Only when this liability is greater than a threshold value is the disease triggered. This affects only a minority of the population towards one extreme of the liability distribution (Figure 6.3). For the reasons outlined above, the distribution of liability in relatives of an affected individual will be shifted from the population mean, thus they are more likely to be above the threshold and suffer from the disease themselves.

Direct evidence for the threshold hypothesis comes from congenital diseases that are more likely to affect one sex compared to the other. For example, pyloric stenosis (see Section 6.2.1) is five times more common

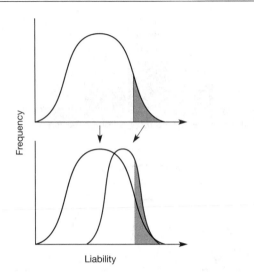

Figure 6.3. The threshold model for discrete diseases. There is a continuous distribution of liability in the population. When this exceeds a threshold the disease results (shaded blue). The relatives of individuals who have exceeded this threshold will have a distribution of liability that regresses towards the population mean but is still shifted (blue line). Thus a greater proportion of relatives will exceed the threshold.

in boys than in girls. The implication is that for some reason the liability threshold is higher in girls compared to boys. If this is the case, then affected girls must have more susceptibility alleles and so relatives of affected girls should be at greater risk than relatives of affected boys. As in the general population, male relatives are at a correspondingly greater risk than female relatives. However, as predicted, both male and female first-degree relatives of female probands are four to five times more likely to be affected than the corresponding sex in the first degree relatives of male probands.

6.3.3 The properties of polygenes

The two-gene example outlined in the preceding section makes a number of implicit and explicit assumptions about the properties of the alleles. It assumes that all alleles are present in equal frequency, that the effect of each allele is additive, that is there is no dominance of one allele of each pair over the other, and finally that the contribution of an allele at one locus is independent of the genotype at the second locus, i.e. that is there is no epistasis. Furthermore the extent to which each allele causes a shift from the population mean in Figure 6.2 was entirely arbitrary. To model the contribution of each allele more realistically it is necessary to describe these properties explicitly. Consider a gene that contains two alleles, A and a. We can define its properties as follows:

1. The displacement, t, is the difference in phenotype between the two homozygotes AA and aa. Because we are considering a normal distribution, the units of t are standard deviations (s.d.). This parameter measures the magnitude of the risk conferred by the risk-allele.

2. If the frequency of allele A is p and the frequency of allele a is q, then the distribution of the alleles in a randomly mating population will conform to the Hardy–Weinberg distribution (Chapter 9), that is $p^2 + 2pq + q^2 = 1$. The contribution to the overall genetic variation in a population made by a risk allele is clearly going to depend on its frequency.

3. Dominance, d, describes the mean phenotypic value of the heterozygote Aa relative to each homozygote. A value of $d = 1$ signifies that the mean phenotype of Aa heterozygotes is the same as the mean of AA homozygotes, that is allele A is dominant over allele a. Conversely a value of $d = 0$ signifies that allele a is dominant over allele A. A value of $d = 0.5$ signifies that the heterozygote has a mean phenotype that is intermediate between the means of the two homozygotes – in this case the alleles are said to be additive.

4. Epistasis describes the situation where the phenotype that arises from alleles at two or more genes is not the arithmetic sum of the contribution made by each individual allele. The simplest example would be a multiplicative interaction, which would be occurring if, for example, two alleles each separately caused a displacement of 3 s.d. but when present together caused a displacement of 9 s.d.

Specifying these parameters allows Figure 6.2 to be re-drawn in a more realistic fashion illustrating the different ways in which the alleles can act. Figure 6.4 shows the results of computer simulations for two different situations that could result in a hypothetical disease with a population incidence of 4%, typical of a complex disease. In both cases alleles at two genes are segregating, each of which has two alleles: A and a at one gene and B and b at the other gene. The A and B alleles act to increase the risk. Disease occurs when the liability exceeds a threshold value of 2.5 s.d. In one case each the risk alleles are rare ($p = 0.01$), dominant ($d = 1$) and each has a large displacement value ($t = 5$ SD). In the other case, the alleles are relatively common ($p = 0.1$), additive ($d = 0.5$) and each risk allele has a small displacement value ($t = 0.5$ SD). Three important lessons emerge from this simulation. Firstly, the observed population prevalence of the disease can be explained by rare alleles each exerting a large effect on the risk or by common alleles each exerting a small effect on the risk. Secondly, in the case of alleles with a large effect there is a very good correlation between phenotype and genotype; 98% of individuals with the risk allele but only 0.13% of aa homozygotes will suffer from the disease. Essentially this case corresponds to Mendelian segregation, which will be readily analysable by the methods used so successfully with single-gene disorders. Such alleles are sometimes described as highly penetrant. In the second

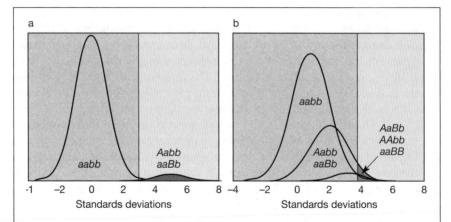

Figure 6.4 Population distribution of liability for a disease in which liability is determined by two loci. In panel a) Alleles *A* and *B* are uncommon ($p=0.01$) but each has a large displacement (t = 5 SD). Most individuals with either risk allele will exceed the threshold value of 2.5 SD. In panel b) *A* and *B* alleles are relatively common ($p=0.1$) but each has a small displacement ($t = 0.5$ SD). Only a small fraction of those carrying the risk alleles will exceed the threshold. (Redrawn with permission from N.J. Risch (2000) *Nature*, **405**, 847-856.)

case the correlation between genotype and phenotype is very poor. The disease risk for individuals with one risk-allele is 6.7%, barely above the risk for the general population. Only one-third of individuals with two risk alleles will be affected. Of individuals with three risk alleles, 69% will be affected, but such genotypes will be rare. Such alleles are sometimes described as having low penetrance. In order to detect the small increase in risk associated with low penetrance alleles it will be necessary to study large numbers of individuals and use powerful statistical methods. This is a general lesson for the study of complex diseases. The third point to note is that unexpectedly, alleles that have an additive effect on the liability scale affect the disease risks in a non-additive fashion and show epistasis with other alleles when their population distribution is taken into account.

6.3.4 Measurement of locus-specific contributions to genetic liability

The proportion of overall heritability that is due to variation at a specific locus is defined as the locus-specific heritability. Theoretical analysis shows that the size of this parameter is dependent both on the size of the displacement parameter, *t*, and the population frequency of the risk allele, *p*. As we shall see, in practice the contribution made by each locus can be estimated either from family-based linkage studies or from population-based association studies. In the case of family studies the parameter defined is the locus-specific λ defined as

$$\frac{\text{Expected proportion of affected sib pairs sharing zero alleles identical by descent (25\%)}}{\text{Observed proportion of affected sib pairs sharing zero alleles identical by descent}}$$

For example we shall see below that a locus called *IDDM1*, which maps to the HLA locus, contributes strongly towards the liability of Type 1 or insulin-dependent diabetes. The observed proportion of affected sib-pairs not identical at this locus is 8%. Since the expected proportion is 25% (see below), λ_s for *IDDM1* is calculated to be 3.6. The overall value of λ_s for diabetes is 15, so *IDDM1* contributes about 25% of the overall genetic liability.

Another parameter that is used to measure the risk due to a single allele is haplotype or genotype relative risk (GRR). This measures the relative risk of an individual with a particular haplotype or genotype relative to the average risk of a member of general population.

6.3.5 Summary of the possible genetic factors contributing to the risk of complex disease

From the preceding analysis we may imagine a variety of scenarios by which genetic factors could influence the risk of complex disease.

- A small minority of cases may be caused by rare, highly penetrant alleles segregating in a Mendelian fashion.

- The alleles involved may be common but each may have a small effect on the risk and so have low penetrance. The genes involved are usually referred to as **polygenes**. Although for simplicity we considered alleles of two genes, the number of polygenes is usually assumed to be large.

- An intermediate situation may exist where one or a small number of alleles may contribute a relatively large effect, but which falls short of the decisive effect of Mendelian alleles. The genes involved in this situation are referred to as **oligogenes**.

Of course in real life we may expect a mixture of some or all of these possibilities, either as different types of alleles in a single affected individual or with different cases in a population having different causes. For example a disease may mainly occur in a population through the combined effect of many polygenes, but a subset of cases may arise through the action of rare alleles of large effect segregating in a Mendelian fashion. Alternatively, a disease may be influenced by a small number of oligogenes, the effect of which is modified by a genetic background of many polygenes. Finally, it is important to remember the influence of the environment, which will modify the effect of the genetic components, possibly in a locus-specific fashion.

6.4 Identifying genes involved in complex disease

Identifying genes contributing to the risk of complex disease has proved to be an immensely difficult task. There have been many reports that claim to have identified such genes, which have subsequently been withdrawn or have failed to be replicated. We saw in Section 6.3 that we may expect a poor correlation between phenotype and genotype for genes that have small displacement values, which only increase the risk by a small amount. To identify these alleles it will be necessary to examine very large numbers of affected individuals and use rigorous statistical methods to reject false positives, without missing the genuine effects of these alleles. Furthermore, there are many sources of error, some quite subtle, which may falsify the results. In this section we shall consider the experimental approaches that may be used and consider some of the statistical and other problems that have complicated this research. Five main approaches have been used in the search for the genes involved in complex disease:

1. Candidate gene
2. Parametric linkage analysis
3. Non-parametric linkage analysis
4. Population association studies
5. Animal models

These approaches are not mutually exclusive – they are usually used in combination to study a particular disease.

6.4.1 Candidate genes

In some cases the nature of the disease will suggest possible candidate genes that can be examined before a comprehensive search of the whole genome is undertaken. For example, type I diabetes mellitus is an autoimmune disease resulting in the destruction of the β cells in the Islets of Langerhans (see Box 6.2) where insulin is produced. Genes involved in cell-mediated immunity are situated at the HLA locus on chromosome 6 (see Box 6.3) and the locus encoding insulin (*INS*) is located on chromosome 11. Both of these loci are good candidates to be involved in susceptibility to type I diabetes. As we shall see in Section 6.4.3.2, population association studies that specifically focused on these loci identified risk alleles in both cases.

In other cases, mapping studies identify a region of the genome within which a susceptibility allele is located. Examination of the region then identifies a likely candidate. A good example of this is the involvement of the *ApoE* locus in susceptibility to Alzheimer's disease. Non-parametric linkage studies suggested a risk allele was located on chromosome 19. The region implicated was known to contain the *ApoE* locus encoding apolipoprotein E. This is found in amyloid plaques, characteristic of the disease, and it binds *in vitro* to the β-amyloid peptide found in amyloid plaques. Clearly the *ApoE* locus was a good candidate to examine further. Population association studies did indeed show that an allele of *ApoE* (*ApoE*-ε4) led to disease susceptibility. We shall consider this example in more detail in Section 6.4.5.

6.4.2 Parametric linkage analysis

This refers to LOD score-based linkage analysis used so successfully to map and clone genes responsible for single-gene disorders. It requires a genetic model to be tested in which parameters for the mode of transmission, dominance, allele frequency, penetrance, genetic heterogeneity, etc. are specified. It is the most powerful means of mapping a disease gene, but it only works when the genetic model is correct. It is not suitable for analysing diseases where more than one gene is involved because of the almost inevitable errors that will arise in specifying all the parameters for two or more genes.

In the context of complex disease, parametric analysis has proved successful for mapping the genes responsible for a subset of cases caused by alleles with a large effect segregating in a Mendelian fashion. Table 6.6 lists some examples.

To illustrate the successful application of parametric analysis we shall consider the mapping of the *BRCA*1 and *BRCA*2 genes, where dominant alleles segregate in Mendelian fashion to cause about 5% of breast cancer cases. A two-stage approach was adopted: firstly it was necessary to demonstrate that a small subset of cases were segregating in a Mendelian fashion. Only then was it possible to map and eventually clone the genes involved. In the first stage a group of 1579 probands were identified. In order to ensure that the collection was truly random with no bias in their selection, the women were chosen as consecutive cases diagnosed in a single breast cancer clinic in California. The history of breast cancer in the families of each

Table 6.6 Examples of loci identified as responsible for subgroups of complex disease

Disease	Restricting criteria	Loci identified
Breast cancer	Age of onset Bilateral tumours Familial clustering	*BRCA*1, *BRCA*2
HNPCC (See Chapter 5)	Familial clustering Age of onset	*MSH1, MLH1, PMS1, PMS2*
FAP (See Chapter 5)	Extreme polyposis	*APC*
Alzheimer's disease	Age of onset Familial clustering	Presenilin II, Presenilin I, *ApoE*, APP
Premature heart disease	Hypercholesterolaemia Age of onset	LDL receptor
Non-insulin dependent diabetes	Onset in young adulthood	*MODY1* Hepatocyte nuclear factor 4α
	Familial clustering	*MODY2* Glucokinase
		MODY3 Hepatocyte nuclear factor 4α
		MODY4 Insulin promoting factor-1
		MODY5 Hepatocyte nuclear factor 1β

proband was obtained. In some families there were multiple cases of breast cancer and in some of these the age of diagnosis was abnormally low. These could be families where a Mendelian allele was segregating. However, the clustering could be occurring simply by chance. An analysis similar to LOD score analysis was carried out to test which possibility was most likely to produce the observed pattern in the whole group of families. Various models were specified including no inheritance, polygenic inheritance and single-gene inheritance with either dominant or recessive alleles. Box 6.1 provides more details of this process.

The model that best fitted the data was a single gene with a dominant allele with a population frequency of 0.0006. This would mean that about 5% of breast cancer would be caused by this allele and it would be carried by about 1 in 800 women. The analysis also allowed the age-related

---

BOX 6.1: TESTING MODELS TO FIT THE DISTRIBUTION OF BREAST CANCER

Model	Hypothesis	H	T2	d	t	q	$-2\ln L + C$	Comparison	χ^2 (df)
1	Unrestricted model	0.00	0.70	0.72	2.74	0.0006	−11692.9	−	−
2	No inheritance of susceptibility	(0)	−	−	−	(0)	−11515.7	2 vs. 1	177.2 (2)
3	Multifactorial heritability without major locus	0.26	−	−	−	(0)	−11660.3	3 vs. 1	32.6 (4)
4	General single locus	(0)	(0.5)	0.77	3.01	0.0006	−11691.9	4 vs. 1	1.0 (2)
5	Dominant single locus	(0)	(0.5)	(1)	2.32	0.0006	−11691.9	5 vs. 4	0.0 (1)
6	Recessive single locus	(0)	(0.5)	(0)	(2)	0.065	−11683.7	6 vs. 4	8.2 (2)

Data from Newman, B. *et al.* (1988) *Proc. Natl Acad. Sci. USA*, **85**, 3044-3048

The observed pattern of breast cancer occurrence is fitted to various models by a computer program called POINTER. Five possible models are defined by parameters for Heritability (*H*), probability of transmission of the susceptible allele from a heterozygous parent (*T2*), dominance (*d*), displacement (*t*) and frequency of the susceptibility allele (*q*). The models are specified by fixing key parameters shown in brackets. For example no inheritance of susceptibility is specified by setting heritability, *H* = 0 and the frequency of the susceptibility allele, *q* = 0 (i.e it doesn't exist). Single-gene models are set by setting *T2* = 0.5 and *H* = 0. POINTER varies the other parameters to find the best fit to the data. To provide a base for comparison POINTER finds the values for all these parameters that will provide the best fit to the observed data without regard to biological plausibility. For example the value *T2* = 0.7 implies a bias in the transmission of the susceptible allele from a heterozygous parent that would break the laws of Mendelian segregation. The likelihood of the observed pattern occurring is then calculated for each model and this likelihood is compared to the likelihood of the unrestricted model. The models specifying no inheritance and a multifactorial model are much less likely than the unrestricted model and so can be rejected. Single gene models fit the data almost as well as the unrestricted model, with the best fit being a single dominant allele with a population frequency of 0.0006.

penetrance to be varied to find the best fit to the observed data. It was clear that breast cancer caused by the allele would have a much earlier age of onset than sporadic cases.

Once it was established that it was likely that there was a subset of families with Mendelian inheritance of breast cancer, families where there were multiple cases of breast cancer were collected and the gene was mapped by testing linkage between mapping markers and the disease. Linkage was found to a marker on 17q21, but to see the linkage it was necessary to rank the families according to age of onset. Only in families where the mean age of onset was less than 47 years was linkage observed; breast cancer in the other families was therefore not caused by an allele of this gene. Mapping the gene allowed it to be cloned using methods similar to those described for other single-gene diseases in Chapter 5. It was given the designation *BRCA*1. In other families, although breast cancer was segregating as an autosomal dominant, there was no linkage to *BRCA*1. This led to the identification and cloning of a second gene, *BRCA*2, where autosomal dominant mutation caused a high risk of cancer. *BRCA*1 mutations also result in a high risk of ovarian cancer while *BRCA*2 mutations result in an elevated risk of male breast cancer, an otherwise rare disease. The possibility that there were two genes that could each have dominant disease-causing alleles was not tested in the original population survey. Together *BRCA*1 and *BRCA*2 account for about two-thirds of the cases where there is family clustering. Attempts to identify highly penetrant alleles at other loci that would account for the other cases of family clustering have failed, and it is likely that these are caused by the combined effect of polygenes.

Breast cancer and other diseases listed in Table 6.6 show that this approach can be very useful in analysing a subset of complex diseases that are due to Mendelian alleles. However, there is a long history of cases where apparent positive linkage between a disease and a marker but this was not repeated in subsequent analyses. Usually these are psychiatric disorders such as schizophrenia, manic depression and alcoholism. There are two reasons why this has occurred. Firstly, it is possible that there are no major loci segregating in a Mendelian fashion responsible for any of these diseases. Recent research using genome scans with affected sib-pairs shows that indeed this is likely to be the case (see Section 6.4.3). Secondly, these diseases cannot be diagnosed with the same certainty as breast cancer or diabetes. It may be that a large number of conditions that are different at a cellular and molecular level give rise to symptoms that fit the same psychiatric profile. Parametric analysis is particularly sensitive to only one or two members of an extended family being wrongly diagnosed. A well-known example occurred in an attempt to map a susceptibility locus for bipolar affective disorder (manic depression) in the Amish community in the USA. The pedigree consisted of 81 subjects with 6 probands suffering from mental illness, five of which were diagnosed with bipolar affective disorder. Initially linkage to *HRAS* and *INS* chromosome 11 was reported with LOD scores of 4.08 and 2.63 respectively. Subsequently two individuals initially categorised as unaffected developed the disease. This reduced the LOD scores to 1.03 and 1.75 for linkage to *HRAS* and *INS* respectively. Further studies on other members of the extended pedigree showed that there was definitely no linkage.

6.4.3 Non-parametric linkage analysis

If a gene is contributing towards the liability of a disease, then the region of the genome within which the gene is situated will be co-inherited from a common ancestor by affected members of a pedigree more frequently than would be expected by chance. Such a statement makes no assumptions about the way the gene is operating, nor any assumptions about how many other genes may also contribute to the risk. Any method that can be used to search for such commonly inherited regions is therefore a **non-parametric** method.

Commonly inherited regions can be recognised by a genome scan following the inheritance of polymorphic microsatellite markers in affected members of a pedigree. If affected members co-inherit the same allele more frequently than would be expected by chance, then the genomic region marked by that allele may contain a gene that contributes towards disease susceptibility. It is important to note that the genome scan is not, in itself, a non-parametric method. It lends itself equally well to mapping genes by parametric methods. Indeed, the linkage analysis used to map the *BRCA1* gene described above was a form of genome scan.

Non-parametric analysis genome scans are usually based on **affected sib-pairs** (ASPs), that is, two sibs in the same family who both suffer from the disease. As well as affected sib-pairs, other members of an affected pedigree can be examined (**affected pedigree member** or APM). Highly polymorphic microsatellite markers, which are evenly spaced throughout the genome, are genotyped. For each autosomal locus the sibs could have zero, one or two alleles in common (Figure 6.5). The expected ratio of these outcomes is 25%, 50% and 25% respectively. If the parental genotypes are known to differ at the locus then we can we specify that sibs that have both

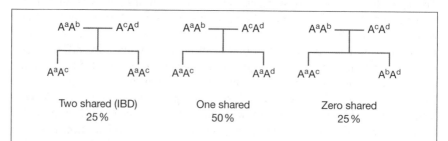

Figure 6.5. Allele sharing. The diagram shows two parents, each heterozygous at a locus A (alleles: superscript a,b,c,d). There are three possible outcomes where two sibs share 2,1 or 0 alleles, which will occur with relative frequencies of 25% : 50% : 25% respectively.

alleles in common are **identical by descent** (IBD). Because the markers are highly polymorphic, it possible to follow the segregation of chromosomes from each parent separately. If this is done, there is a 50% chance that two sibs will be IBD if the alleles segregate randomly.

If the parental genotypes are not available we can only say that they are **identical by state** (IBS). This distinction is important because two sibs could be IBS because one or both parents were homozygous. If a number of closely linked markers can be scored, IBD can be inferred in the absence of parental genotypes, because the parents would be unlikely to be homozygous for a number of different markers. If such markers are not available, IBS values must be corrected for the average population heterozygosity at the locus concerned.

In a genome scan of affected sib-pairs, a set of families containing two affected sibs is assembled. Typical datasets in early studies were composed of about 100-200 families, but as we shall see, this number is insufficient to reliably detect weakly acting alleles, i.e. alleles with low λ or GRR values. DNA samples are collected from the affected sibs along with their parents. The microsatellite loci are genotyped using PCR with fluorescent tagged primers. Using primers tagged with different colours it is possible to analyse up to 18 loci in the same lane and up to 500 on the same gel (Figure 6.6). About 300 primers are used in the initial scan, which will cover the genome at 10cM intervals. At each locus there will be two alleles in each subject (except for sex-linked loci in boys) distinguished by the length of the amplified PCR products. For each locus the genotype of the parents and both sibs can thus be specified. Using the combined data from all the families the IBD frequencies at each locus can be determined and each tested to see whether it deviates from the value expected from random segregation. If the inheritance of chromosomes from each parent is analysed separately the random expectation is that 50% of the sibs will be IBD. So the method is attempting to identify loci where, over the whole dataset, there is a significant excess of sib pairs that are IBD.

Statistical issues are critically important in this field. The problem is to reject false positives without needlessly rejecting weak but genuine effects. Tests for significance are carried out in two ways. Firstly, a modified form of LOD score analysis can be used that tests for linkage between the mapping marker and a hypothetical disease gene. This produces a maximum likelihood score (MLS). Secondly the probability, P, that a deviation as great as that observed could have arisen by chance can be calculated. The difference between these two parameters is that an LOD score measures the ratio of two probabilities – that the observed results would have occurred if two markers are linked compared to the likelihood that they would have occurred if the markers were unlinked. A P-value refers to the probability of the observed deviation from 1:1 occurring by chance. Each genome scan involves hundreds of independent tests. The traditional threshold of significance is the point at which the expected result will only occur by chance in 1 in 20 tests. If hundreds of tests are carried out then clearly such a result

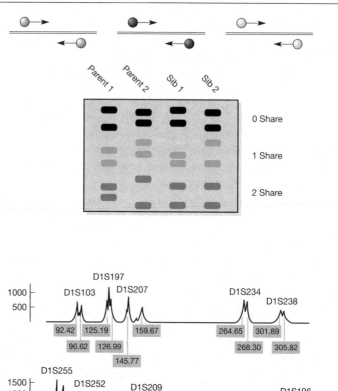

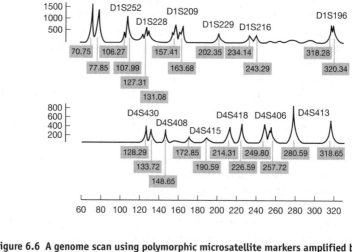

Figure 6.6 A genome scan using polymorphic microsatellite markers amplified by PCR.
(top) Each locus is amplified by PCR primers that hybridise to unique sequences that flank
the microsatellite. The number of repeat units determines the size of the microsatellite and
hence the position it migrates to on the gel. Different primers are tagged with fluorescent
dyes so it is possible to run the products of several reactions in one lane. A single band
indicates homozygosity for the allele migrating to that position. In this example three loci
are genotyped denoted by blue, black and grey. In practice 18 different loci can be analysed
in one lane and up to 500 on a single gel. (bottom) The gels are analysed by an Applied
Biosystems DNA sequencer using ABI Prism automated genotyping software. Shown here is
part of the output from the diabetes genome scan.

will be achieved by chance on several occasions. It is necessary therefore to modify the traditional thresholds of significance to take account of this multiple testing. As we saw in Section 6.4.1 there is a long history where reports of linkage have proved to be false. So replication of a finding of linkage in an independent study is critical before a linkage is accepted as being demonstrated. The guidelines for genome scans based on ASPs, used to define the degree of confidence in an observation, are as follows:

- Suggestive linkage: expected to occur by chance about once per genome scan. This corresponds to an MLS score of 2.2 and a value of $P = 7$ x 10^{-4}.

- Significant linkage: expected to occur by chance about once per 20 genome scans. This corresponds to an MLS score of 3.6 and a value of $P = 2$ x 10^{-5}.

- Highly significant linkage: expected to occur by chance about once per 1000 genome scans. This corresponds to an MLS score of 5.4 and a value of $P = 3$ x10^{-7}.

- Confirmed linkage: significant linkage in one study replicated in a second independent study. Because the replication doesn't involve multiple testing a value of $P = 0.01$ is sufficient to confirm linkage.

Because large numbers of sib-pairs are required to reliably detect alleles with a weak effect, genome scans are often carried out in stages to minimise the amount of work necessary without jeopardising the chances of spotting genuine effects. In the first stage regions are identified showing suggestive linkage. These regions are then re-examined in a second collection of families and using more densely spaced mapping markers.

Table 6.7 illustrates how a genome scan was applied to search for alleles leading to susceptibility to type 1 diabetes or insulin-dependent diabetes mellitus (IDDM). The different types of diabetes are described in Box 6.2. Table 6.7 shows that a clear deviation from a 1:1 random segregation ratio is evident for alleles linked to HLA locus (see Box 6.3). The MLS score exceeds the level of significance by a large margin. This locus had already been identified by population association studies (see Section 6.4.4.2) and shows that the method will detect alleles that have a significant effect, but which is less than that of a Mendelian locus.

This and other studies with type 1 diabetes also suggested a number of other regions contributing to the overall genetic liability, but with a much smaller effect (λ values between 1 and 2). The problem is that regions identified in different studies are not the same and the record of replication from one study to another is poor. Theoretical calculations show that sib-pair analysis with a few hundred sib-pairs will reliably detect loci with λ or GRR values of 4 or more but that thousands of sib-pairs are required to detect alleles with lower values (see Table 6.8, p. 211). Moreover, computer simulations show that studies involving 100–200 sib-pairs can be expected to detect linkage that is not replicated in a second study unless the number of sib-pairs in the second study is increased about 5-fold.

Table 6.7. A genome scan for IDDM. A set of 96 families from the UK where two sibs suffered from IDDM were genotyped at 290 marker loci (see Figure 6.4). Because the markers were highly polymorphic it was possible to distinguish between maternally and paternally derived chromosomes and so determine for each chromosome whether the sibs were IBD. In this situation a significant deviation from a 1:1 ratio suggests linkage between the marker and IDDM. The value of MLS for significance was set at 2.3. (this study was carried out before the guidelines outlined above were established). A total of 20 chromosome regions showed some evidence of linkage, although only some are expected to be genuine. The table below shows data for four of these where there is good evidence of linkage to IDDM. Two of the regions correspond to *IDDM1* (the HLA locus) and *IDDM2* (the insulin locus) which had previously been implicated by other studies (see text). The effect of the *IDDM1* genotype on other loci was analysed. This increased the power to detect linkage not apparent in the total dataset (FGF3) or increased the level of significance (ESR). Biologically this suggests epistatic interactions between *IDDM1* and each of these other two loci (designated *IDDM4* and *IDDM5*). Two additional sets of families were used to analyse further cases where there was some positive evidence for linkage. Moreover, markers closely linked to FGF3 also showed linkage to IDDM. Taking these into account the MLS at FGF3 increased to 3.4. Interestingly, in these other two family sets, there was no evidence for linkage at *IDDM2* even though the first dataset and other evidence (see text) had convincingly implicated this locus.

Chromosome	Marker	Locus	IBD		MLS	Data used
			1 share	0 share		
6p	TNFa	*IDDM1*	97	35	7.3	All
6q	ESR	*IDDM5*	95	59	1.8	All
			61	30	2.5	Sibs sharing two alleles at *IDDM1*
11q	INS	*IDDM2*	107	67	2.1	All
11p	FGF3	*IDDM4*	71	60	-	All
			38	21	1.8	Sibs sharing zero/one alleles at *IDDM1*

BOX 6.2: DIABETES MELLITUS

Diabetes mellitus is a disease characterised by the inability of body cells to take up glucose, leading to high levels of glucose in blood and urine, but intracellular starvation. In order to gain energy cells break down fats. The end product of fat breakdown is acetyl CoA which is metabolised to acetoacetone, β hydroxybutyrate and acetone, collectively known as ketone bodies. These accumulate in the blood stream (**ketonemia**) and are excreted in the urine (**ketonuria**). The acetone contaminates the breath of diabetics with a characteristic odour. The build-up of ketone bodies is known as **ketosis**. The uptake of glucose into cells is promoted by the hormone insulin produced in the islets of Langerhans in the pancreas.

There are two different diseases, both of which result in elevated blood glucose levels:

1. **Type 1** or **insulin-dependent diabetes mellitus** (IDDM). This disease is characterised by lack of insulin, and can usually be treated by insulin injections. In most cases the lack of insulin is caused by the autoimmune destruction of the insulin-producing β cells in the islets of Langerhans. Typically the disease shows rapid juvenile onset – the median age of diagnosis is 12 years. However, a build-up of autoantibodies can be detected much earlier. Although the immediate symptoms can be treated with insulin, in the long term IDDM sufferers may face complications which lead to blindness, kidney failure, amputation of lower limbs and heart attacks. It affects about 4 in 1000 Caucasian children of European descent. There are striking ethnic differences in its frequency, the condition being rare in Mongoloid and Negroid populations.

2. **Type 2** or **non-insulin dependent diabetes mellitus** (NIDDM). As its name implies, this disease is not caused by a lack of insulin, but is characterised by high blood and urine glucose levels that are resistant to insulin. It typically affects people over 40, and progresses more slowly than IDDM. There are clearly predisposing genetic factors that are apparently unmasked by environmental factors. Insulin resistance is common in obese people and may result from prolonged exposure to high glucose levels resulting from sugar-rich diets, or there may be genetic factors that predispose both to diabetes and obesity. NIDDM accounts for about 85% of diabetes cases and rates are much higher than IDDM in developed societies, probably reflecting the effects of affluence on diet. A subtype of this disease is MODY (Maturity onset diabetes of the young). This disease shows earlier onset than normal diabetes, affecting teenage children and young adults. It often shows a monogenic segregation (see Table 6.6).

Diabetes is a major source of ill health. It may lead to blindness, renal failure, loss of lower limbs, heart attacks and strokes. In children of European-Caucasian descent, IDDM is the second most common cause of chronic ill health after asthma. In the United States, there are more than 14 million cases of diabetes. According to the National Diabetes Research Commission the consequences each year of diabetes in the United States include:

- 15 000–39 000 new cases of blindness
- 13 000 cases of end-stage renal disease
- 54 000 amputations, mostly of lower limbs
- 162 000 deaths from heart attacks, strokes, etc.

Both types of diabetes show strong evidence of a genetic component, but there is also a strong environmental component, as evidenced by the continued rise in its incidence. The environmental factors are still not fully characterised. A childhood

Continued ▶

infection with the Rubella virus has been shown to predispose to IDDM. However, inbred strains of mice that show a high disposition to a similar disease are more likely to succumb if they are raised in a germ-free environment. This suggests that an overly hygienic environment in childhood may increase the risk, an observation that has been made in other autoimmune diseases. The population incidence of Type 1 diabetes in the Western world is 0.4%. The frequency in sibs of affected individuals is 6%. We can therefore calculate that $\lambda_s = 15$. The concordance in identical twins is 36%. Clearly, genetic factors influence the risk of disease. Note, however, that the twin concordance is very much lower than 100%, emphasising the importance of environmental factors.

BOX 6.3: THE HLA LOCUS

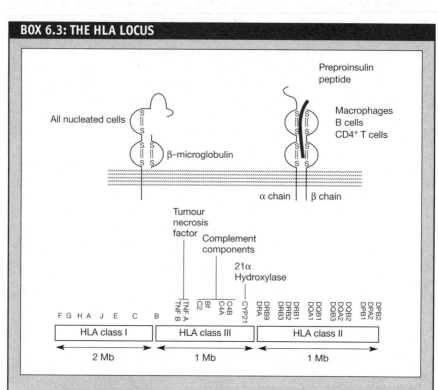

When organs are transplanted from one human to another the transplanted organ is rejected because the host recognises the organ as foreign. The rejection is controlled by the major histocompatibility antigens displayed on the cell surface. Many of these are encoded by a cluster of genes called the HLA locus (human leucocyte antigens). The HLA locus occupies 4Mb of DNA on chromosome 6 and contains over 100 genes. The region is divided into three classes of genes: class 1, class II and class III. Organs are only accepted as self if they are identical to the host at class I and class II genes, which are highly polymorphic. The proteins encoded by these genes are displayed on the cell surface of all cells. During the neonatal period, they induce tolerance in cytotoxic T cells in the thymus. The extracellular domains of both class I and class II proteins are part of the immunoglobulin superfamily. Class III genes encode proteins and peptides with diverse functions of the immune system such as the production of cytokines like tumour necrosis factor, and complement fixation.

As well as marking self and non-self, class I and class II molecules are responsible for this presentation of foreign antigens to T-helper cells. Class I molecules present intracellular antigens from all cells to T cells, which allows the recognition of intracellular infective agents such as viruses. This results in the stimulation of cytotoxic T cells, which kill the infected cells. Class II molecules present extracellular antigens in dendritic cells, B cells and macrophages, so triggering the humoral or antibody response. Class II molecules are heterodimers of α and β chains, encoded by pairs of closely linked genes. For example DQA1 and DQB1 encode DQα1 and DQβ1 that form a heterodimer. The two chains come together so that a groove is formed between them in which the foreign antigen is located and presented to receptors on the T-helper cell. In the diagram the DQα1 and DQβ2 chains are shown presenting a fragment of the insulin precursor chain to T-helper cells.

Thus the usefulness of genome scans of affected sib-pairs is limited to the detection of alleles that exert a reasonably large effect, equivalent to those we classified as oligogenes in Section 6.2. Such studies have now been carried out on a number of common diseases including Type 1 and type 2 diabetes, multiple sclerosis, Alzheimer's disease, rheumatoid arthritis, Crohn's disease, manic depression, schizophrenia and autism. These studies confirmed linkage to the HLA locus in autoimmune diseases such as Type 1 diabetes, multiple sclerosis and rheumatoid arthritis. In the case of type 1 diabetes, Alzheimer's disease and Crohn's disease, alleles were detected where the evidence is convincing and replicated in other studies. We shall consider Alzheimer's disease later in this chapter. However, apart from these examples, the success of this technique has been limited. In each study a number of regions showing evidence of linkage were detected, but the λ values were low and in general were not replicated in other studies. However, there is an important negative value to all of these studies. Although the power to detect weakly acting alleles was limited given the number of sib-pairs used, the power to detect more strongly acting alleles was high. In other words if alleles do exist that have a major effect on the development of these diseases they would have been detected. The fact that no such alleles came to light is good evidence that none exist. Clearly this will have to be taken into account in further studies attempting to identify the genes involved.

6.4.4 Population association studies

In contrast to linkage-based studies, which work with family pedigrees, association studies are based on populations. An attempt is made to find an allele that is associated with the disease. That is, affected individuals are more likely to carry a particular allele than unaffected members of a control group. This can come about for two fundamentally different reasons. The most obvious reason is that the allele may actually affect a biologically relevant locus. However, association may also be due to **linkage disequilibrium** (LD), which can occur because an ancestral mutation, responsible for the disease in a population, occurred on a chromosome carrying the associated allele at a closely linked site. We shall examine these alternatives in more detail below.

6.4.4.1 The problem of population stratification in association studies

Population association studies are subject to a particular error if there is a mixture of ethnic subgroups present in the test population, as is found in the USA or the UK, because of the inward migration of a variety of different ethnic groups. If a disease is more common in one ethnic subgroup, then any allele that is more common in that subgroup will show a positive association with the disease, whether or not it is actually involved the aetiology of the disease. This problem is called **population stratification**. There are a number of possible solutions to this problem.

1. The most obvious way is to match each case of the disease with an unaffected control matched for ethnicity and other possible confounding factors such as age and sex. This is known as a **case control study**. Although straightforward in principle, the divisions of ethnic subgroups in a population are not always clear, especially the degree of mixing that may have occurred. In practice, case control studies have a poor track record, with associations being reported that have not been replicated.

2. The controls can be drawn from unaffected family members who are necessarily matched for ethnicity. The transmission disequilibrium test described in Section 6.4.4.2 is a commonly used test that is based on this principle. The main problem with this is that it restricts the data that can be used, as it relies on the availability of parents who are heterozygous for the allele being examined. This reduces the statistical power to detect weak associations. An advantage of TDT is that it tests linkage as well as association, so it both identifies and simultaneously maps the putative allele.

3. Association studies can be carried out in isolated populations that have descended from a few founder individuals and so are expected to be more homogeneous. The populations of Iceland, Finland and Sardinia are examples of populations that are used in such surveys. See Section (6.4.4.6)

4. Most association studies in the future will be carried out using panels of SNP markers, which will contain a large number of markers (Section 6.4.4.5). If an apparent association with one SNP is actually due to population stratification, then other SNP markers in the panel should also show a positive association. These are other markers that also happen to have a higher frequency in the ethnic group that has a higher incidence of the disease. Thus when one marker is apparently showing a positive correlation, the other markers in the dataset can be used as an internal control for population stratification.

Historically association studies have been based on examining candidate loci which may be plausibly involved in the aetiology of the disease. Examples of this are associations between alleles at the HLA locus and autoimmune diseases such as Type 1 diabetes, muscular dystrophy, ankylosing spondylitis and rheumatoid arthritis. The HLA locus is responsible for encoding cell surface antigens that are used to distinguish self and non-self so it not surprising that variation in these alleles could be a factor in the risk of developing an autoimmune disease (see Box 6.2).

In the case of type 1 diabetes, groups of diabetic patients and control groups matched for age and ethnicity were examined to see if alleles at any HLA loci were significantly more common in affected individuals, suggesting that those alleles lead to a susceptibility to diabetes. Conversely, one can ask whether any alleles are significantly more common in healthy controls, suggesting that those alleles exert a protective effect. The first results suggested that *DR3* and *DR4* alleles of a Class II gene called *DRB1* lead to susceptibility, while *DR2* alleles are protective. One example of many studies was carried out in Alberta, Canada. In Type 1 diabetic patients 96% had either a *DR3* or a *DR4* allele and 38% had both alleles. In contrast in the general population only 45% had one allele and 3% had both. Clearly the possession of one or both of these alleles is an important risk factor. However, it is not sufficient. Of the 3% in the general population who had both risk alleles, only 10% would develop diabetes. The allele *DR2* at this locus exerts a dominant protective effect; only 1 in 20 diabetic patients had this allele. Subsequent research showed that the actual risk locus is not *DRB1* but the nearby *DQB1*. The susceptibility alleles were in linkage disequilibrium with the risk alleles at this locus. The protective allele encoded an amino acid, asparate, at position 57 of the polypeptide chain, which is called DQβ1. All the susceptibility alleles encoded an amino acid other than aspartate at this residue. DQβ1 forms a heterodimer with DQα1 and presents processed antigens to T-lymphocytes. It is thought that the DQα1 proteins encoded by susceptible alleles promote the recognition of an autoantigen (the protein or part of a protein recognised by the body's immune system) whose identity is not yet known.

A second locus which is biologically relevant to diabetes is the insulin locus (*INS*) on chromosome 11. When it was cloned in 1981 it was examined to see if any alleles were associated with diabetes. Restriction mapping revealed a length polymorphism that was caused by a minisatellite or VNTR 5' to the insulin coding region. Although the number of repeats was highly polymorphic, they could be broadly divided into three classes:

1. Class I is composed of 26–63 repeat units.
2. Class II is composed of 80 repeat units. This class is rare in Caucasian populations.
3. Class III is composed of 141–209 repeat units.

Class I homozygotes showed a positive association with type 1 diabetes in Danish, British and USA populations. However, it was puzzling that there appeared to be no linkage between the *INS* VNTR and diabetes in family pedigrees. The reason for this is that the class I allele is frequent in the population so parents could be homozygous. This would distort linkage studies because all progeny would necessarily inherit the allele whether or not they suffered from diabetes. This problem was solved by restricting analysis to where one of the parents was a class I/class III heterozygote. Linkage was revealed when this was done in one pedigree that previously failed to show linkage (Figure 6.7).

This type of analysis is called a **transmission disequilibrium test** (TDT). It formally examines whether the putative risk allele is inherited by affected offspring more often than the non-risk allele from a heterozygote parent. As noted in the previous section it provides an important control for spurious associations between an allele and a disease due to population stratification.

How the VNTR polymorphism affects the risk is still not certain. Insulin is produced in the thymus during the neonatal period, allowing it to be recognised as a native protein. The VNTR polymorphism is located in the promoter region. It affects the level of insulin production in the thymus, thus it may influence the successful development of tolerance to insulin by the immune system.

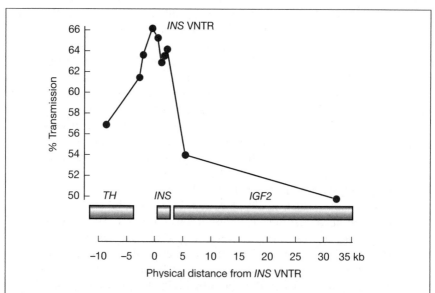

Figure 6.7 Mapping the *IDDM2* locus using the transmission disequilibrium test.
Polymorphisms at loci in the *INS* region were genotyped in 446 families with two diabetic offspring. Alleles associated with IDDM were transmitted from heterozygous parents significantly more frequently than 50% – that is the alleles and IDDM were in linkage disequilibrium. The degree of disequilibrium rises to a peak at or very close to the *INS* VNTR. (Redrawn with permission from Bennet, S.T. and Todd, J.A. (1996) *Ann. Rev. Genet.*, **30**, 343-370.)

6.4.4.3 Linkage disequilibrium (LD)

Linkage disequilibrium (LD) refers to a situation where a particular combination of alleles at two closely linked loci occurs more often than would be expected by chance from their population frequencies. It can occur when most cases of a disease in a population are caused by a mutation in a common ancestor. The particular alleles at closely linked loci on the chromosome where the mutation occurred will tend to be co-inherited by individuals who have inherited the mutation (Figure 6.8). The mutation in the ancestor is said to be a **founder mutation**. Recombination will eventually erode the association, so the degree of LD will depend on the number of generations since the founding mutation and the distance between the mapping marker and the disease gene. Thus LD provides a means of mapping genes that treats affected individuals as members of giant, but incomplete extended pedigrees extending back over many generations. LD has been extremely effective in mapping genes involved in monogenic disorders. For example, LD around the Huntington's disease and cystic fibrosis loci were very important in mapping the position of these genes (see Section 5.5.2.).

LD rarely results in simple gradient of linkage away from the disease locus (Figure 6.9). Some markers may show high levels of LD, whereas nearby markers, which may even be physically closer, show little or no LD. The reason is that present-day haplotypes are dependent on population history and evolution after each of the limited number of recombination events

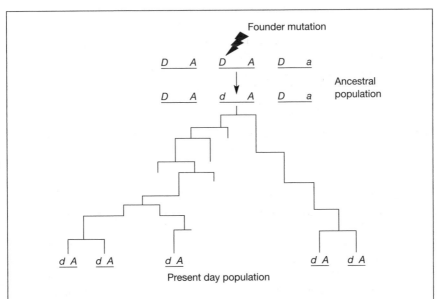

Figure 6.8 Linkage disequilibrium. A polymorphic mapping locus with alleles A and a is located near a disease locus. The diagram shows an ancestral population of chromosomes in which a founder mutation produces an allele, d, that causes susceptibility to the disease. Because it occurred on a chromosome that happened to carry allele A, the disease allele d will be co-inherited with allele A. On a population-wide basis the frequency of the haplotype dA will significantly exceed the frequency of haplotype da, i.e. alleles d and A will be associated.

that have occurred between the two linked loci. Linkage disequilibrium mapping should not be confused with linkage mapping. Both rely on the tendency of linked markers to be co-inherited and not separated by recombination. However, linkage studies are carried on single families or single extended pedigrees and the linkage detected will extend over a relatively large distance (up to 20 cM). LD studies are carried out on whole populations. The extent of LD detected depends on the number of generations since the founder mutation and population history and thus can be complex to predict. For common alleles that arose in the ancestral human population in Africa it could be as low as 3–10 kb. In isolated populations it can be much higher. The reasons for this difference will be discussed below.

LD will only work if affected individuals in a population have inherited the same founder mutation. If more than one mutation has affected a particular disease locus then the mutations could show LD with different alleles at the linked loci. In this situation no particular allele or haplotype will be associated with the disease. This is one reason why LD studies are often carried out in isolated populations that have expanded from a relatively small number of founders so that there is more chance that only one founder mutation is responsible for the disease.

6.4.4.4 Linkage or association?

So far we have discussed how non-parametric linkage analysis and population association studies can be used to identify and map alleles contributing to complex disease. Which one is likely to be the most useful? Because both have their different sources of error, replication in both types of analysis will add confidence to a positive result. So in one sense, the answer is that both should be used. However, theoretical analysis shows that population association studies are in principle more powerful. **Power** has a particular meaning in this context – it defines the probability that a study will identify the involvement of a risk allele, with a given locus-specific λ or GRR, to a defined statistical threshold. So for example in the genome scan for diabetes alleles described in Table 6.7, the authors were able to calculate that the study had a 95% chance of identifying another allele with the same or greater locus-specific λ as the HLA locus with an MLS score of 2.3 or more.

Power calculations can be performed to ask how many subjects need to be studied to identify an allele with a given effect. Examples of such calculations comparing linkage and association are given in Table 6.8, which shows the results for different values for GRR and allele frequency. Allele sharing should identify alleles with GRR values of 4 or more with between 200 and 300 families depending on allele frequency. However, unrealistically large numbers are required with GRR values of 2 or less. For example, 67 816 affected sib-pairs would be required to identify an allele with a GRR value of 1.5 with an allele frequency of 0.5. These calculations explain the lack of reproducibility between allele sharing studies carried out to date, which were based on a few hundred affected sib-pairs. Such studies would have an approximately 20% chance of detecting an allele with a GRR value of 2.0. If a number of such alleles exist, it is inevitable that each study will detect some of them, but that linkage will not be detected to these alleles in a second study.

	Allele sharing			Association by TDT test		
GRR	Allele frequency	Probability of allele sharing	No. of families required	Probability that disease allele transmitted	No of cases	
					Single	Sib-pairs
4	0.1	0.597	185	0.8	150	48
	0.5	0.576	297		103	61
2	0.1	0.518	5382	0.67	695	264
	0.5	0.526	2498		340	180
1.5	0.1	0.505	67816	0.6	2218	941
	0.5	0.512	17997		949	484

Source: Risch, N. and Merikangas, K. (1996) *Science*, **273**, 1516–1517.

Table 6.8 shows that population association studies should detect alleles with a weak effect with a manageable number of cases. Moreover, there is no need to use families with two or more affected sibs, which makes the study much easier to carry out on a large scale. However, if affected sib-pairs are used, fewer families are used and the results will be more robust since the other family members can be used as controls for population stratification. The calculations in Table 6.8 also illustrate that allele frequency affects the numbers required to identify an allele. Generally fewer cases are required to identify less common alleles ($P = 0.1$ compared to $P = 0.5$), but there are exceptions, e.g. allele sharing where GRR = 4.

6.4.4.5 Large-scale population surveys based on SNP maps

In principle, LD can be used on a genome-wide basis to map the actual loci contributing to the genetic risk. Affected individuals are examined for a common haplotype that suggests they have commonly inherited a particular chromosomal region. The calculations in Table 6.8 assumed that degree of LD was 1.0 between the marker and susceptibility alleles. Because this won't be the case in practice, larger samples than those indicated will need to be used.

The *ApoE* locus – already been demonstrated to be involved in susceptibility to Alzheimer's disease – was used to demonstrate the validity of this approach (Figure 6.9). A peak of LD around the *ApoE* locus clearly stands out from the background signal in surrounding areas. When amplified, this peak shows that three SNPs spanning 60 kb show highly significant LD. This example suggests that in practice LD would have been readily detected with markers spaced 10–30 kb apart, which would represent one SNP per gene. However, note that in both cases some markers within the region identified failed to show any LD with the disease. As noted above, this is a common feature of LD mapping. If the study had been carried out with less densely spaced SNP markers the peaks of LD could have been missed. The example also illustrates another general point: LD studies with SNPs identify regions containing only one or a small number of genes. In this example the region showing LD contains only two genes, *ApoC* and *ApoE*. If the involvement of *ApoE* was unknown, such a study would have rapidly narrowed the search for the locus responsible. So once a susceptibility locus has been mapped with LD, identifying the actual locus responsible should be relatively straightforward. This contrasts with the situation from linkage studies where the region identified is usually a megabase or more in size and contains many genes.

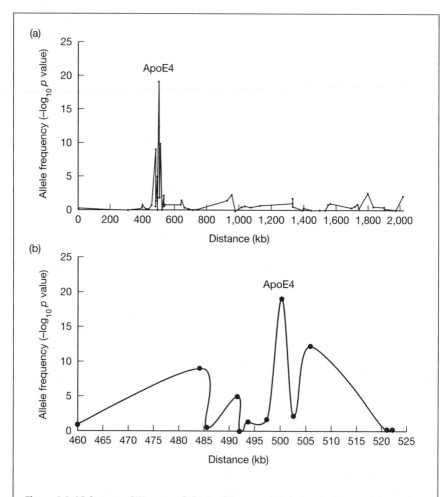

Figure 6.9 LD between SNPs around the *ApoE* locus and Alzheimer's disease. (a) Data for a large number of SNPs over a 2 Mb region around the *AopE* locus. (b) An expansion of the peak around the *ApoE* locus. (Redrawn with permission from Roses, A.D. (2000) Nature, **405**, 857–865.)

There is still considerable debate as to how many SNP markers will need to be used in a genome-wide search using LD. Theoretical calculations suggest markers spaced every 3–10 kb will be required if the allele is common (>1%). This is because the origin of common polymorphisms is likely to have been in the ancestral human population before it expanded out of Africa. Recombination in the 5000 generations since then would limit LD to this small interval. The use of isolated populations may be useful to increase LD. However, if the allele is common then the founding population must be very small (Section 6.4.4.6). Recent results have suggested that although LD in isolated populations may be useful for mapping rare monogenic disorders it does not offer significant advantages for mapping more common alleles involved in complex disorders.

Others argue that more widely spread markers will be sufficient to detect the presence of a susceptibility allele in the first place. Once detected it can be mapped with great precision with more densely spaced markers. Proponents of such a view cite examples such as the *ApoE* locus described above. An interesting development suggests that LD may extend for a much greater length in European compared to other populations. A recent study found that LD around 19 randomly selected genomic loci was much greater in 44 unrelated individuals of North European descent (from Utah) compared to 96 individuals from Nigeria (Yorubans). The 'half length' of LD was found to average 60 kb in the European population and in 50% of the cases was higher than 80 kb. In contrast the half length of the LD was only 5 kb in the Nigerians in keeping with theoretical predictions. The long-range LD in the European population was confirmed by an independent study of 48 Southern Swedes. The most likely explanation for the increased LD in Europeans is that there was a small founder group or severe population bottleneck between 800 and 1600 generations ago (27 000–53 000 years ago) in which the effective ancestral population size was between 50 and 1000 individuals depending on how many generations the bottleneck lasted. This finding raises the possibility of a two-tiered search for disease susceptibility loci. Genome scans using relatively few SNPs in European populations may identify genomic regions which contain susceptibility loci – fine structure mapping can then be carried out using LD in populations such as the Yorubans. Although this is an attractive scenario, the study also highlighted two general problems with LD mapping.

1. Although the average range of LD was high in Europeans, there was great variation between the individual loci. In one locus (*PCI*) the half length of LD was only 6 kb, in another (*WASL*) the half length of LD was 155 kb. This variation may explain why different studies, investigating only one or a small number of loci, have come to different conclusions about the extent of LD.

2. The degree of LD to a locus shown by closely linked SNPs can vary widely. This is very similar to the variation seen in the *ApoE* study described above.

Both of these uncertainties will make it difficult to standardise the design of LD studies to be sure that any relevant loci have been detected. Genuine associations may slip through the net of genome scans with widely spaced markers.

It may be possible to restrict SNPs to only those affecting coding sequences, most of which have been identified in the human genome project (see Chapter 4). These SNPs may actually be the causative mutations, or at the very least be very closely linked to the causative mutations. In such a survey only about 100 000 SNPs will need to be used. However, there is no a *priori* reason why the mutations should only affect coding sequences. Regulatory sequences could equally well be affected and these could extend considerable distances from the coding sequences. A second important issue is that multiple risk alleles at a disease locus may have arisen on independent occasions. Different cases of a disease may have been influenced by different risk alleles at the same locus, so in a population survey the disease may not appear to be associated with any one of them. Again this problem is less likely to occur in isolated populations, which are likely to be more homogeneous.

Even if it proves necessary to use very closely spaced SNPs, the development of SNP maps of the human genome with 2 million markers and technology for high throughput genotyping makes it feasible to undertake genome scans searching for LD between SNPs and the disease. Large-scale programmes of population screening are being undertaken involving collaborations between National Health Services, medical charities and commercial organisations. For example, in the UK the NHS, MRC and the Wellcome Trust are planning to undertake surveys correlating the disease histories of 500 000 individuals with their SNP genotypes. Data generated will be sold to pharmaceutical companies. A similar project is underway in Iceland to exploit the special advantages of this isolated population (see Section 6.4.4.6).

6.4.4.6 Small isolated populations

Isolated populations that have expanded from a small group of founders offer several advantages for the identification of disease alleles. The Finnish and Icelandic populations have been used extensively for this purpose. Finland was colonised in two waves 4000 and 2000 years ago. The size of the founding population is unknown, but in the 12th century the population was only 50 000, whereas today it is 5 million. Since the 16th century, starting in the reign of the Swedish King Gustavus of Vasa (1523–1560) records of every member of the population have been kept. In the last few centuries there have been migrations from the populous south-east to northern and western regions. This has resulted in isolated communities founded by very small numbers of individuals. Iceland was settled in the 9th and 10th centuries and there has been little or no immigration since. These settlers, mainly from Scandinavia, were the ancestors of the present population of 275 000. The Norse culture meant there are complete genealogical records from the very beginning. The advantages of using populations such as the Finnish and Icelandic populations are summarised as follows:

- They tend to be more uniform, lessening the chance of population stratification that always complicates population association studies.

- A common culture and religion makes the environment more uniform, making the contribution of genetic factors easier to detect. It also increases the chances that mating is truly random so that allele frequencies conform to the Hardy–Weinberg distribution.

- Often there is also a common medical culture. So criteria for diagnosis and treatment will be more uniform. For example, in Finland all doctors are trained in just five medical schools that share a common tradition.

- The good genealogical records allow apparently unrelated individuals to be traced to a common ancestor. If they share a haplotype that indicates they have co-inherited a chromosomal region from that common ancestor, then that region is likely to contain the disease allele. Because relatively few generations have elapsed since the common ancestor, LD will extend over a large distance, making it possible to detect with a small number of affected individuals. This is only likely to apply to autosomal recessive diseases or late acting dominant diseases that have no selective disadvantage, because they act after an affected individual has reproduced. Mutations giving rise to dominant and sex-linked disorders will die out after a few generations due to selection. Most cases will therefore be due to recent mutations that have arisen *de novo*.

- As the number of founders was small, allele frequencies may not reflect the overall genetic structure of the general population world-wide. Moreover, during the initial stages of settlement the small population will lead to genetic drift that will exaggerate the initial effect of a small founding group. Population bottlenecks caused by famines and epidemics may also further exaggerate this effect. Some disease alleles may be more frequent, leading to a greater prevalence of certain diseases, while other diseases common elsewhere may be absent. In terms of complex diseases this decreases the chances of heterogeneity, i.e. the chances that an apparently identical disease in a population has more than one genetic factor in its aetiology.

- For common alleles that occurred before the expansion out of Africa, LD may extend over a greater distance. This is because among the founding group there is likely to be only one individual with the susceptible allele. All the alleles at closely linked polymorphic loci will therefore be in LD with the susceptibility allele. Erosion of LD will start afresh as the population expands. The chances that there will be a single founder carrying the susceptibility allele depend on the number of founders and the allele frequency. It is more likely if there were a small number of founders and the frequency of the allele is rare. Computer simulations show that if the allele is common (>5%) then the founding population must be very small for there to be a useful degree of LD. This has been confirmed in practice by pilot studies.

A commercial organisation called deCode Genetics will carry out a survey in Iceland to take advantage of these special features of isolated populations. A database will be created that will assemble and correlate information on the health record of individuals from the healthcare system, their genealogical history and genotypic data obtained from a high through-put genotyping facility. This required special legislation in the Icelandic parliament and the agreement of the Icelandic population. For obvious reasons this enterprise caused considerable controversy both in Iceland and in the wider human genetics community. Over 90% of the Icelandic people have agreed to take part, although critics argue that this figure reflects a requirement to opt out rather than a positive choice to opt in.

Isolated populations have a proven track-record in identifying genes affected in Mendelian disorders. It is not clear whether they will help in complex disorders, because LD with common risk alleles may not be significantly greater than in outbred populations. If very large-scale studies are necessary, it will be much easier to assemble a collection of experimental subjects in larger communities such as the USA and the UK. Moreover, there is a concern that the alleles identified may be peculiar to that population and not relevant to wider humanity.

6.4.5 Genetic susceptibility to Alzheimer's disease

The analysis of the aetiology of Alzheimer's disease (AD) is a good illustration of how the various techniques described in this chapter have been used in concert to study a major complex disease and provides good examples of the types of difficulty that are encountered in the process.

AD is the most common form of senile dementia. In America the population incidence is 0.1% at age 60 rising to 10% by age 80 and to over 30% by age 90. Cases occurring before age 60 are classified as early onset. The more common form is usually referred to as late onset Alzheimer's disease or LOAD. Pathologically both forms are characterised by the presence of neurofibrillary tangles in the cerbral cortex and senile or amyloid plaques in the brain. Neurone degeneration and death occurs in the hippocampus and parts of the cerebral cortex, regions of the brain that function in memory and learning.

Neurofibrillary tangles contain a highly polymerised form of a cytoskeletal protein called tau. The amyloid plaques contain aggregates of an insoluble 42-residue peptide called $A^\beta 42$. The $A^\beta 42$ peptide is released by proteolytic cleavage from a precursor protein called amyloid precursor protein (APP) encoded by the *APP* gene on chromosome 21. APP is anchored in the cell membrane by a single transmembrane domain, resulting in a large extracellular domain and a small cytoplasmic tail. This cytoplasmic tail in APP is cleaved by three different proteases called the α, β and γ-secretases to release different forms of the A^β peptide. Cleavage with the α-secretase releases a 40-residue form called $A^\beta 40$. Cleavage with either the β or γ-secretases results in the release of the insoluble and toxic $A^\beta 42$ form. In a healthy brain, 90% of the A^β peptide is in the $A^\beta 40$ form.

A number of non-genetic risk factors have been proposed including head trauma, exposure to aluminium and a family history of Down's syndrome. Vascular disease is a strong contributory factor. Genetically the disease appears to cluster in families, but the genetic epidemiology is complicated by a number of factors.

- The disease symptoms are mimicked by a number of other forms of dementia.

- Because of its late age of onset, first degree relatives of affected individuals are either too young to show symptoms (children) or have died long ago so that information is only available by anecdotal family evidence or medical records that may not have reliably distinguished between AD and other forms of dementia.

- Identical twin studies are complicated by competing causes of death in the twin who is not the proband.

- Extended families apparently showing an autosomal dominant pattern of inheritance are probably distorted by bias of ascertainment. That is they are only noticed and reported because they are unusual and they may not reflect a true picture of the normal pattern of inheritance.

Despite these difficulties it is clear that there is a strong genetic component to the liability of AD. About 5% of cases occur in families, showing an autosomal dominant form of inheritance. These families are more common where the proband is classified as early onset; however, only 30% of early onset cases show autosomal dominant inheritance. In the remaining sporadic cases the relative risk to first degree relatives is 3.6, from which it can be calculated that about 34% of the overall risk of AD is due to hereditary factors. Twin studies also support a genetic component – the concordance in monozygotic twins is 80% compared to 35% in dizygotic twins. This suggests a high heritability, but for the reasons outlined above the results of twin studies are treated cautiously.

Analysis of the rare early onset forms of AD has led to the identification of three loci that cause a small minority of all AD cases. Because AD segregated in Mendelian manner in affected families the loci affected could be mapped by parametric analysis. The first gene to be identified was the *APP* gene on chromosome 21, which encodes the precursor of the $A\beta42$ peptide found in senile plaques. Disease-causing mutations cluster around the site of cleavage by the β and γ-secretases and result in a five- to eight-fold increase the formation of both $A\beta40$ and $A\beta42$.

Two more loci were subsequently identified. Both encode related proteins, called presenilin I and presenilin II, encoded by the *PS1* and *PS2* genes on chromosomes 14 and 2 respectively. The gene encoding presenilin I is located on chromosome 14 and presenilin II is located on chromosome I. The latter is responsible for a cluster of AD families in the Volga basin region of Germany. Presenilin I and II are proteins with multiple transmembrane domains. For a long time their function was unknown but it has recently been shown that presenilin I is actually the γ-secretase responsible for the release of the disease-inducing $A^{\beta}42$ peptide.

Thus it is possible to explain how both types of dominant mutation leading to familial Alzheimer's disease may be operating. Although the mutations are responsible for less than 1% of Alzheimer's disease, they inform us about disease-causing mechanisms that are relevant to the more common sporadic form, because in both familial and sporadic forms there is an increase in the production of the $A^{\beta}42$ peptide. Moreover, they suggest that inhibitors of the γ-secretase could potentially be used as a treatment for Alzheimer's disease. Inhibitors of the γ-secretase activity *in vitro* are already known. But they cannot be used *in vivo* because they do not cross cell membranes. Nevertheless, γ-secretase is clearly a very promising target for further drug screens.

The first progress in dissecting the genetic factors in the much more common sporadic cases came when a set of 32 families where 87 out of 293 members suffered from Alzheimer's disease were analysed using the APM method (Section 6.4.3). This detected linkage between AD and markers on chromosome 19. Located within this region was the gene for apoliprotein E (*ApoE*), a plasma protein involved in cholesterol transport. *ApoE* was a candidate because the protein was detected in both the senile plaques and neurofibrillary tangles that characterise AD. Moreover, *in vitro* it has an affinity for $A^{\beta}42$ peptide. So a population association study was undertaken. There are three major alleles at this locus: *ApoE**ε2 (population prevalence 6%), *ApoE**ε3 (78%) and *ApoE**ε4 (16%). The alleles differ from each other according to the identity of amino acid residues at two positions in the polypeptide chain. *ApoE**ε4 was found to increase the risk and decrease the age of onset in a dose-dependent fashion (Table 6.9). This association has been repeated in numerous other studies.

The *ApoE**ε4 allele is thought to be responsible for the genetic component of the liability in about 40–50% of AD cases. Clearly, there are further alleles that contribute to the susceptibility. Further genome scans initially identified a locus on chromosome 12 between 12p11.23 and 12q13.12q. Although confirmed in a further study the exact location varied. Furthermore another study failed to find any linkage at all. The number of cases used in these studies was small and this probably explains the inconsistent results, as discussed above. Strong evidence of involvement to a locus on chromosome 10 was obtained with independent studies with larger numbers of cases. In the first study, a two-stage non-parametric genome scan was

Table 6.9 Dosage-dependent risk from *ApoE**ε4 allele. The *ApoE* status of 95 AD patients from 42 families was ascertained.

ApoE*E4 dose	% affected	Relative risk	Age of onset
0	20	1.00	84.3
1	46.6	2.84	75.5
2	91.3	8.07	68.4

Source: Corder, E.H. *et al.* (1993) *Science*, **261**, 921–923

carried out using ASPs with late onset Alzheimer's disease (LOAD), defined as families where the mean age of diagnosis was over 65 years. In the first stage 292 ASPs were examined with markers spaced 20 cM apart. Four chromosomal regions were detected which met the criteria for suggestive linkage, including a region on chromosome 10 (10q24). The second stage used an additional 168 families and examined more densely spaced markers in the region on chromosome 10. Convincing evidence for linkage to 10q24 was found. The linkage was detected in both sets of families analysed independently, so the linkage has been replicated. In the two stages together, a total of 429 ASPs were used and the MLS for the whole dataset was 3.83, which would be expected to occur by chance only once in 100 genome scans. This MLS value is higher than the value for the *ApoE* locus, which also showed linkage in the same dataset.

A second, partially independent approach also detected linkage to 10q24. High levels of the Aβ42 peptide are found in the bloodstream of both LOAD patients and their close relatives. Thus the level of Aβ42 forms a quantitative trait that can be analysed independently. This approach is very useful in complex disease, because the genetic components that control one aspect of the disease may be easier to identify than when their effect is not conflated with other biological factors that contribute to the cause of the disease. In this case the same families were used as in the first stage of the genome scan described above. However, the analysis could use multigenerational families and included individuals with high Aβ42 levels who were not diagnosed with AD. Thus the analysis is more powerful than the genome scan with ASPs. The peaks showing suggestive linkage to AD were tested for linkage to plasma Aβ42 levels. Linkage, with an MLS of 3.83, was observed to exactly the same region of chromosome 10 (10q24) that was identified in the genome scan. This not only provides independent confirmation of the linkage but suggests that the biological action of the locus is to elevate plasma Aβ42 levels.

Biochemical research suggests that the Aβ42 peptide is degraded and cleared by a protein called IDE (insulin degrading enzyme). The gene encoding IDE is located on chromosome 10 and IDE shows linkage to AD in parametric and non-parametric analysis in a group of 435 families with 1426 AD patients. However, although IDE is close to the locus at 10q24 identified in the studies described above, the peak of linkage is about 40 cM from the IDE locus. So, frustratingly, it is not clear whether there are two closely linked AD loci on chromosome 10 or whether the three studies have identified the same locus.

The successful analysis of the genetic components of AD illustrates the different approaches to identifying risk alleles for complex disease described in this chapter.

- A few cases are due to an autosomal dominant mutation that can be mapped by parametric analysis.

- Non-parametric linkage analysis was used to detect linkage to chromosome 19 and chromosome 10.

- Biological knowledge was used to identify *ApoE* and IDE as candidate loci.

- Population association studies were used to identify *ApoE**ε4 as a risk allele.

- Breaking down AD into component factors allowed plasma Aβ42 levels to be mapped as a quantitative trait.

- *ApoE* was used as proof-of-principle for genome scans based on LD of SNPs (see Section 6.4.4.5)

6.5 Conclusions: complex diseases are complex

The use of densely spaced SNP maps in population association studies using large numbers of affected individuals will enable searches of unprecedented power for susceptibility alleles. The identification of all human genes by the human genome project and increasing knowledge of their function further enhances the chances of identifying susceptibility alleles. Thus, with good reason, there is great optimism that progress will be made in the next few years. However, there have been many false dawns in this field and caution is still appropriate. We have seen in this chapter that the aetiology of common diseases is complex and the many difficulties that arise at a result. In diseases such as breast cancer and Alzheimer's disease autosomal dominant mutations have been characterised that are highly penetrant. However, these only account for a small fraction of the overall incidence of these diseases. Indeed, they only account for a minority of the cases where there is apparently autosomal dominant inheritance within a family. The genetic risk associated with most complex diseases is unlikely to be due to a few loci where risk alleles each have a large effect. Even where such alleles with relatively large effects have been identified, such as type 1 diabetes, only a small percentage of the at-risk individuals actually suffer from the disease. Partly this is due to environmental variation experienced by at-risk genotypes and partly to the moderating effect of the rest of the genetic background. In most cases there are likely to be many loci where alleles individually elevate the risk by a small amount. Furthermore, there will be complex gene–gene and gene–environment interactions. At best we may expect a risk profile composed of the cumulative risk contributed by several loci. If this does indeed come about, then there are considerable social, legal and ethical implications concerning the ownership of the information and the way that it is used. These will be considered in more detail in Chapter 11.

6.6 Summary

- Evidence for genetic involvement in common diseases comes from twin concordance studies, the increased risk to relatives of affected individuals and adoption studies.

- Identifying these factors would help to identify those at risk, define environmental risk factors and lead to novel and more rational methods of treatment.

- The heritability of a trait measures the proportion of phenotypic variability that can be ascribed to genetic factors. Even if the heritability of a trait is high, it may be still be strongly influenced by environmental factors.

- The inheritance of these diseases is similar to quantitative traits studied in model organisms that show a Gaussian distribution in the population. This inheritance is explained by postulating that the trait is controlled by the action of a number of polygenes and influenced by environmental factors. Since most diseases are not quantitative in nature it is proposed that the disease occurs when combined genetic and environmental liability exceeds a certain threshold. In order to model the action of polygenes it is necessary to specify parameters for each locus such as dominance, additivity, population frequency and displacement.

- Because the risk of a disease is the combined effect of many polygenes, modified by the environment, these diseases are generally known as complex or multifactorial diseases.

- Alleles may make a large contribution to overall liability and be relatively uncommon or make a small contribution and be common. Thus there are several different genetic architectures that would explain the inheritance of complex diseases.

- Risk alleles may be identified and mapped by five different methods.
  - Candidate genes based on the biology of the disease.
  - Parametric linkage analysis, only suitable for identifying a single gene responsible for a subset of cases where the disease segregates in a Mendelian fashion.
  - Non-parametric analysis, usually based on genome scans with affected sib-pairs (ASPs).
  - Population association studies based either on mapping an allele in linkage disequilibrium with the risk allele or searching for alleles that are directly responsible for the increased risk.
  - Animal studies.

- Genome scans with affected sib-pairs have had disappointing results. Theoretical analysis has shown that if the contribution of the allele is small, then very large numbers of sib-pairs will need to be included in the study.

- The focus has now switched to population studies using single nucleotide polymorphisms (SNPs).

- All of the above techniques have been used to study Alzheimer's disease.
  - A minority of cases are caused by the segregation of one of three autosomal dominant alleles at the following loci: *APP*, *PS1* and *PS2*.
  - The *ApoE4* allele is a major susceptibility locus on chromosome 19.
  - Recently a further major locus (possibly two loci) has been mapped on chromosome 10.

Further reading

General

CARDON, L.R. and Bell, J.I. (2001) Association study designs for complex diseases. *Nature Reviews Genetics*, **2**, 91–99.

Provides a guide to the complexities of setting up a properly designed population study.

LANDER, E. and KRUGLYAK, L. (1995) Genetic dissection of complex traits – guidelines for interpreting and reporting linkage results. *Nature Genetics*, **11**, 241–247.

Sets out and explains the reasoning behind the generally accepted guidelines for reporting linkage.

LANDER, E.S. and SCHORK, N. (1994) Genetic dissection of complex traits. *Science*, **265**, 2037–2047.

Classic review that formally sets out the different strategies for identifying susceptibility loci.

PELTONEN, L., PALOTIE, A., and LANGE, K. (2000) Use of population isolates for mapping complex traits. *Nature Reviews Genetics*, **1**, 182–190.

Review of the special properties, benefits and disadvantages of isolated populations.

RISCH, N.J. (2000) Searching for genetic determinants in the new millennium. *Nature*, **405**, 847–856.

Excellent all-round review. Explains the mathematical basis of the polygenic model of complex disease.

RISCH, N. and MERIKANGAS, K. (1996) The future of genetic studies of complex human diseases. *Science*, **273**, 1516–1517.

Classic paper, which demonstrates that, in theory, population association studies are more powerful than linkage studies for identifying susceptibility loci. Sets out the mathematics used to calculate the figures shown in Table 6.1.

ROSES, A.D. (2000) Pharmacogenetics and the practice of medicine. *Nature*, **405**, 857–865.

Provides an optimistic view of the power of LD mapping using SNPs to identify susceptibility loci.

SCHORK, N.J., CARDON, L.R. and XU, X.P. (1998) The future of genetic epidemiology. *Trends in Genetics*, **14**, 266–272.

Alzheimer's disease

BERTRAM, L., BLACKER, D., MULLIN, K. *et al.* (2000) Evidence for genetic linkage of Alzheimer's disease to chromosome 10q. *Science*, **290**, 2302–2303.

CHAPMAN, P.F., FALINSKA, A.M., KNEVETT, S.G. and RANSEY, M.F. (2001) Genes, models and Alzheimer's disease. *Trends in Genetics*, **17**, 254–261.

CORDER, E.H., SAUNDERS, A.M., STRITTMATTER, W.J., *et al.* (1993) Gene dose of apolipoprotein E type 4 allele and the risk of Alzheimer's disease in late onset families. *Science*, **261**, 921–923.

MYERS, A., HOLMANS, P., MARSHALL, H. *et al.* (2000) Susceptibility locus for Alzheimer's disease on chromosome 10. *Science*, **290**, 2304–5.

PERICAK-VANCE, M.A., BEBOUT, J.L., GASKELL, P.C. *et al.* (1991) Linkage studies in familial Alzheimer's disease – evidence for chromosome-19 linkage. *American Journal of Human Genetics*, **48**, 1034–1050.

ROKS, G., and VAN DUIJN, C.M. (2000) Genetic epidemiology of Alzheimer's disease. In *Analysis of Multifactorial Disease*, Bishop, T., Sham, P. (eds), pp. 85–100. Bios Scientific Press, Oxford.

STRITTMATTER, W.J., SAUNDERS, A.M., SCHMECHEL, D. *et al.* (1993) Apoliprotein E: High avidity binding to beta-amyloid and increased frequency of type 4 allele in late onset familial Alzhimer's disease. *Proceedings of the National Academy of Sciences USA*, **90**, 1977–1981.

DE STROOPER, B. (2000) Closing in on γ secretase. *Nature*, **405**, 627–629.

Describes the evidence that leads to the conclusion that presenilin I is the γ secretase.

Hereditary breast cancer

HALL, J.M., LEE, M.K., NEWMAN, B. *et al.* (1990) Linakge of early-onset familial breast cancer to chromosome 17q21. *Science*, **250**, 1685–1688.

NEWMAN, B., AUSTIN, M.A., LEE, M. *et al.* (1988) Inheritance of human breast cancer: evidence for autosomal dominant transmission in high-risk families. *Proceedings of the National Academy of Sciences USA*, **85**, 3044–3048.

Diabetes

BENNETT, S.T. and TODD, J.A. (1996) Human type 1 diabetes and the insulin gene: Principles of mapping polygenes. *Annual Review of Genetics*, **30**, 343–370.

Provides a classic example of the use of the Transmission Disequilibrium Test.

DAVIES, J.L., KAWAGUCHI, Y., BENNETT, S.T., *et al.* (1994) A genome-wide search for human diabetes 1 type susceptibility genes. *Nature*, **371**, 130–135.

One of the first examples of a complete genome scan using affected sib-pairs. Displays the data set more comprehensively than later examples, which makes it easier to understand how the method works in practice.

FIELD, L.L. (2000) Type 1 diabetes. In *Analysis of Multifactorial Disease*, Bishop, T., Sham, P. (eds), pp. 149–175. Bios Scientific Press, Oxford.

General Review

Linkage disequilibrium

KRUGLYAK, L. (1999) Prospects for whole genome linkage disequilibrium mapping of common diseases. *Nature Genetics*, **22**, 139–144.

Reports computer simulations showing that LD is unlikely to extend for more than 3 kb either side of common alleles that arose before the expansion out of Africa.

REICH, D.E., CARGILL, M., BOLK, S. *et al.* (2001) Linkage disequilibrium in the human genome. *Nature*, **411**, 199–204.

The study that compared LD in Europeans and Yorubans.

WEISS, K.M. and TERWILLIGER, J.D. (2000) How many diseases does it take to map a gene with SNPs? *Nature Genetics*, **26**, 151–157.

Provides a sceptical view of the power of LD mapping with SNPs to reliably identify susceptibility loci.

CHAPTER SEVEN

Gene therapy

Key topics

- Somatic and germline gene therapy
- Gene replacement and gene addition
- *In vivo, ex vivo* and *in vitro* gene therapy
- Transgenic animal models
- Vehicles for gene transfer
  - Retrovirus
  - Adenovirus
  - Adeno-associated virus
  - Liposomes and lipoplexes
  - Naked DNA
- Gene therapy for cystic fibrosis
- Gene therapy for Duchenne muscular dystrophy
- Gene therapy for bleeding disorders
- Gene therapy for severe combined immunodeficiency syndrome
- Gene therapy for non-heritable disorders
  - Cancer
  - DNA vaccines

7.1 Introduction

In theory, genetic diseases may be treated by the introduction of the wild-type gene into the cells affected by a mutation. Such an approach is called gene therapy. Following the successful isolation of the genes affected in many common monogenic disorders, there is considerable expectation that gene therapy will, for the first time, offer the prospect of a cure for genetic diseases. Moreover, the introduction of DNA molecules and oligonucleotides into cells can be used to provide novel treatments for many kinds of non-hereditary diseases such as cancer. In this chapter various approaches to gene therapy are described and progress that has been made in their application reviewed. It will be seen that experiments *in vitro* and with animal models demonstrate that, in principle, gene therapy is possible. Nevertheless, the results of the first clinical trials demonstrate that successful application to human patients for the treatment of hereditary diseases is still a long way off; however, gene therapy to treat other types of diseases such as cancer is giving more promising results.

7.2 Types of gene therapy

7.2.1 Somatic and germline gene therapy

There are two strategies for correcting an inherited disease by gene therapy (Figure 7.1). In **somatic gene therapy** the genetic defect is corrected only in the somatic cells of a person affected by the disease. In **germline gene therapy** a genetic modification is made to a gamete, fertilised egg or embryo before the germline has split off from the cells that will make the rest of the body. The crucial difference between these two strategies is that in the first, any genetic changes are restricted to the lifetime of the person treated, while in germline therapy any change is passed on to subsequent generations. Somatic gene therapy poses few ethical problems, but germline gene therapy represents a fundamentally new type of human activity whose consequences need to be thought through carefully before any experiments are attempted. Because of this, such experiments are currently prohibited in most countries.

7.2.2 *In vivo* and *ex vivo* gene therapy

One strategy to introduce the **transgene** (the exogenous gene) is to isolate cells so that they can be manipulated *in vitro*. This would allow the cells that have received the transgene to be selected and perhaps cultured *in vitro*. These cell clones would then be reintroduced into the patient's body. Such a strategy is called *ex vivo* **gene therapy** (Figure 7.1). Alternatively the transgene could be directly introduced into the cells affected by the disease; this is called *in vivo* **gene therapy**.

Ex vivo gene therapy is suitable for genetic defects that affect blood cells. All blood cells are derived from pluripotent stem cells in the bone marrow. They can be isolated, cultured *in vitro* and successfully reintroduced *in vivo*. As described below, *ex vivo* therapy was used for the treatment of adenosine deaminase (ADA) deficiency, the only gene therapy experiment that has so far resulted in any sign of clinical improvement. Clinical trials using *ex vivo* approaches are currently underway for Gaucher's disease, Fanconi's anaemia, Hurler's syndrome and chronic granulomatous disease. In principle, the major group of monogenic disorders that could be treated by *ex vivo* therapy are the haemoglobinopathies. Gene therapy for this group of diseases is complicated by the mechanisms that control the expression of both the α-globin and β-globin gene cluster. Expression of genes in these clusters requires the action of LCRs that oper-

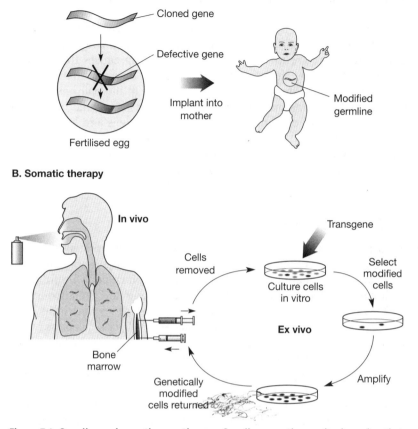

Figure 7.1 Germline and somatic gene therapy. Germline gene therapy (top) requires that the transgene is introduced into the fertilised egg or into the embryo before the germline separates from the soma. As a result, the baby is born with modified somatic and germline cells and will pass the modification on to the next generation. In somatic gene therapy the change is limited to the cells of the subject's body and is not passed on to the next generation. Somatic gene therapy can be *ex vivo* or *in vivo*. *Ex vivo* therapy involves removing cells from the patient, in this case blood cell progenitors from the bone marrow, and modifying them *in vitro*. After the modification, the cells may be cultured to increase their number before reintroduction into the patient. *In vivo* therapy involves direct introduction of the transgene into the cells of the subject's body. In this case a transgene is introduced into the cells that line the lung by means of an aerosol.

ate at a considerable distance from the genes controlled (see Chapter 5). Increased understanding of these processes has led to recent advances in this area (see Further reading).

Clearly, *ex vivo* approaches for diseases such as CF and muscular dystrophy will not be practical. The cells of the lung epithelia divide only very slowly so it is not possible to culture them *in vitro*. Even if it were, it is not easy to imagine how the lungs of affected individuals could be readily repopulated by cells manipulated *in vitro*. The accessibility of such cells thus becomes a major factor in determining whether gene therapy is possible. In the case of diseases such as CF it will be necessary to manipulate the cells *in vivo*. This requires that the transgene is delivered to those cells affected by the disease. Thus DMD would be treated by introducing the dystrophin gene into muscle cells, CF by introducing the *CFTR* gene into the cells of the airway, and so on. In both diseases these are the cells whose malfunction causes the major life-threatening effects. Nevertheless, in both examples there are many other cell types that fail to function normally and contribute to the pathophysiology of the disease. CF patients may suffer, *inter alia*, from pancreatic insufficiency, gastrointestinal malformation, defects in the vas deferens and liver problems. DMD patients may also have heart and central nervous system defects. Even if the defects were successfully corrected in the tissues primarily affected by the disease, the disease may not be fully cured. Furthermore, many somatic cell types may not be easily accessible for treatment.

7.2.3 Gene addition or replacement

The most perfect form of gene therapy would be to completely replace the defective gene with a functioning transgene. This would allow therapy for all types of mutation including dominant ones. Replacement would involve homologous recombination between the transgene and the chromosomal copy. This is possible as there are many situations where geneticists rely on such recombination, for example in the generation of gene deletions in trans-genic mice. However, gene replacement requires procedures to select a small minority of cells that have the alteration required. The selection procedures make use of sensitivity or resistance to drugs. *In vivo* gene therapy would need to be carried out without such selection. Although gene replacement has been shown to occur at a very low frequency in experiments with mouse embryonic stem cells, such procedures in human cells will not be technically feasible in the foreseeable future. Germline therapy will require accurate gene replacement, because this is the only way to be sure that the gene will be expressed in a completely normal fashion. This is one of the powerful practi-cal reasons that precludes germline therapy for the present.

For somatic gene therapy, this leaves gene addition rather than replace-ment, where the transgene works alongside the mutated gene in the same cell. This should allow correction of the lack of gene function that occurs in reces-sive conditions. However, it cannot correct a dominant mutation because the mutant protein will still be produced and exert its detrimental effect. It is theo-retically possible to use antisense strategies to inhibit a gene affected by a dominant mutation. This relies on producing an RNA molecule, or introducing an oligonucleotide, that is complementary to the mRNA originating from the gene. The mRNA and this **antisense mRNA** form a double-stranded hybrid that cannot be translated and is specifically degraded. Targeting to a mutant mRNA can be achieved by using an oligonucleotide whose sequence is com-plementary to mRNA from the mutant version of the gene.

The transgene can be integrated into a chromosome or can exist as an independent copy (**episome**). The advantage of the former is that the trans-gene will be stable in cell division and be inherited by progeny cells. The disadvantage is that, as noted above, the site of integration cannot be controlled. It is possible that the new gene will insert within a functioning gene, resulting in its inactivation. Alternatively it may activate an oncogene by integrating next to it. As a consequence the cell may become transformed into a cancer cell. This is not a theoretical speculation. Some retroviruses are known to be oncogenic by this mechanism.

7.2.4 Animal models

Development of gene therapy techniques will require a large amount of experimentation. For obvious reasons it is difficult to do most of this on human subjects. Animal models are extremely useful to demonstrate the principle and develop the technology. Some diseases have natural counterparts in animals. For example, *mdx* in mice and *xmd* in dogs (golden retrievers) are X-linked mutations that cause a disease similar to DMD. Alternatively, transgenic mice can be generated with the homologue of the human disease gene deleted. Usually these mice suffer from similar symptoms and can therefore be used to develop gene therapies. It is important to realise that there will also be important differences and the results may not always be directly transferable from the animal model.

Table 7.1 Advantages and disadvantages of the various approaches used to transfer genes.

	Advantages	Disadvantages
Retroviruses	Integration results in stable modification of target cell Infect replicating cells Suitable for *ex vivo* treatment	Uncontrolled integration may have oncogenic consequences Cannot infect non-dividing cells, may be overcome by use of related Lentiviruses Provoke immune response Expression of transgene becomes attenuated in long term
Adenoviruses	Infect non-dividing cells, so suitable for CF and DMD Non-integration avoids safety hazard of uncontrolled integration Efficient gene transfer, at least *in vitro*	Expression of transferred gene is transient Provoke strong immune reaction
Adeno-associated virus	Non-pathogenic Broad tropism Long-lasting expression of transgene	Difficult to raise high titres in packaging cell lines Small insert size cup to 4.5 kb
Lipoplexes	Non-immunogenic, so safe Can carry large DNA molecules	Inefficient delivery Transient expression
Naked DNA	Non-immunogenic Simple to prepare and administer to muscle cells	Inefficient gene transfer Limited in cell types that can be treated Expression transient

7.3 Methods of transferring transgenes into target cells

Gene therapy relies upon methods of introducing transgenes into target cells. This means a **vehicle** is needed for its delivery. Sometimes the term 'vector' is used in this context. In this account, the word 'vector' is reserved for its normal meaning of a DNA molecule to which a gene may be spliced to mediate its expression and/or replication in the host cell. Sometimes the use of the two words will overlap, for example a retroviral genome into which a transgene has been spliced is a vector, but the **virion** (see below) containing the recombinant genome is a vehicle. The utility of the vehicle will depend on a number of factors:

- efficiency of delivery to the target cell;

- specificity of delivery for the target cell (**tropism**);

- whether the target cell needs to be dividing;

- whether the vehicle will provoke an immune response;

- the size of the DNA that can be carried;

- stability and longevity of the gene in the target cell;

- expression of the introduced gene.

Various approaches can be taken to transfer the gene to the target cell. The advantages and disadvantages of these are summarised in Table 7.1 and are discussed in more detail in the following sections.

7.3.1 Virus-based vehicles for gene therapy

7.3.1.1 Retrovirus-based gene transfer

Retroviruses (Figure 7.2) have two identical RNA genomes. Upon infection the RNA is copied into DNA by a virally encoded reverse transcriptase. This DNA copy is then integrated into the host genome in a process that depends upon the **long terminal repeats** (LTRs) at each end of the genome (Figure 7.3). The retroviral genes are transcribed from a promoter within the 5' LTR regions. The resulting RNA molecules are packaged into the virus particles. This packaging process requires a sequence called ψ. As well as the *pol* gene encoding reverse transcriptase, the retroviral genome contains the *gag* gene that encodes a viral core protein and the *env* gene that encodes a viral envelope protein. The *env* protein recognises receptors on target cells, facilitating infection.

Figure 7.3 shows how retroviruses may be used as gene delivery systems. Because retroviruses can be oncogenic if they integrate upstream of a cellular oncogene, it is essential that any retrovirus-based gene therapy product is not contaminated by replication-competent viruses. This is achieved by using a packaging cell line in which essential retroviral genes are divided between two different DNA molecules, neither of which contains the ψ packaging sequence necessary for incorporation into the virion (the mature

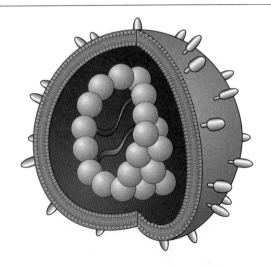

Figure 7.2 Retrovirus virion. The mature virus particle found outside the cell is called a virion. The retrovirus virion is composed of an outer lipid envelope into which are inserted envelope proteins that recognise specific proteins on the outside of target cells. Within the outer lipid layer is an inner nucleocapsid, a hollow protein structure made up of the *gag* and *pol* proteins surrounding two RNA genomes.

virus particle). A third DNA molecule carries the gene to be transferred together with a functioning ψ sequence sandwiched between LTRs. This packaging cell line produces retrovirus particles containing the gene of interest that can infect the target cell. Once inside the target cell, the pol protein contained within the virus particle will ensure that the gene is integrated into the host genome. The LTRs are necessary for integration and also act to express the gene once integrated. Because multiple recombination events will be required to produce a functioning retrovirus in the packaging cell line, it is highly unlikely that the preparation will be contaminated. Nevertheless, each batch produced for clinical testing is carefully examined for any replication-competent virus particles.

The advantages of retrovirus-mediated gene transfer are its efficiency compared with other methods of gene transfer, together with the stability of the introduced gene. It is generally used for *ex vivo* gene therapy, which is based on replicating cells. Many gene therapy trials are currently in progress using retroviral vehicles for gene transfer. Nevertheless, there are significant problems with the use of retroviruses for gene therapy.

1. Retroviruses can only infect proliferating cells because the viral genome integrates into the host chromosome and can only gain access to the chromosomes when the nuclear envelope disappears at mitosis. This prevents the use of retroviral vehicles for the treatment of genetic diseases such as DMD or CF where the cells most affected are non-dividing or only dividing slowly.

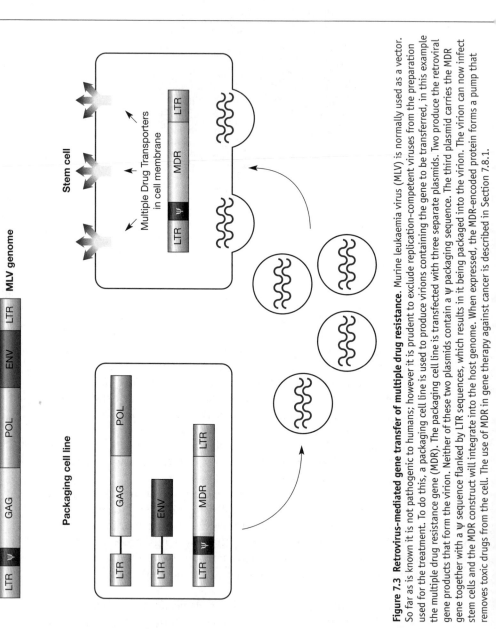

Figure 7.3 Retrovirus-mediated gene transfer of multiple drug resistance. Murine leukaemia virus (MLV) is normally used as a vector. So far as is known it is not pathogenic to humans; however it is prudent to exclude replication-competent viruses from the preparation used for the treatment. To do this, a packaging cell line is used to produce virions containing the gene to be transferred, in this example the multiple drug resistance gene (MDR). The packaging cell line is transfected with three separate plasmids. Two produce the retroviral gene products that form the virion. Neither of these two plasmids contain a ψ packaging sequence. The third plasmid carries the MDR gene together with a ψ sequence flanked by LTR sequences, which results in it being packaged into the virion. The virion can now infect stem cells and the MDR construct will integrate into the host genome. When expressed, the MDR-encoded protein forms a pump that removes toxic drugs from the cell. The use of MDR in gene therapy against cancer is described in Section 7.8.1.

2. Retrovirus-mediated gene transfer is limited in the size of DNA molecule that can be accommodated, which mostly limits its use to cDNA.
3. Retrovirus-mediated gene transfer does not specifically infect one cell type. This results in a loss of efficiency as most of the genes do not reach the target cells. Moreover, when it is used to alter the characteristics of a particular cell type, such as a cancer cell, modifying the wrong cell type may be actively counter-productive. One approach for improving target specificity is to modify the viral env protein that interacts with receptors on the exterior of the host cell. This can be done by fusing a protein hormone to the env protein so that the virus may be targeted to cells displaying the receptor for that hormone. Another strategy is to replace the env protein with that of another virus. For example, in one experiment the mouse leukaemia virus (MLV) env protein was replaced with the env protein of human vesicular stomatitis virus (HSV). The engineered MLV virus only infected cells displaying the HSV receptor and no longer infected the normal target cells of MLV.
4. Retroviruses integrate at random sites in the genome. If they integrate near a cellular oncogene they may cause its activation through transcriptional activation.

Lentiviruses, such as HIV, are related to retroviruses but have evolved the capacity to replicate in non-dividing cells. Thus in principle, viruses such as HIV could be used for gene therapy in those cases where the target cell is non-dividing. Naturally there would be considerable concern about the prospect of using HIV as a vehicle for gene therapy because it causes AIDS. Moreover, HIV only infects T cells because the Env protein on the viral surface binds the CD4 protein during infection. So its natural tropism would prevent it infecting the targets of most gene therapy regimes. The tropism has been changed by substituting the HIV *env* gene for the *env* gene of human vesicular stomatitis as described above. To lessen the risk still further, HIV vectors have been engineered to remove all viral genes except those necessary for infection of non-dividing cells. Such vectors contain only 25% of the original HIV genome. Nevertheless, there is a fear that recombination *in vivo* with related viruses could reconstitute an infectious virus.

7.3.1.2 Adenovirus-based gene transfer

Adenoviruses (Figure 7.4) are double-stranded DNA viruses that naturally infect the non-dividing cells of the respiratory and gastrointestinal tracts. They make an attractive vehicle for gene therapy because they have evolved to evade host immune mechanisms and to be highly efficient at infecting their target cells. Furthermore, once inside the target cell they replicate as episomes and so avoid the dangers of uncontrolled integration. There are at least 50 different **serotypes**. Serotypes 3 and 5 show a high degree of tropism for the respiratory tract and thus these strains have been used as vehicles for CF gene therapy. The ability of adenoviruses to infect non-proliferating cells has resulted in attempts to use them for gene therapy of DMD, where it is necessary to treat non-dividing muscle cells.

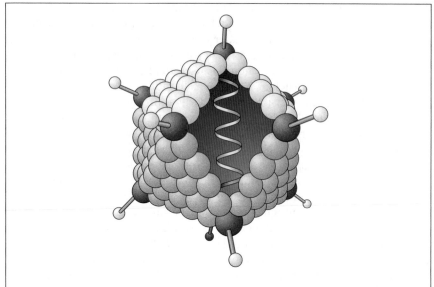

Figure 7.4 An adenovirus virion. This is an icosahedral structure consisting of a capsid made up of 12 different proteins enclosing a single double-stranded DNA molecule 35 kb in size.

Of course it is necessary to disable the adenovirus to prevent it replicating and killing the host cell. Gene expression can be divided into early and late phases. DNA replication and expression of the genes whose products constitute the virion depend upon prior expression of early genes. The infective cycle can therefore be halted by deleting E1, an essential early gene (Figure 7.5). At the same time this makes room for the insertion of foreign DNA, up to a size of about 6 kb.

Adenoviruses are efficient at delivering DNA to their target cells. However, a major drawback is that clinical trials have shown that even disabled forms provoke a strong immune response. This limits the dose that can be safely administered. Moreover, expression is transient so that the therapy would need to be repeated. Repeated therapy results in the immune response becoming more severe. Recently, adenovirus vehicles with deletions of other early genes such as E2, E3 and E4 (Figure 7.5) have been constructed to further reduce late gene expression which might be responsible for the immune response.

The dangers of adenovirus vectors were highlighted by the tragic death in 1999 of a patient undergoing gene therapy with an adenovirus vector. James Gelsinger was an 18-year-old who was taking part in a Phase 1 trial to treat an inherited deficiency of ornithine transcarbamylase (OTC). Phase 1 trials are designed to determine whether the treatment is safe, but are not concerned with efficacy. The death occurred even though the trial was using an adenovirus vector in which both the E1 and E4 genes had been deleted, which should have reduced the harmful immune response to the

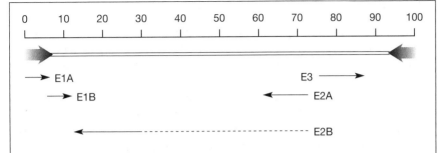

Figure 7.5 Early gene transcription in adenoviruses. The adenovirus genome is 36 kb in size conventionally represented by a map with 100 units. Each end of the molecule consists of a 100 bp sequence that is an inverted repeat of the sequence at the other end. During the first 6–8 hours after infection, mRNA molecules representing about 25% of the genome are derived from scattered regions of the genome called E1 to E4. Each region gives rise to multiple transcripts by differential splicing but all have the same 5′ end. The resulting overlapping mRNA molecules give rise to proteins that are subsets of one another. In some cases one mRNA molecule is read in overlapping reading frames so that the resulting proteins have common amino-terminal sequences but different carboxy-terminal sequences. The very first region to be transcribed is E1A; the E1A protein is required for the expression of all other adenovirus genes. Early gene transcription allows DNA replication followed by late RNA transcription that encode the capsid proteins of the virion.

virus. This event was a severe setback to gene therapy trials, especially to the use of adenovirus vectors. It also highlights the ethical issues associated with gene therapy, which will be discussed in Chapter 11.

7.3.1.3 Adeno-associated viruses

Adeno-associated viruses (AAV) are naturally replication-defective viruses that depend on other helper viruses, such as adenovirus, to provide essential functions. AAV-based systems offer a number of advantages over other virus systems: they are not naturally pathogenic, they are tropic for airway cells, they infect non-dividing cells, they show longer-lasting expression (up to 6 months in animal models) and they can show chromosome-specific integration in chromosome 19. The genome of AAV is a single-stranded DNA molecule containing two genes, *rep* and *cap*. The Rep protein mediates viral integration into the specific site on chromosome 19. Cap is the structural protein of the capsid. These genes are flanked by two inverted terminal repeats that are essential to pack the DNA into the virus capsid. For gene therapy the *cap* and *rep* genes are replaced by the transgene. This construct is introduced into a packaging cell line that supplies the Cap and Rep proteins *in trans*. A disadvantage of this arrangement is that the specific targeting to chromosome 19 is lost as the Rep protein is not present. In animal models there has been long-lasting expression of transgenes from such tissues as muscle, liver and brain. Recently, an AAV vector was used in a promising human trial to correct haemophilia A (see below).

7.3.2 Non-viral methods of gene transfer

7.3.2.1 Liposome- and lipoplex-mediated gene transfer

In aqueous solution DNA molecules are negatively charged. The outside of most cells are also negatively charged, so DNA molecules and cell surfaces may be expected to be mutually repellent, which is thought to limit the efficiency of DNA uptake by cells. This problem may be overcome by encapsulation of the DNA in **liposomes**. Liposomes are spheres consisting of lipid molecules surrounding an aqueous interior. The lipid molecules used contain hydrophobic and hydrophilic domains. They form a bilayer in which the hydrophilic domains face outwards towards the surrounding aqueous environment and inwards towards the water-filled interior. DNA molecules can be encapsulated within liposomes. However, liposomes are typically 0.025–0.1 μm in size but plasmids are typically over 2 μm in size. Thus only a small number of plasmids can be fitted into each liposome and as a result liposomes are very inefficient vehicles for gene transfer.

This problem was overcome with specially formulated lipids where the hydrophilic domain is positively charged (cationic), which attracts both the negatively charged DNA and the negatively charged cell surface. These cationic lipids were very efficient at encapsulating DNA. The resulting lipid–DNA complexes are much more complicated than liposomes. The plasmids are encapsulated in tube-like structures, which in some ways resemble the lipid envelope that surrounds the protein capsid of some viruses. Because these structures are so different from liposomes they were given a new name, **lipoplexes**. The main advantage of lipoplex-mediated gene transfer is that lipoplexes are non-immunogenic so they are safer than virus-based methods. In addition, lipoplexes are easy to prepare and are not limited in the size of DNA molecule that can be carried. The main disadvantage is that the efficiency of transfer is lower than virus-based methods. Nevertheless recent trials have shown that in some cases they can be as efficient as virus-based vehicles.

7.3.2.2 Naked DNA

During the course of experiments designed to test the efficiency of various liposome/lipoplex formulations, naked DNA was used as a negative control since, theoretically, naked DNA should not be taken up by cells. Surprisingly it was found that naked DNA injected directly into the muscle tissue of mice was taken up and expressed by muscle cells. This is thought to occur by DNA entering the cell through small lesions in the cell membrane. This opens up a novel route for the introduction of genes into muscle cells. One possibility is to use muscle cells to produce proteins whose action is not cell-limited, such as insulin or blood clotting factors. As discussed below, another idea is the use of naked DNA to express proteins that would act as vaccines against infectious diseases or even cancers.

Soon after the *CTFR* gene was cloned it was demonstrated that the wild-type gene could correct the chloride conductance defect of a CF cell line *in vitro*, demonstrating that in principle it should be possible to treat the disease by gene therapy (see Chapter 5). The life-threatening consequences of the disease affect the epithelial cells that line the lung airways and thus these are the cells into which the wild-type gene would have to be introduced. Clearly, it will not be possible to treat the cells by *ex vivo* therapy, so an *in vivo* method of treatment will need to be developed.

The first experiments were carried out with an adenovirus vector using transgenic mice lacking *CFTR* function as a model (CF transgenic mice). They showed that the gene could be delivered to the target cells in the airway and correct the chloride ion transport defect, for 6 months after introduction of the wild-type gene in some experiments. The scene was set for gene therapy trials with human CF patients. There are several methodological problems that need to be considered before describing the results.

Firstly, in which part of the airway is expression most important? *CFTR* expression is highest in the submucosal gland, which may secrete water on to the airway surface. However, *CFTR* is also expressed in the epithelia that line the alveoli, so CFTR function may be important there as well. It may be that restoration of function in airway progenitor cells may be necessary for successful treatment. At present, gene delivery systems are based on aerosols, which can only access epithelial cells and not submucosal cells or progenitor cells

Secondly, what proportion of cells need to be corrected for relief of the clinical symptoms? *In vitro* experiments with monolayers of CF epithelial cells suggest that restoration of chloride ion conductance in only 6% of the cells is sufficient because chloride ions can move laterally from cell to cell through **gap junctions**, which connect adjacent cells. However, sodium ion absorption is cell-limited; thus most cells will need a functioning CFTR protein if regulation of sodium ion absorption is important.

Thirdly, how should preliminary trials be conducted? If the gene is delivered to cells deep in the lungs it will be difficult to gain access to these cells to measure whether the gene has been delivered and whether there is any evidence of *CFTR* function. Alternatively, the gene could be delivered to the nose where it is possible to sample cells to determine whether the *CFTR* gene has been delivered and, if it has, to monitor its expression and function. However no clinical outcome can be observed because the nasal passages are not involved in the pathophysiology of CF. A compromise may be the maxillary sinus, which can be accessed to monitor the success of gene delivery and function but which is also clinically relevant.

7.4.1 Adenovirus trials

There have been a number of clinical trials based on adenovirus vehicles, but with only limited success in gene delivery and function. *CFTR* expression was variable and transitory. Increasing the dose of the virus to improve the outcome results in a strong immune response in the form of inflammation and fever. The problem may be that the cells that need to be treated are the columnar ciliary cells that line the airway. Previous *in vitro* experiments have been carried out with the underlying basal cells. The columnar cells are much more resistant to infection by adenoviruses. Perhaps this should not be surprising. Adenoviruses were chosen as vehicles because they have evolved to infect the lung. While this may be true, it is probably equally true that columnar cells have also evolved to be resistant to such infection.

7.4.2 Lipoplexes

Trials where aerosols of lipoplex-encapsulated DNA were sprayed into the nasal cavity have had some very limited success. Gene transfer was monitored by the presence of *CFTR* DNA and mRNA, and *CFTR* expression was monitored by chloride ion conductance and sodium ion absorbance. Both transfer and expression were detected, but the levels were low and did not persist for any significant length of time. The levels were comparable to those observed with adenovirus vehicles but, importantly, without the inflammatory response. These trials were conducted using the nasal epithelium, which is not medically relevant to CF and is probably an easier target than the usually infected lower airways typical of CF patients. Moreover, although the degree to which it is necessary to remediate *CFTR* function is not known, the levels of *CFTR* expression observed are not likely to prove sufficient for clinical improvement.

Lipoplexes are taken up into the cell by **endocytosis**: the cell membrane invaginates to form an intracellular vesicle. These vesicles fuse to form structures called **endosomes**. A problem with lipoplexes is that they remain trapped in the endosome and so the DNA does not reach the nucleus. A suggestion to overcome this problem is to encapsulate an adenovirus vector in a lipoplex. This may combine the benefits of both vehicles: the negatively charged lipoplex provides an efficient means of access into the cell, while the adenovirus can escape from the endosome and deliver the DNA to the nucleus.

7.4.3 Future directions

Although it is exciting that clinical trials for CF have been undertaken with some positive results, clinically effective treatment is still a long way off. Far more efficient ways have to be found to deliver and express the *CFTR* gene. One exciting development is the construction of human artificial chromosomes based on an α-satellite sequence as a centromere. This may provide a vector for the safe and permanent maintenance of the *CFTR* gene. Since such a vector will be capable of carrying a large DNA insert, it might be possible to transfer the entire *CFTR* chromosomal locus instead of *CFTR* cDNA. This may enable the pattern of expression to match the natural system more faithfully.

A second way forward may be treatment *in utero*, where immune tolerance to adenovirus may be greater. It may also avoid the early and possibly irreversible damage to the lung that occurs in CF patients. One intriguing report shows that in a CF transgenic mouse, transient expression of *CFTR in utero* results in a permanent reversal of the CF phenotype. The *CFTR*-deleted mice used in the study suffer from abnormalities of the gastrointestinal tract that mimic meconium ileus, a form of intestinal blockage seen in 5–10% of newborn CF patients. Transient expression completely prevented this abnormality so *CFTR* expression may only be required briefly during development, but not subsequently. If this is true, it may be necessary to reassess current strategies for gene therapy of CF. However, some caution would be appropriate before such a conclusion is drawn. At present it is not known how relevant this model will be to the human airway. Previous experience has shown that studies with the CF mouse model do not always transfer to the human situation.

7.5 Gene therapy for Duchenne muscular dystrophy

Gene therapy for DMD has been successfully achieved in transgenic mice, but has not reached the stage where human trials can be attempted. Several problems specific to DMD need to be solved. Firstly, ways have to be found to access muscle cells to effect gene transfer. Secondly, the extreme size of the dystrophin gene (14 kb) makes it difficult to incorporate the wild-type gene into adenovirus vehicles, which have an upper limit of 6 kb.

The *mdx* mouse shows similar defects to the human disease. Human dystrophin cDNA has been transferred to muscle cells using an adenovirus vector either directly injected into muscle tissue or injected into the heart and allowed to spread around the body with the bloodstream. The size problem has been partially solved by making use of a mini-dystrophin gene, which originated in a patient with BMD. The deletion responsible resulted in a reduction in the size of the cDNA to 6.3 kb. The reading frame was maintained and the resulting dystrophin protein retained considerable function, even though it was only half the size of the wild-type protein. Expression of this mini-gene in *mdx* mice results in expression of dystrophin, which is correctly located within the muscle cells and completely corrects the pathological phenotype. Retroviral vehicles have also been used. Because they only infect replicating cells, the gene was transferred to embryonic **myoblasts**. Myoblasts can be reimplanted *in vivo*, but unfortunately expression of the dystrophin gene is transient when this is done.

One approach to gene therapy for DMD is based on the observation that even in patients with a severe form of the disease a small amount of normal or nearly normal dystrophin protein is often detected. This is thought to arise as follows. Most severe DMD mutations destroy the reading frame of the dystrophin-encoding mRNA. During RNA processing, mistakes can occur that result in an exon carrying the mutation being omitted. In a small number of mRNA molecules this can restore the reading frame and result in the production of dystrophin. The splicing process can be redirected using synthetic oligonucleotides. Experiments with cells *in vitro* have shown that it is possible to restore the reading frame of a mutated dystrophin mRNA using such an approach.

7.6 Gene therapy for bleeding disorders

Haemophilia A and B are caused by a lack, or very low levels, of the blood clotting factors VIII and IX, respectively. They make attractive targets for gene therapy for two reasons:

1   For many years both forms of haemophilia have been successfully treated by partially purified serum preparations. It is thus well established that the diseases can be treated by augmentation of clotting factor levels in serum. Moreover, significant clinical improvement is seen with very low levels of the factors. In the treatment of haemophilia B, the standard therapeutic objective is only to maintain serum levels above 1% of normal levels. Thus even if the therapy is inefficient, there is still the promise that it will be effective.

2   Unlike CFTR or dystrophin, the action of clotting factors is not limited to the cell in which they are synthesised. They can thus be made in a wide variety of tissues, simplifying the problem of gene delivery and control of gene expression.

Following successful experiments in mice and canine models of haemophilia there have been two phase 1 clinical trials that showed some limited success in treating both types of haemophilia in humans. In the first, the gene for factor IX on an AAV vector was introduced to three haemophilia B patients via intramuscular injection. As it was a phase 1 trial, it was designed to test safety. The first objective was to establish what virus titre was safe by gradually increasing the dose in successive batches of patients. The report was concerned with the first batch of test subjects, who received a very low virus titre. Despite this, significant positive effects were observed. After treatment, the gene was shown to be present in the muscle cells by Southern blots and PCR, and immunohistochemistry showed the muscle cells were producing factor IX. In one patient serum levels of factor IX were measurably increased and clinical efficacy was demonstrated by a reduction of bleeding episodes requiring the use of external clotting preparations. Significantly, in a second patient there was also evidence of clinical improvement, even though there was no evidence that serum levels of factor IX had increased. This implies that an increase in factor IX that is too low to measure is still effective. Alternatively, it is possible that low concentrations of factor IX may be sequestered by binding to collagen IV, for which it is known to have an affinity. Low-level expression from the transgene may have saturated this binding and released some factor IV to act as a clotting factor. Importantly, there was no evidence that the injection of the AAV had any harmful effect.

In the second study an *ex vivo* approach was used to treat haemophilia A. Skin fibroblasts were removed by biopsy from each patient and Factor VIII cDNA cloned into plasmids was introduced by electroporation. Cells that had been successfully transfected were selected and grown *in vitro* to increase cell number. The genetically modified clones were then reintroduced into the peritoneal cavity by laproscopy. The advantage of this method is that it avoids the use of viruses, with their attendant safety problems, while at the same time being efficient because only modified cells are grown and reintroduced into the patient. Four of the six patients showed a detectable increase in serum factor VIII levels and two had greater than 1% of normal activity. The number of bleeding episodes and the use of exogenous factor VIII were reduced in all patients showing an increase in factor VIII levels. Although the improvement was transitory, increased serum levels of factor VIII persisted for 10 months in one patient and six months in another. The reason for the eventual decline is not understood. It could be that the fibroblasts had reached the end of their reproductive span, which is limited in all cells (see Section 2.2.4), because they were allowed to proliferate *in vitro* as part of the experimental procedure.

7.7 Gene therapy for Severe Combined Immunodeficiency Syndrome

Severe Combined Immunodeficiency Syndrome (SCIDS) is a group of diseases with heterogeneous causes. One form, caused by a lack of adenosine deaminase (ADA), was the first in which gene therapy was attempted. Two ADA-deficient children were treated by the *ex vivo* treatment of isolated lymphocytes. Adenosine deaminase was introduced via a retroviral-based vehicle together with the enzyme itself encapsulated by polyethylene glycol. The treatment, which was repeated every 6 weeks, did result in a marked clinical improvement, but it is unclear whether this was due to the introduced DNA or the ADA protein, especially as no ADA gene expression could be detected in one of the children.

More recently, another form of the disease, SCID-XI, has been successfully treated. SCID-XI is a sex-linked disorder affecting the gene encoding a subunit called γC that is common to five different interleukin receptors. As a result of the mutation, signals for the survival, growth and differentiation of early lymphoid progenitor cells are blocked, affecting the production of T-cells, B-cells and natural killer (NK) cells. Haematopoietic stem cells from patients were transduced *ex vivo* with the gene on a Moloney retrovirus-derived vector. Counts of T-cells, B-cells and NK cells were all restored to normal levels and remained so for at least 10 months. The patients were apparently completely cured, the first time such a claim could be made for gene therapy.

7.8 Gene therapy for non-heritable disorders

Gene therapy can also be used for non-heritable disorders, and these applications are a lot nearer introduction into medical practice than gene therapy for heritable diseases. In reality, they form the bulk of gene therapy trials in progress.

7.8.1 Cancer therapies

Many gene therapy protocols are being investigated as possible treatments for cancers. One idea is to deliver a gene to a target cell whose product will activate a prodrug, which then kills the manipulated cell but does not affect the rest of the cells in the body. For example, the thymidine kinase gene from herpes simplex virus can be delivered to brain tumour cells using a retroviral vehicle (Figure 7.6).

The engineered retrovirus is injected directly into the brain tumour and the patient is then treated systemically with ganciclovir. This drug is an analogue of guanosine. It is phosphorylated by thymidine kinase to the triphosphate form, which can be incoporated into DNA. Incorporation inhibits further DNA replication and kills the cells. Thus ganciclovir is cytotoxic but only when phosphorylated by thymidine kinase. Although only a small proportion of the cells express thymidine kinase, a larger number of surrounding cells are killed. This is known as the bystander effect. The reason why this happens is not completely understood, but it is thought that the toxin generated in one cell can spread to adjacent cells through gap junctions.

Retroviruses can only replicate in dividing cells but brain cells are non-dividing. The use of retroviruses thus targets the treatment to tumour cells, which will be the only dividing cells in the brain. When this treatment was used in phase I clinical trials a further unexpected benefit was observed: secondary tumours elsewhere in the body showed signs of regression. It is thought that lysis of cells in the primary tumour released a high concentra-

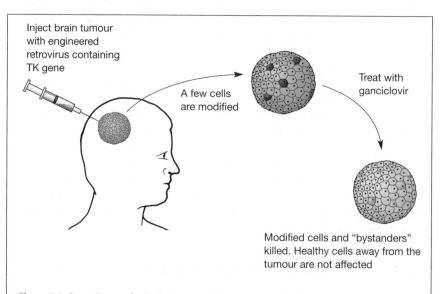

Inject brain tumour with engineered retrovirus containing TK gene

A few cells are modified

Treat with ganciclovir

Modified cells and "bystanders" killed. Healthy cells away from the tumour are not affected

Figure 7.6 Gene therapy for brain tumours. The tumour is directly injected with a retrovirus containing the mouse thymidine kinase (TK) gene and a few cells take up the vector, shown in blue. These cells convert the prodrug ganciclovir into an active form and are killed (grey cells). Because of the bystander effect surrounding cells are also killed.

tion of tumour-specific antigens, which provoked an increased immune response against the secondary tumours.

Another possible use for gene therapy against cancer is to protect blood haemopoietic stem cells from the effects of chemotherapy. The usefulness of chemotherapy is often limited by two factors. Firstly, resistant lines arise that express elevated levels of the multiple drug resistance (MDR) protein, which pumps toxic drugs out of the cell. Secondly, haemopoietic stem cells naturally express very low levels of MDR and are killed by low concentrations of toxic drugs. This leads to a condition known as leucopenia, a lack of white blood cells, and the patient becomes immunocompromised. The problem of resistance could be overcome by increasing the initial concentration of drugs to kill the cancer cells before resistant lines arose. However, such a course of action would exacerbate the second problem and kill the patient. A solution to this impasse is to use gene therapy to increase the expression of MDR in haemopoietic cells. The way a recombinant retrovirus can be used to elevate the level of MDR expression is described in Figure 7.3.

Haemopoietic stem cells can be efficiently transduced by retroviruses using *ex vivo* therapy and can then repopulate bone marrow after reimplantation. Haemopoietic stem cells express a surface antigen called CD34 and thus may be selected using an anti-CD34 antibody. There are sufficient CD34+ cells in peripheral blood for the treatment so there is no need for painful bone marrow sampling. They are induced to grow *in vitro* using a cocktail of growth factors that stimulate their proliferation and are then transduced with the retrovirus carrying the MDR gene. Meanwhile the patient is subjected to aggressive chemotherapy for 3–4 days with drugs such as taxol. The manipulated stem cells are then reimplanted along with unmanipulated marrow cells. In phase I clinical trials of this treatment one patient showed MDR transfer in CD34+ cells 3 months after treatment.

There is much research directed towards correcting the somatically acquired mutations that give rise to cancer. Dominantly acting oncogenes such as activated *ras* may be inhibited by antisense mRNA that prevents translation of the oncogene mRNA. Over half of cancers have mutations in tumour suppressors such as RB and p53 (see Chapter 5). When wild-type genes are introduced into cell lines from tumours carrying these mutations the tumour characteristics disappear. If these genes could be introduced into tumour cells *in vivo* then cells could either reacquire normal control of cell regulation or, if the genomes are irreversibly damaged, p53 may induce apoptosis (programmed cell death). The p53 protein functions as a dimer. Some p53 mutant proteins interfere with the assembly of normal subunits into dimers. Such mutations are said to be **dominant negative mutations**. Clearly, introducing a wild-type copy will not restore p53 function and it is necessary to turn off the mutant gene. This may be achieved by antisense oligonucleotides that bind to the normal mRNA and prevent translation.

One important defence against cancer is the immune system. Several trials have been designed to recruit the immune system to help fight the cancer. One idea of how this could be achieved is to programme cancer cells to express non-self HLA markers on the cell surface. This should result in the cells being destroyed by cellular immunity mechanisms. In one experiment, lipoplexes containing the gene encoding the major histocompatibility antigen HLA-B7 were injected into malignant melanoma tumours. In one-third of the patients the tumour shrank or even disappeared. Encouragingly, there were signs that even tumours that were not injected started to regress. This is important because malignant melanoma spreads through the body and it would be difficult to treat all tumours separately.

A second way to enlist the immune system to fight a cancer is to increase its sensitivity to cancer cells. The immune system is thought to be an important natural line of defence against cancer. It is capable of recognising cancer cells and killing them, but in an established cancer is clearly failing to do this. The idea of stimulating the immune system to fight cancer is an old one. In the early part of the twentieth century an American surgeon called William Coley noticed that the tumours of some patients regressed after a severe bacterial infection. From this he developed a cancer treatment based on deliberate bacterial infection. As might be expected, this did not turn out to be very successful because the infection could not be controlled. Coley modified the treatment by injecting a preparation of dead bacterial cells called Coley's toxins. This treatment was effective in some cases but was never consistently successful and after a while it fell from use. The reason for the original observation was later shown to be that the bacterial infection stimulated the production of a **cytokine** called tumour necrosis factor (TNF) that stimulated cell-mediated immunity. The production of recombinant TNF led to hopes that a modern version of the old bacterial infection treatment could be used against cancer, this time using TNF to stimulate the immune system. Unfortunately, TNF has severe side-effects, such as the induction of fevers and general malaise; once again the treatment failed and for very similar reasons.

The latest version of this treatment is to target the production of cytokines to the cancer cells themselves. This is done by targeting genes encoding cytokines to the cancer cells. This stimulates the immune system, but only in the vicinity of the tumour. The attraction of this approach is that it is not necessary to target every cancer cell or even every tumour where there are multiple secondary tumours. By stimulating the immune system to destroy a few cancer cells, its interest in all cells carrying the same tumour antigen is aroused.

7.8.2 DNA vaccines

A different area that is attracting a lot of interest is the use of naked DNA as a novel type of vaccine. The idea here is to induce muscle cells to produce antigens from infectious agents or tumour-specific antigens. A particular advantage of this strategy is that as the antigen is expressed by cells it may trigger cellular as well as humoral immunity. The principle has been shown to be effective for a variety of infectious diseases in mice. Clinical trials are under way in humans for influenza vaccination using this strategy. It is possible that trials for HIV, herpes and malaria may start in the near future. In the longer term, it is hoped to use proteins commonly found on the surface of cells in some cancers to produce vaccines against these cancers.

7.9 Summary

- Gene therapy seeks to treat genetic disease using the unmutated version of the gene affected.

- Two strategies can be employed: somatic gene therapy, which limits any genetic change to the lifetime of the person treated, and germline therapy, which affects gametes and any changes will thus be inherited by future generations. Germline therapy is currently prohibited for ethical reasons.

- Gene therapy can act by replacing the mutated gene or adding a wild-type version to work alongside it. Gene replacement requires techniques that are not yet available, so gene therapy of inherited diseases is limited to recessive mutations.

- Cells may be modified by *ex vivo* techniques, where cells are removed from the body, genetically modified and then reintroduced. *Ex vivo* therapy is suitable for modification of blood stem cells.

- Diseases such as DMD and CF must be treated by *in vivo* therapy, where the gene is introduced to the cells within the body of the patient.

- Various vehicles are used to introduce genes into cells:
  - Retroviruses introduce genes into replicating cells and are used for *ex vivo* therapy.
  - Adenoviruses are tropic for airway cells and are efficient at carrying genes into cells, but provoke a strong immune reaction in the patient.
  - Lipoplexes encapsulate the DNA into cationic lipids. They do not induce an immune reaction, but may be less efficient than other methods.
  - DNA can be directly injected into muscle cells. This is being used for DNA-based vaccines.

- The first trials for CF gene therapy have successfully demonstrated that genes can be delivered to cells *in vivo*. However, expression was transient and not sufficiently strong to cause a clinical improvement. Adenovirus-based therapy provoked a strong immune reaction.

- Gene therapy for DMD has been demonstrated in a mouse model but so far there have been no clinical trials.

- There has been promising progress in gene therapy for haemophilia A and B. The haemophilia A therapy was based on an *ex vivo* approach in which naked factor VIII DNA was introduced into cultured fibroblasts by electroporation. The haemophilia B therapy used adenoassociated virus (AAV). In both cases a measurable expression of the transgene and some clinical improvement was observed.

- Two forms of Severe Combined Immunodeficiency Syndrome (SCIDs) have been treated by gene therapy. Replacement of the adenosine deaminase gene was the first gene therapy trial attempted. The results were ambiguous. Treatment of SCID-XI was based on an *ex vivo* approach in which the gene was introduced into haematopoietic stem cells on a retroviral vector. This apparently effected a complete cure, the first time this has been achieved for any condition using gene therapy.

- Most gene therapy research is currently focused on the treatment of non-inherited diseases. Various ways of using gene therapy to combat cancer have been tried and some have shown encouraging results in clinical trials.

Further reading

General

BANK, A. (1996) Human somatic cell gene therapy. *Bioessays*, **18**, 999–1007.

MOUNTAIN, A. (2000) Gene therapy: the first decade. *Trends in Biotechnology*, **18**, 119–128.

SOMIA, N. and VERMA, I.M. (2000) Gene therapy: trials and tribulations. *Nature Reviews Genetics*, **1**, 91–99.

VARIOUS AUTHORS (1997) Special report: making gene therapy work. *Scientific American*, **276** (6), 79–103.

Cystic fibrosis

BOUCHER, R.C. (1996) Current status of CF gene therapy. *Trends in Genetics*, **12**, 81–84.

ENNIST, D.L. (1999) Gene therapy for lung disease. *Trends In Pharmacological Sciences*, **20**, 260–266.

WAGNER, J.A. and GARDNER, P. (1997) Towards cystic fibrosis gene therapy. *Annual Review of Medicine*, **48**, 203–216.

Duchenne muscular dystrophy

DICKSON, G. (1996) Gene therapy of Duchenne muscular dystrophy. *Chemistry and Industry*, **8**, 294–297 (15 April 1996).

FASSATI, A., MURPHY, S. and DICKSON, G. (1997) Gene therapy of Duchenne muscular dystrophy. *Advances in Genetics*, **35**, 117–149.

Haemoglobinopathies

MAY, C., RIVELLA, S., CALLEGARI, J. *et al.* (2000) Therapeutic haemoglobin synthesis in beta-thalassaemic mice expressing lentivirus-encoded human beta-globin. *Nature*, **406**, 82–86.

WEATHERALL, D.J. (2001) Phenotype-genotype relationships in monogenic disease: lessons from the thalassaemias. *Nature Reviews Genetics*, **2**, 245–255.

Reviews the complex control exerted over α and β globin production and points out the formidable difficulties that will have to be overcome for successful gene therapy.

Haemophilia
KAY, M.A., MANNO, C.S., RAGNI, M.V. *et al.* (2000) Evidence for gene transfer and expression of factor IX in haemophilia B patients treated with an AAV vector. *Nature Genetics*, **24**, 257–261.

ROTH, D.A., TAWA, N.E., O'BRIEN, J.M. *et al.* (2001) Nonviral transfer of the gene encoding coagulation factor VIII in patients with severe hemophilia A. *New England Journal of Medicine*, **344**, 1735–1742.

Severe combined immune deficiency syndrome (SCIDS)
CAVAZZANA-CALVO, M., HACEIN-BEY, S., BASILE, C.D. *et al.* (2000) Gene therapy of human severe combined immunodeficiency (SCID)-X1 disease. *Science*, **288**, 699–672.

Gene therapy for cancer
VILE, R.G. (1996) Gene therapy for treating cancer: hope or hype? *Chemistry and Industry*, **8**, 285–289.

Genetic testing

Key topics

- When gene testing is used
- Types of mutation
- General principles of testing
- Testing for mutations in the *CFTR* gene
- Testing for mutations in the dystrophin gene
- Testing for uncharacterised mutations
  - Single-stranded conformational polymorphisms
  - Denaturing gradient gel electrophoresis
  - Mismatch cleavage detection
  - Protein truncation test
  - DNA chips
  - DNA sequencing
  - Chromosome tracking

8.1 Introduction

The advances in gene isolation and characterisation described in this book enhance enormously the ability to diagnose genetic disease by the more traditional methods based on a physician's skill supported by biochemical tests. This chapter is concerned with the techniques that can be used routinely to detect mutations in genes of interest. The application of these techniques often raises significant ethical problems; these are considered in Chapter 11.

There a number of different situations where gene testing might be carried out.

- Where there is a known risk of a particular disease, embryos may be tested with a view to termination; alternatively, prospective parents may wish to know their carrier status in order to make reproductive decisions. The optimum situation in embryo testing is where the nature of the mutation has previously been defined, thus allowing its presence to be specifically tested in the embryo. In an autosomal recessive disorder this will require two mutations to be identified. In the absence of such information the chromosome(s) harbouring the mutation may be tracked by linked polymorphic markers.

- Neonatal population screening for genetic diseases where there is a clear benefit derived from early diagnosis and there is a simple and reliable test available. Neonatal screening for phenylketonuria is common using the Guthrie 6-day blood-spot test. A small sample of blood is taken from the baby and preserved by drying on to a card. Phenylalanine levels in the blood spot are then assayed. However, the same blood spot can provide material for other tests, such as screening for the ΔF508 *CFTR* mutation in European populations.

- To confirm or rule out a diagnosis of genetic disease made on the basis of medical symptoms.

- To diagnose the status of an individual, normally an adult, at risk from a late-onset genetic disease such as HD or one of the inherited cancer predispositions.

- To determine the status of genes predisposing to complex diseases. This area is likely to grow in importance as tests are developed for susceptibility to diseases such as Type 1 diabetes mellitus.

- Where a candidate gene involved in a genetic disease has been identified, genetic testing is used to search for a correlation between the presence of a mutation and the onset of disease symptoms. This is often the final step in the hunt for a disease gene.

8.1.1 DNA testing has been revolutionised in the last few years

A few years ago work in a DNA diagnostic laboratory would mainly consist of tracking disease chromosomes through a family using RFLPs analysed by Southern blotting. Southern blotting is laborious, involves the use of radioisotopes, with the attendant safety problems, and requires significant amounts of biological material. RFLP analysis required extensive work to establish the phase of mutations and linked RFLP polymorphisms segregating in a family. Many meioses were uninformative because the RFLPs were not heterozygous, and if the linked markers were not sufficiently close to the mutation the diagnosis could be falsified by recombination.

Techniques used to screen for mutations have been revolutionised in the last few years by two developments. Firstly, cloning the disease genes has allowed tests to be directed towards the gene sequences themselves rather than tracking the inheritance of mutant chromosomes through families. Secondly, PCR is used to amplify sequences. This generates enough product to be analysed by conventional agarose or polyacrylamide electrophoresis. However, increasingly the technology used for automated DNA sequencing is applied to analysing PCR products. Apart from increasing throughput and accuracy, a big advantage of this is that it allows the quantification necessary to detect heterozygous deletions.

8.1.2 Three different challenges to the gene tester

In some cases the nature of the mutation likely to affect a gene is known. Firstly, where only one or a small number of different mutations in a population are responsible for a disease; typically, this results from founder mutations. Secondly, where the disease commonly occurs because of a particular type of mutation. The obvious example of this would be trinucleotide repeat expansion (TRE) mutations, but we may also consider cases such as DMD, where deletions are responsible for 60% of cases.

In other cases such information will not be available. This is inevitable when a gene is first cloned, during the attempt to correlate the presence of mutations with disease symptoms. There are also many genes where a large number of different mutations have been catalogued. Sometimes these mutations are unique to a particular family or individual and are known as **private mutations**. Testing for the presence of mutations in these genes is a formidable technical challenge. Often the genes may be large, and it is possible that the aetiological mutation may be a single base change.

We may thus summarise three different problems that a gene tester will be called upon to solve.

1. The disease is caused by one or a small number of well-characterised mutations. This will be approached by using tests designed to detect specific mutations. These tests are described in Section 8.2.
2. The gene is cloned but there are many possible mutations. This requires tests that can detect differences between normal and mutant genes without necessarily defining the nature of the mutation. This is a more difficult challenge requiring a different approach that is described in Section 8.2.
3. The mutation cannot be characterised, but the chromosome harbouring the mutation can be tracked through a family. This would have been the normal situation 10 years ago; it is less common now but the need still arises on occasion. An example is given in Section 8.2.

8.1.3 Nature of mutations

Before we consider techniques for detecting mutations, it is necessary to consider the different types of mutation that may occur (the nomenclature used to describe mutations is shown in Box 8.1):

BOX 8.1: NOMENCLATURE USED TO DESCRIBE MUTATIONS

A formal description of the recommended nomenclature can be obtained from http://www.interscience.wiley.com/jpages/1059-7794/nomenclature.html

Amino acid substitution

Amino acid	Code	Amino acid	Code	Amino acid	Code
Alanine	A	Glycine	G	Proline	P
Arginine	R	Histidine	H	Serine	S
Asparagine	N	Isoleucine	I	Threonine	T
Aspartic acid	D	Leucine	L	Tryptophan	W
Cysteine	C	Lysine	K	Tyrosine	Y
Glutamine	Q	Methionine	M	Valine	V
Glutamic acid	E	Phenylalanine	F	Nonsense	X

Each amino acid is represented by a single-letter code, shown in the table above. The amino acid substitution is represented by a number representing the residue in the polypeptide flanked by the single-letter code of the original and replacing amino acid, e.g.

R553X   A truncating mutation in which arginine at position 553 in the polypeptide is replaced by a stop codon.

N1303K   Asparagine at 1303 is replaced by lysine.

Nucleotide substitutions

These are described by a number representing the position of the nucleotide in the cDNA or genomic DNA (prefaced by c or g if it is not clear which is being used), followed by a letter (or letters) representing the original nucleotide (G, A, T or C) and a chevron before the letter (or letters) representing the replacing nucleotide. The A of the ATG codon is designated by +1. The nucleotide immediately preceding the initiator codon is designated by −1. Zero is not used. If the change occurs in an intron, the change is specified by indicating the number of nucleotides to the nearest nucleotide in the cDNA. Changes in introns may also be specified by reference to intron number using the abbreviation IVS. Nucleotide changes are usually only specified when they occur in introns, because changes in exons, which result in mutations rather than polymorphisms, will cause a change in the amino acid sequence, the nomenclature for which is described above.

621 +1 G>T   G is replaced by T at the first nucleotide in the intron following nucleotide 621 in the cDNA

1717-1G>A   G is replaced by A at the last nucleotide in the intron preceding nucleotide 1717 in the cDNA

IVS4 +1G>T   G is replaced by T at the first nucleotide in intron 4

Deletions and insertions

Deletions are represented by 'del'; insertions are represented by 'ins'. This is preceded by the number indicating the first nucleotide affected by the deletion, or the nucleotide before the insertion, and followed by the nucleotide(s) that are inserted or deleted, e.g.

1609delCA   The C and the A nucleotides starting at position 1609 are deleted
394 delT   The nucleotide T at position 394 in the cDNA is deleted
3905 insT   T is inserted after nucleotide 3905

Formerly, the greek letter Δ was used to indicate a deletion. This is no longer used because it cannot be represented in HTML files. However, the practice is continued with mutations such as the CFΔF508 (the phenylalanine at position 508 is deleted) and CFΔ1507 (the nucleotide at position 1507 is deleted), because they have become so well known by their original names. Often these mutations are represented by spelling out the Greek letter, e.g. deltaF508.

- large-scale chromosomal changes such as deletions and inversions, which may be detected by cytogenetic examination;

- small-scale deletions;

- TREs: this type of mutation is easy to identify and proved very useful in the hunt for genes such as the HD gene;

- mutations affecting RNA production and processing, such as promoter or splice site mutations;

- protein-truncating mutations, such as nonsense and frameshift mutations;

- base substitutions that result in a change in the amino acid sequence of the encoded protein.

Some of these mutations, such as protein-truncating mutations or deletions, will quite obviously affect gene function. Others may be more problematic. For example, as discussed in Chapter 9, apparently neutral polymorphisms are common, so it is sometimes difficult to know whether a base change has functional consequences. Sometimes the nature of the change may be significant, for example it may affect a highly conserved amino acid or a splice acceptor site.

8.1.4 Material used for tests

Material used for tests may be recovered from a number of different sources. An important advantage of PCR-based tests is that they can be used with very small amounts of material.

- A single cell from an eight-cell embryo produced by *in vitro* fertilisation; this allows preimplantation testing. Parents known to be at risk of having an affected child produce an embryo by *in vitro* fertilisation. This is then tested for the presence of the genetic disease and only implanted if it is shown not to be affected.

- **Chorionic villous sampling**. The chorion is derived from the zygote but is not part of the developing embryo. It is a membrane with projections or villi that surround the embryo. Chorionic villous sampling is carried out by introducing a catheter through the vagina until it touches the chorion. The test can be safely carried out in the tenth to eleventh week of pregnancy. It thus has an advantage over amniocentesis because, combined with PCR amplification of the sample, it allows an earlier termination of pregnancy should this be indicated by the results of the genetic test.

- **Amniocentesis**. Sampling of cells for genetic testing from the amniotic fluid surrounding the developing foetus. The procedure involves inserting a large tube through the mother's abdomen. These cells are cultured *in vitro* for about 3 weeks to increase numbers. The procedure cannot be attempted before the 16th week of pregnancy.

- Blood from the Guthrie card. DNA is stable so the card can be readily posted to different locations or retrieved for tests later.

- Buccal cells recovered from a mouth wash.

- Blood sample, most commonly used for adult tests.

Genetic tests are usually carried out on material amplified by PCR. Both mRNA and DNA can be used as templates. Genomic DNA can be recovered from any convenient source, such as a chorionic villous sample, a blood sample or mouth wash. It contains potentially relevant information in sequences such as promoters that are not expressed in mRNA. The main disadvantage of genomic DNA is that many genes are large with a complex organisation of introns and exons. If the structure of the gene is known, the exons can be amplified by specific primers. DNA samples are usually used to characterise the presence of specific mutations.

Because the introns have been removed by processing, cDNA produced using mRNA as a template is a much simpler target to analyse when the structure of the gene is not known. This is always the case when a gene is first isolated by positional cloning. In this situation it is necessary to prove that the gene is responsible for the condition by showing a correlation between the presence of a mutation and the occurrence of the disease. This requires that the gene from affected and normal individuals is scanned for differences, using techniques described in Section 8.3. The expression of a particular mRNA will often be restricted to a particular tissue and may be present in very low quantities. Thus in many cases it is not practicable to use mRNA.

This section considers the general types of test that can be used to detect specific mutations and then illustrates how these are used, with examples drawn from the diagnosis of CF, DMD and a TRE mutation.

8.2.1 General principles

Tests for particular mutations are now mostly based on PCR. These tests are designed according to the particular type of mutation being tested (see above). One advantage of PCR is that more than one pair of primers can be included in a single PCR reaction, allowing more than one mutation to be screened in each reaction. Such a reaction is referred to as a **multiplex PCR** reaction. Specific mutations may be recognised by **allele-specific oligonucleotides** (ASOs), which anneal either to a mutant or to wild-type sequence but not both. An important principle with PCR is that tests whose interpretation depends on the presence or absence of an amplification product must have a positive control to ensure that the PCR reaction does not fail for technical reasons, such as an inactive polymerase or impurities in the DNA sample.

General approaches for recognising specific mutations are listed below.

- Large deletions may be recognised by using PCR primers in which either one or both primers anneal with the region deleted. Deleted DNA used as a template fails to produce an amplification band. In sex-linked diseases such as DMD the absence of a band can be readily detected in affected males. In female carriers of X-linked deletions, or heterozygotes with deletions affecting autosomal genes, there will be a change in the quantity of the amplified band. This can now be detected using automated sequencing technology because there will be a reduction in the size of the peak in the trace of the output.

- Small deletions may be recognised by amplifying a region of 100 nucleotides that spans the site of the mutation. A deletion is recognised by a band shift in the product. Even differences as small as one nucleotide can be resolved by careful electrophoresis.

- Point mutations, which include both base substitutions and deletions/insertions of only one or two nucleotides, can be recognised by PCR. The ARMS test (Amplification Refractory Mutation System) is commonly used. In its original form two PCR reactions are carried out in parallel and the products run in adjacent lanes during electrophoresis. One primer that anneals at a site away from the site of the mutation is common to both reactions. In one reaction the other primer anneals to the sequence of the mutant allele. In the second reaction the other primer anneals to the sequence of the wild-type allele. Thus a band is produced in one or other of the lanes according to whether the template carries the wild-type or mutant allele. Generally the ARMS test is designed in a multiplex form that allows a number of mutations to be simultaneously screened. This requires careful design of the primers to result in well-separated bands upon electrophoresis of the amplification products. Usually this test is now carried out in a modified form. Only a primer that anneals to the mutant allele is included, so that the presence of a band indicates that the mutation is present. Positive controls for the PCR reaction are included in the multiplex test. Usually two controls are used that produce bands that are either smaller or larger than any of the bands that will be produced by mutant alleles. These controls test that the PCR reaction is functioning properly, so that the absence of a mutant band can be safely interpreted as indicating that the mutation is not present.

- Sometimes a mutation will either create or destroy a restriction site. This may be simply detected by amplifying a region around the site and attempting to digest the product with the appropriate restriction enzyme.

- TREs can be screened using PCR products that anneal either side of the expansion site. Replication slippage during amplification often results in a ladder effect. This can be very useful for determining the repeat number precisely. Large amplification will sometimes result in the distance between the two primers becoming too large for the PCR reaction to work efficiently. Such cases need to be investigated by Southern blotting.

- ASOs can also be used in hybridisation assays to detect a mutation. Usually the ASOs are attached to the membrane. This is then hybridised to DNA amplified by PCR from the region being tested. A number of different point mutations can be simultaneously screened by having a number of ASOs arranged in an array. DNA chips are an extension of this idea, where the array is miniaturised to allow a very large number of ASOs to be incorporated. In the extreme, an ASO for every single nucleotide position can be included, allowing every possible point mutation to be detected.

8.2.2 Test for mutations in the CFTR gene

In people of European extraction, the ΔF508 mutation is the most common CF allele. The proportion of CF alleles that are ΔF508 shows a gradient across Europe, increasing from about 50% in southern European countries to 85% in Denmark and the Basque region of Spain (see Figure 9.8). The average figure is about 70%. Twenty other mutations are also relatively common, having a combined frequency of about 15% (depending on country). The remaining 15% of mutations are individually uncommon, but over 800 have been characterised. The distribution of CF alleles has consequences for genetic testing for CF. Assuming a frequency for the ΔF508 allele of 70% and a combined frequency of 15% for the other common alleles, the frequency of the various classes of CF-affected individuals can be calculated as follows:

49%	ΔF508/ΔF508
21%	ΔF508/common mutation not ΔF508
21%	ΔF508/rare mutation
4.5%	common mutation not ΔF508/rare mutation
2.25%	common mutation not ΔF508/common mutation not ΔF508
2.25%	rare mutation/rare mutation.

These figures will vary in different European countries as allele frequencies change. The important point is that over 90% of CF patients will carry at least one ΔF508 allele. An initial test for the ΔF508 mutation will therefore identify most individuals who are either affected by CF or who are carriers. Testing for ΔF508 together with the 20 common mutations will identify 70% of CF cases in Europeans (more in some countries). However, detecting **compound heterozygotes** between ΔF508 and a rare mutation will be difficult. This is an important complicating factor in embryo testing, neonatal screening or carrier screening in adults. In other populations, different mutations may be more common and testing for *CFTR* mutations must therefore be designed to take account of the genetic structure of the local population or the ethnic origin of the subject.

The ΔF508 mutation may be detected by a simple and robust PCR test that amplifies a 100 bp region spanning the site of the mutation. The 3 bp deletion results in the ΔF508 allele migrating slightly faster than the wild-type allele. As a result, all three possible genotypes may be readily distinguished from each other (Figure 8.1). A multiplex ARMS test can be used to recognise the other 20 most common CF mutations found in European populations. Figure 8.2 shows the result of a multiplex test using a commercial testing kit.

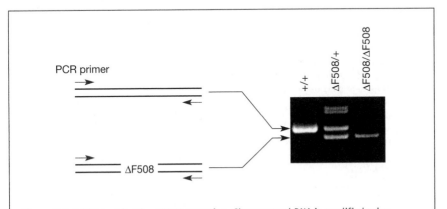

Figure 8.1 A PCR test for the ΔF508 mutation. Chromosomal DNA is amplified using primers that anneal either side of the site of the ΔF508 mutation. The amplification products are separated on a denaturing polyacrylamide gel and visualised by ethidium bromide fluorescence. The wild-type allele is 3 bp larger than the mutant allele so migrates slightly more slowly. The three possible diploid genotypes are clearly distinguishable. The two additional bands in the heterozygote lane are heteroduplexes between wild-type and mutant DNA strands. (Data courtesy of the North Trent Molecular Genetic Laboratory, UK.)

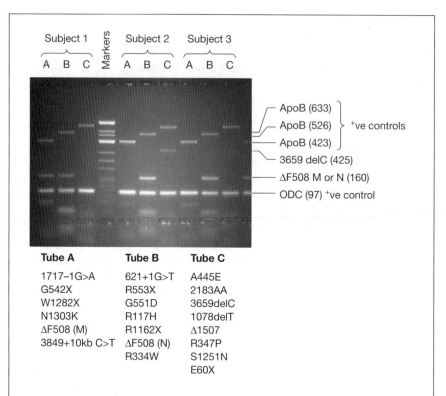

Tube A

1717–1G>A
G542X
W1282X
N1303K
ΔF508 (M)
3849+10kb C>T

Tube B

621+1G>T
R553X
G551D
R117H
R1162X
ΔF508 (N)
R334W

Tube C

A445E
2183AA
3659delC
1078delT
Δ1507
R347P
S1251N
E60X

Figure 8.2 Multiplex PCR ARMs test for CF alleles. A commercial testing kit called Elucigene CF20, marketed by Cellmark Diagnostics, can detect the presence of the 20 most common CF mutations. Three multiplex PCR reactions are carried out in parallel for each subject. Each tube contains a number of separate primer pairs. Each pair is designed to anneal a particular allele to produce a band whose size is well separated from any other possible bands from the same tube. Therefore a band is only produced if the allele is present. Because the absence of the band is used as evidence that the allele is absent, it is essential to have positive controls that show the PCR reaction is functioning properly. Each tube also contains primers that will amplify two control sequences, one larger and one smaller than any of the expected products. The diagram illustrates tests on three different subjects. Subject 1 is producing bands with the primers targeted at the wild allele (N) and mutant(M) alleles at the ΔF508 site. This subject is therefore a ΔF508 heterozygote. Subject 2 is producing a band in tube B corresponding to that expected from the 369delC allele. This subject is therefore heterozygous with respect to this allele. Subject 3 only amplifies the wild-type allele at the ΔF508 site, and produces no other bands. Therefore, this subject does not contain any mutations at the CF locus. (Data courtesy of Richard Kirk, North Trent Molecular Genetics Laboratory, UK.)

Sometimes the base change creates or destroys a restriction site, which forms the basis of a very simple test for the presence of the mutation. An example is given in Figure 8.3 for the G551D and R553X mutations, which are the third and sixth most common CF mutations worldwide. Both mutations affect the sequence GTCAAC, which is the target site for the restriction enzyme *Hinc*II. Thus the presence of one or

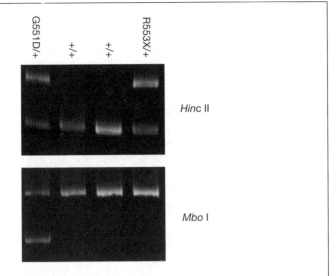

Figure 8.3 Diagnostic restriction enzyme assays for the *CFTR* G551D and R553X mutations.
The wild-type sequence, GGTCAAC is a target for *Hinc*II (underlined). This is altered to GATCAAC
in G551D, which contains a target for *Mbo*I (underlined) but is no longer digested by *Hinc*II. In
R553X the wild-type sequence is altered to GGTCAAC, which is not a target for either enzyme. In
the test, a fragment containing the mutation site is amplified by PCR and digested separately
with each enzyme. If the individual contains either mutation, the *Hinc*II digest fails and so a
heterozygote produces two bands (the wild-type allele is still digested). When digested with
*Mbo*I only a G551D heterozygote produces two bands because one allele is a target for this
enzyme. (Data courtesy of the North Trent Molecular Genetic Laboratory, UK.)

the other is indicated by loss of this restriction site. In addition, the
G551D mutation creates an *Mbo*I site, which can therefore be used as a
further test to distinguish between the two mutations when loss of the
*Hinc*II site is observed.

These and other similar tests provide a simple means for identifying
80–90% of CF mutations. The remaining mutations provide a much sterner
challenge because it is not practicable to develop individual tests for each of the
hundreds of mutations that have been recorded. Instead it is necessary to use
techniques that scan the gene to detect differences from the wild type without
defining the precise nature of these differences. These techniques are reviewed
below and some examples of the detection of rare CF mutations are included.

Neonatal screening for CF can be carried out by assaying for levels
of immunoreactive trypsin in the Guthrie blood-spot sample. An abnormally high level may indicate CF but it could be due to other causes.
Such babies would therefore be subjected to DNA tests. Because most
CF cases would have at least one ΔF508 allele, screening for this mutation would detect nearly all cases. This test can also be carried out on
the Guthrie blood-spot sample. A diagnosis of CF would be confirmed
by continued elevation of immunoreactive trypsin levels after 28 days
and by measurement of sweat electrolytes.

8.2.3 Duchenne muscular dystrophy

Two-thirds of DMD cases are caused by large deletions that remove one or more exons. The deletions cluster in two regions: at the 5' end, affecting exons 3–8, and towards the 3' end, affecting exons 44–60 (predominantly 44–50). Multiplex PCR can be used to detect loss of these exons. In this test PCR is used with a mixture of different primer pairs. Each primer pair is designed to amplify a single exon using genomic DNA as a template. The products of the PCR show a band for every exon present; a missing band is therefore diagnostic of a deletion. Figure 8.4 shows an example of this multiplex PCR test used to screen for deletions of exons in the 3' deletion hotspot.

The remaining cases of DMD not caused by deletions are mostly nonsense or frameshift mutations that cause chain termination; these may be detected by the protein-truncation test described below.

8.2.4 Trinucleotide repeat expansions

Each TRE is a separate event; however each mutation affects the same site and so can be readily detected by a PCR reaction based on primers that anneal either side of the expansion site. The size of the PCR product indicates whether an expansion event has occurred. The extent of the

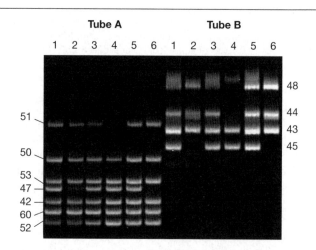

Figure 8.4 Multiplex test for deletions in the dystrophin gene. Each test consists of two tubes (A and B) containing pairs of PCR primers that amplify specific exons of the dystrophin gene, as shown at the sides. Individual number 6 has a deletion of exons 45 and 47, as judged by the missing bands in tubes B and A respectively. This probably results from a single deletion that spans exon 46, which was not tested. Individual 2 has a deletion of exon 45; the band for exon 47 is weak in this individual but was clearly visible in the original photograph. Individual number 4 apparently has a deletion of exons 44, 48 and 51. These exons are non-contiguous and so this result would be regarded with suspicion and the sample retested. The most likely explanation is that the PCR reaction failed to amplify the three largest bands in this sample. (Data courtesy of the North Trent Molecular Genetic Laboratory, UK.)

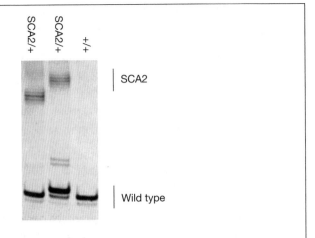

Figure 8.5 TREs in spinocerebellar ataxia type 2 (SCA2). The CAG repeat sequence was amplified using PCR primers that anneal to flanking regions. SCA2 heterozygotes have an extra band corresponding to the amplified allele. The samples have been separated on a polyacrylamide gel and visualised by silver staining. (Data courtesy of the North Trent Molecular Genetic Laboratory, UK.)

expansion may be important in the prognosis of age of onset and severity; this information is also revealed by the expansion. Figure 8.5 shows an example, where spinocerebellar ataxia type 2 is diagnosed by such a test.

8.3 Scanning genes for unknown mutations

Section 8.2 considered some case studies where the nature of the likely mutation is known and it is possible to design specific tests to identify them. However, many genes are affected by a large number of different mutations; 10–20% of CF alleles are rare mutations. Genes such as *BRCA1* and *BRCA2* are affected by a large number of different mutations. It is necessary to use techniques that can scan a sample and detect differences from the wild-type gene. One ever- present problem in this type of analysis is deciding whether a change detected is a 'polymorphism' or a 'mutation', i.e. whether the change has deleterious consequences for gene function.

8.3.1 Single-stranded conformational polymorphisms

Single-stranded nucleic acid molecules form extensive secondary structure in solution because of intramolecular base pairing. This secondary structure has a large effect on the rate of migration during electrophoresis under non-denaturing conditions. Even a single base change can radically alter the secondary structure and hence the rate of migration of a nucleic acid fragment. This forms the basis of the single-stranded conformational polymorphism (SSCP) technique, which allows the presence of a base change to be detected without defining the nucleotide affected. PCR is used to amplify the gene in short sections of 100–300 bp and the products are separated under non-denaturing conditions by polyacrylamide gel electrophoresis. An example is shown in Figure 8.6 where a rare mutation in the *CFTR* gene was detected. This technique is very simple and cheap and detects 70–95% of base changes in fragments.

8.3.2 Denaturing gradient gel electrophoresis

Partially denatured DNA molecules migrate at a much slower rate during gel electrophoresis. If a molecule is subjected to electrophoresis in a gradient of denaturing agents, such as heat or chemical denaturants, it will migrate to the point where it starts to denature and then effectively stop migrating. This is called denaturing gradient gel electrophoresis (DGGE). The point at which denaturation starts is extremely sensitive to sequence and will be altered by even a single base change. Mutations may therefore be detected by a bandshift in PCR products from the test DNA compared with wild type (Figure 8.7). This method detects about 99% of mutations

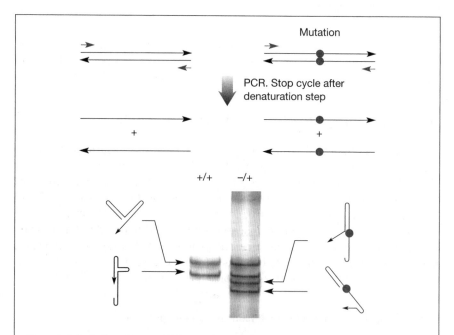

Figure 8.6 Detection of a rare *CFTR* mutation using SSCP. Reference wild-type DNA and test DNA samples are amplified by PCR and the PCR reaction stopped at the denaturation stage of the cycle. The products are loaded on to a native polyacrylamide gel that allows the single-stranded molecules to form secondary structure, which affects their rate of migration. Because each strand takes up a different structure, so a mutant DNA molecule produces two bands in a different position and a heterozygote will produce four bands (two wild type, two mutant). (Data courtesy of the North Trent Molecular Genetic Laboratory, UK.)

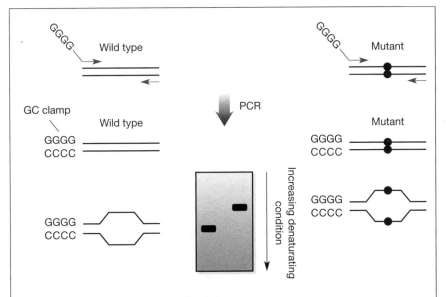

Figure 8.7 An example of DGGE. Reference wild-type and test DNA are amplified by PCR. The 5' end of the upstream primer is extended with a series of C residues, resulting in a 'GC clamp' in the amplified product. The products are separated in a polyacrylamide gel with a gradient of increasingly denaturing conditions. When the duplex DNA molecules start the migration, their motion is dramatically slowed. This point is sensitive to a base-pair change in sequence, so mutant and wild-type DNA molecules migrate to different positions. The GC clamp is slow to denature because of the three hydrogen bonds in a GC base pair. This stabilises one end of the molecule and has been found empirically to increase the discrimination between mutant and wild-type DNA molecules.

and is simple to operate once the conditions are fully established for each DNA fragment tested. However, designing the right conditions is not easy, and the conditions for each fragment have to be developed separately.

8.3.3 Mismatch cleavage detection

If a wild-type DNA molecule is hybridised to a mutant sequence, at the site of the mutation there will be a mismatch in the double-stranded duplex. This mismatch may be cleaved chemically using agents such as piperidine or osmium tetroxide (chemical mismatch cleavage, CMC) or by enzymes involved in mismatch repair (enzyme mismatch cleavage, EMC). DNA is amplified from wild type and DNA to be tested. The products are separately denatured and then allowed to form heteroduplexes, before being subjected to CMC or EMC. The products are fractionated by DGGE; a mutant DNA strand is detected by the appearance of a smaller fragment (Figure 8.8). The use of chemical cleavage agents is problematic because they are highly toxic and carcinogenic. Enzymic cleavage is now well developed and is claimed to detect 99% of mutations. However, as yet it is not widely used.

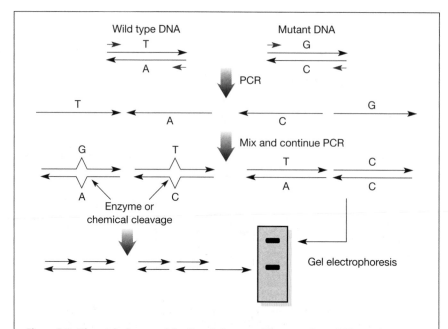

Figure 8.8 Mismatch cleavage detection. Reference wild-type and test DNA samples are amplified separately by PCR for a number of cycles, then mixed and subjected to further rounds of PCR amplification, which allow heteroduplexes to form if a mutation is present in the test DNA. Chemicals (such as osmium tetroxide or piperidine) or mismatch repair enzymes are used to cleave the DNA at the site of any mismatches. Cleaved molecules are smaller and therefore migrate faster on the gel.

8.3.4 Protein truncation test

In many genes the majority of disease-causing mutations result in trunca-tion of the protein product. The presence of such a mutation may be detected by using the product of a PCR reaction as a template for an *in vitro* transcription/translation reaction. The truncated protein from a mutant gene is detected by gel electrophoresis. The advantage of this test compared with those described above is that it will only detect changes that are biologically significant. The problem with the other methods is that many biologically neutral polymorphisms will be detected and there is no simple way of distinguishing these from disease-causing mutations. This test is likely to be the method of choice for screening for mutations in tumour-suppressors such as *BRCA1*, *BRCA2*, and *APC* genes, where over 90–95% of mutations are chain terminating. An example of its use in detecting a mutation in *BRCA2* is shown in Figure 8.9.

8.3.5 DNA chips

DNA chips are a novel and still developing technology that allows the identity of every nucleotide in a test DNA sequence to be examined in a

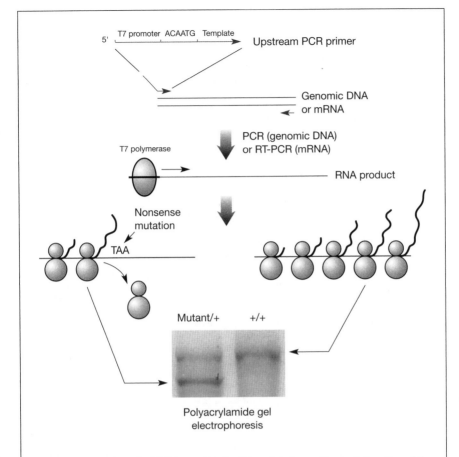

Figure 8.9 Detection of a *BRCA2* mutation by the protein truncation test. A section of the *BRCA2* gene was amplified by PCR. The design of the upstream primer is important. It has a 5' extension that contains the promoter sequence for the phage T7 RNA polymerase followed by a consensus eukaryote protein translation initiation sequence. The primer is designed so that the reading frame of the amplified *BRCA2* sequence will be correctly aligned with the ATG of the translation initiation sequence. The amplified product can be transcribed *in vitro* using phage T7 polymerase and the resulting RNA translated *in vitro* using ^{35}S-labelled methionine to visualise the product by autoradiography. If the *BRCA2* sequence contains a nonsense mutation, translation is halted prematurely and the product is truncated. This results in a faster rate of migration in the polyacrylamide gel used to fractionate the products of the *in vitro* translation. Generally, in large genes such as *BRCA1* an exon will be amplified from genomic DNA rather than mRNA. (Data courtesy of the North Trent Molecular Genetic Laboratory, UK.)

single operation. The general methods by which DNA chips or oligonu-cleotide microarrays are constructed and used were discussed in Section 4.9.1. Figure 8.10 provides an example of a chip that was constructed to detect mutations in exon 11 of the *BRCA1* gene. DNA chips can be designed to recognise specific mutations rather than examine every base pair. This may make interpretation of the results simpler because it would

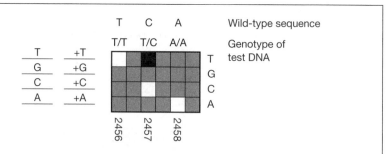

Figure 8.10 Detection of a heterozygous base substitution in the *BRCA1* gene using a DNA chip. The left-hand part of the diagram shows the organisation of an oligonucleotide array that interrogates a single base position in the test DNA. Each oligonucleotide is 20 nucleotides long; the sequence matches the wild-type sequence except that there is a substitution or insertion at residue 11 as shown. Wild-type and test RNA sequences were fluorescently labelled with different coloured dyes and then hybridised separately to the chip. The composite image for three consecutive positions is shown on the right. The oligonucleotides are attached to the chip in square cells. Cells shown in white hybridised to wild-type and test sequences with equal intensities, blue-coloured cells hybridised to neither target and the black cell hybridised to the test RNA but not to the wild-type RNA. The test RNA sequence hybridised to the cells containing T, C and A in the oligonucleotides corresponding to the wild-type sequence, but it also hybridised to the cell that contained T at position 2457, showing that one allele contained a T at this position. The individual from which the test RNA was derived is therefore heterozygous for a base substitution at position 2457. Quantitative analysis of the hybridisation signal also detected the mutation, as the wild-type DNA produced a stronger signal than the test DNA at position 2457C. RNA was used because it was found to give superior results compared with DNA. The RNA was generated from PCR-amplified DNA by *in vitro* transcription. As well as the oligonucleotides shown, oligonucleotides with various deletions were also included in each array.

only identify mutations that are definitely known to be disease-causing, rather than identifying all base changes, many of which may be neutral polymorphisms. Chips have already been constructed that screen for known mutations in the *CFTR* gene, the β-globin gene and the mitochondrial genome. Despite the potential of DNA chips, they are not yet used routinely in gene testing.

8.3.6 DNA sequencing

The advances in DNA sequencing technology, developed for the human genome project, have made it simple and routine to re-sequence genes from subjects suspected of having a mutation. Figure 8.11 provides an example of a mutation in the *BRCA1* gene being detected by automated DNA sequencing.

8.3.6.1 Chromosome tracking

When the mutation causing a genetic disease segregating in a family cannot be identified, it is necessary to track the chromosome carrying the mutation in order to diagnose the status of an embryo. This involves genotyping linked polymorphic markers and establishing the phase of markers and

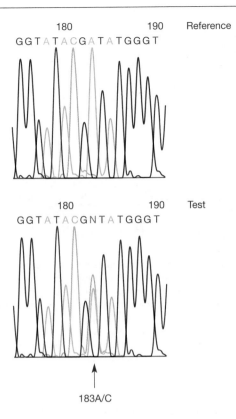

Figure 8.11 Detecting a mutation in the *BRCA1* gene by DNA sequencing. The output from an ABI 3770 prism sequencing apparatus. Peaks coresponding to G, A, T and C are coloured black, grey, dark blue and light blue, respectively (in reality these colours are black, green, red and blue, respectively). Printing limitations prevent the depiction of these colours in this figure). The reference sample shows a clear green peak at position 183, showing an A is present at this position. The test sample shows both a light blue (C) and grey (A) peak, indicating that the subject is an A/C heterozygote at this position. Note how the size of the peaks fluctuates along the length of the gel, reflecting the efficiency of chain termination at each position. However, the peak fluctuations are reproducible in the test and reference samples. For example, in both cases there is a high dark blue (T) peak at position 179 but a low red peak at position 177. At position 183 in the test sample the expected high grey peak has reduced in size by approximately 50%. Thus, a comparison of peak height provides information as well as peak order. In this case the reduction of the grey peak at position 183 is additional evidence that a mutation has occurred. (Data courtesy of Ann Dalton, North Trent Molecular Genetics Laboratory, UK.)

mutation. Highly polymorphic microsatellites are normally used. An example is shown in Figure 8.12, where a polymorphic microsatellite was used to track the inheritance of a chromosome carrying a Huntington's disease (HD) mutation. Although a precise test for an HD-causing mutation is available, it was not used in this case because it would have led to the diagnosis of the mother of the subject, who did not want to know her status.

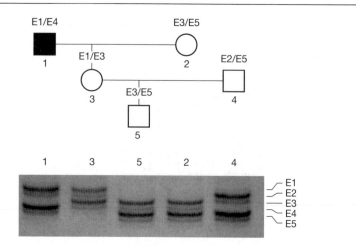

Figure 8.12 Chromosome tracking to exclude Huntington's disease. The subject (labelled 5 in the pedigree) wishes to know his HD status because his maternal grandfather (1) suffered from the disease. However, his mother (3) does not wish to know her status (so far she is disease-free). The subject must have inherited one copy of chromosome 4, which carries the HD locus, from one of his maternal grandparents. If it could be shown that the chromosome was inherited from his maternal grandmother the risk of HD can be excluded without testing the status of the mother. Of course if the subject inherited a copy of chromosome 4 from his grandfather then the risk that he will suffer from the disease is increased from 1 in 4 to 1 in 2. The test used a polymorphic microsatellite at D4S127 that is closely linked to the HD locus, so recombination is unlikely to confound the results. The alleles revealed by this test allow the copies of chromosome 4 to be tracked through the pedigree. The position of bands produced from five alleles (E1–E5) are marked at the side of the figure. The genotypes, deduced from the data, are written above each member of the pedigree. The subject is E3/E5. The E5 chromosome was inherited from his father who has no family history of the disease. The E3 chromosome was inherited from his mother, who in turn inherited it from her mother. Therefore, the copy of chromosome 4 he inherited from his maternal grandparents comes from the grandmother, not the grandfather. We can conclude that he is not at risk of the disease. (Data courtesy of Steve Evans, North Trent Molecular Genetics Laboratory, UK.)

8.4 Summary

- Gene testing has been revolutionised by the isolation of disease genes and the amplification of small amounts of material by PCR.

- Gene testing is carried out in a number of different situations:

  ○ To diagnose embryos known to be at risk of genetic disease.
  ○ Neonatal screening.
  ○ As an aid to medical diagnosis.
  ○ To diagnose the status of an adult at risk of a late-onset disease.
  ○ To determine susceptibility to a complex disease.
  ○ To confirm the identity of a candidate gene by showing a correlation between disease and mutation.

- Some mutations, such as protein-truncating mutations, will clearly ablate protein function. Others, such as base substitutions, may be neutral polymorphisms.

- DNA or RNA may be used for the tests. DNA is easy to obtain, but the extremely large size and complex organisation of some genes may make searching for unknown mutations difficult. RNA from the right tissue may be more difficult to obtain but it will represent the gene sequence in a more manageable form.

- PCR tests have been developed to recognise the different types of mutations, e.g. large deletions, small deletions and point mutations. One commonly used test is the ARMS test, which uses ASOs as primers to test for the presence of several specific mutations in two parallel PCR assays.

- The *CFTR* ΔF508 mutation may be detected by a simple PCR test. A multiplex PCR test can be used to detect other common mutations.

- About 60% of DMD cases are caused by deletions in the dystrophin gene. These may be detected by a multiplex PCR test.

- TREs can be detected by amplifying the expansion site using PCR primers that anneal to flanking regions.

- Unknown mutations may be detected by a number of tests that attempt to find differences between the test sample and a reference wild-type sequence.

- SSCP relies on the fact that the rate of migration of single-stranded DNA through a native gel is highly sensitive to secondary structure, which in turn depends on sequence.

- DGGE is based on the difference in migration rate of duplex and partially denatured DNA. The point at which a DNA molecule starts to denature is dependent on its exact sequence.

- Mismatch cleavage uses chemicals or enzymes to cleave a heteroduplex between test and reference DNA at the site of a sequence mismatch.

- The protein truncation test is based on *in vitro* transcription/translation. If the reference sequence contains a chain-terminating mutation the polypeptide produced will be smaller than the reference.

- DNA chips are, in principle, capable of detecting all possible changes in a gene sequence in a single operation. They are still at an experimental stage.

- Direct DNA sequencing for genetic testing is now routine. There remains the problem of deciding whether a change detected is a disease-causing mutation or a neutral polymorphism.

- When a family requires genetic testing and counselling for a genetic disease in which the aetiological mutation has not been identified, it is necessary to track the chromosome carrying the mutation. This is now normally done with linked microsatellite markers amplified by PCR.

Further reading

General

ANTONARAKIS, S.E. and THE NOMENCLATURE WORKING GROUP (1998) Recommendations for a nomenclature system for human gene mutations. *Human Mutation*, **11**, 1–3. Available online at http://www.interscience. wiley.com/jpages 1059-7794/nomenclature.html

ENG, C. AND VIJG, J. (1997) Genetic testing: the problems and the promise. *Nature Biotechnology*, **15**, 422–426.

Cystic fibrosis

FERRIE, R.M., MARTIN, M.J., ROBERTSON, N.H. *et al.* (1992) Development, multiplexing, and application of ARMS tests for common mutations in the CFTR gene. *American Journal of Human Genetics*, **51**, 251–262.

Methods for unknown mutations

GROMPE, M. (1993) The rapid detection of unknown mutations in nucleic acids. *Nature Genetics*, **5**, 111–116.

JONSSON, J.J. and WEISSMAN, M.S. (1995) From mutation mapping to phenotype cloning. *Proceedings of the National Academy of Sciences USA*, **92**, 83–85.

MASHAL, R.D., KOONTZ, J. and SKLAR, J. (1995) Detection of mutations by cleavage of DNA heteroduplexes with bacteriophage resolvases. *Nature Genetics*, **9**, 177–183.

YOUIL, R., KEMPER, B.W. and COTTON, R.G.H. (1995) Screening for mutations by enzyme mismatch cleavage with T7 endonuclease VII. *Proceedings of the National Academy of Sciences USA*, **92**, 87–91.

Detection of *BRCA1* mutations using a DNA chip

HARCIA, J.G. BRODY, L.C., CHEE, M.S., FODOR, S.P.A. and COLLINS, F.S. (1996) Detection of heterozygous mutations in *BRCA1* using high density oligonucleotide arrays and two-colour fluorescence analysis. *Nature Genetics*, **14**, 441–447.

CHAPTER NINE

Human population genetics and evolution

Key topics

- The Hardy–Weinberg distribution
- Selection
- Drift
- Migration
- Effective population size
- Population bottlenecks and founder effects
- Phylogenetic trees
- Polymorphisms in protein-encoding genes
- Mitochondrial polymorphisms
- Amplification of DNA from Neanderthal fossils
- Y chromosome polymorphisms
- Out-of-Africa and multiregional hypotheses of human origins
- Population variation in disease frequencies

9.1 Introduction

Apart from monozygotic twins, the genetic constitution of each individual is different. What sets us apart from each other is not that we have different genes – we all have copies of the same genes – but rather these genes may take different **allelic** forms. A gene or locus is said to be **polymorphic** if there is more than one allelic form in a population and the rarer allele still occurs at a significant frequency – usually taken to be more than 1%. In some cases, different alleles may produce different phenotypes, as in mutations responsible for monogenic disorders; in others the variation may not have any discernible effect on the phenotype and are said to be **neutral**. The study of these polymorphisms within and between populations is important for two reasons. Firstly, they can be used to reconstruct recent human evolution and the histories of population movement and growth since the emergence of anatomically modern humans. Secondly, as we saw in Chapter 6, the genetic structure of human populations has a direct bearing on the design and interpretation of experiments that use population-based methods to search for risk alleles of common complex diseases.

9.1.1 Types of variation studied

Polymorphisms of nuclear genes encoding proteins were the first type of variation to be studied in human populations. Subsequently other types of variation have become increasingly important. As in the case of genetic maps, neutral molecular polymorphisms such as RFLPs, *Alu* elements, microsatellites and minisatellites have been extensively used. Polymorphisms affecting the mitochondrial DNA (mtDNA) and the Y chromosome have proved particularly important because they are transmitted without recombination. Thus when two molecules are compared each retains a record of all the changes that have occurred since they diverged from a common ancestor. The human genome project has provided a novel source of variation in the form of single nucleotide polymorphisms (SNPs). These provide a detailed picture of the variation in individuals and populations. We shall see that rather than each SNP being informative on its own, the real benefit lies in studying linked haplotypes. We shall consider the studies using each of these forms of variation and consider how they have increased our understanding of human evolution. In the case of SNPs we shall also consider issues that affect their usefulness as tools to search for risk alleles involved in complex disease.

9.1.2 The Out-of-Africa and multiregional theories of modern human evolution

The evolution of anatomically modern humans has generated a fierce controversy. The fossil record shows that the lineage that gave rise to modern *Homo sapiens* started with *Homo habilis* in Africa approximately 2.5 million years ago (mya). This was followed by *Homo erectus* who lived 1.3 mya to 300 thousand years ago (kya), and who spread from Africa to Europe and Asia. The first members of the species characterised as *Homo sapiens* are found in Africa dated approximately 400 kya and had spread to Europe by 300 kya. These first human populations are recognisably different from modern humans and are thus called **archaic**. In Europe archaic humans formed a subspecies called *H. sapiens neanderthalensis* or Neanderthal man. The last Neanderthals disappeared from Europe 30 kya. Anatomically modern humans are first found in Africa just over 100 kya. Their remains have been found in Israel dated 100 kya, but there is no evidence of modern humans in the rest of the world for another 30,000 years. Remains of modern humans are found in China dated 67 kya, Australia 55 kya, Europe 40 kya and North America 35–15 kya.

The controversy centres on the evolution of anatomically modern humans from *Homo erectus* and archaic human populations. Many palaeontologists recognise similarities in the bone structure of modern populations of humans in a particular region and the archaic and *H. erectus* populations that lived in the same region previously. This gave rise to the **multiregional** hypothesis that posited that *H. erectus* populations evolved into modern humans in each separate region. In order to avoid the problem of multiple independent origins of favourable characteristics, which would be inherently unlikely, it is proposed that there was gene flow between the populations so that a favourable trait that arose in one part of the world could spread to all human populations. However, this gene flow would have to be very limited, otherwise the special characteristics of each regional population would be lost.

Not all palaeontologists agree with the multiregional hypothesis. There is also evidence from the fossil record for a contrasting view that modern humans originated in Africa, from where they spread to displace archaic human populations. The initial population of modern humans was only about 10 000 individuals and probably displaced the resident archaic human populations without interbreeding. This Out-of-Africa hypothesis is supported by many separate lines of genetic evidence. We shall review in this chapter the genetic evidence that leads to this conclusion.

9.2 Basic principles in human population genetics

9.2.1 The Hardy–Weinberg distribution

Consider an autosomal gene that has two alleles, say A and a, with the A allele being dominant over the a allele. These alleles can combine in different ways to produce three different genotypes: AA, Aa and aa respectively. In a population at equilibrium, that is where the gene frequencies are not changing, the distribution of these genotypes is described by the Hardy–Weinberg equilibrium which states that:

$$p^2 + 2pq + q^2 = 1$$

where p = the frequency of the A allele
q = the frequency of the a allele
$2pq$ = the frequency of Aa heterozygotes.

This distribution is only valid if a number of conditions are met:

- Mating is random.

- There is no selection for or against any of the genotypes.

- There is no migration.

In human populations these conditions are rarely completely satisfied; in particular mating is rarely random. Humans tend to mate with partners from similar religious, cultural and socioeconomic backgrounds. We shall consider the effects of migration and selection below. Nevertheless, in the absence of obvious factors, such as a clear selection for one of the genotypes or large-scale migration, allele frequencies do generally conform to that described by the Hardy–Weinberg distribution.

The usefulness of the Hardy–Weinberg distribution is that if we measure the frequency of one of the genotypes, it allows the other two to be calculated. For example the average frequency of cystic fibrosis (CF) in European populations is about 1 in 2000. Because CF is an autosomal recessive condition, the value of q^2 in the Hardy–Weinberg distribution is 1/2000. This allows the following simple calculation:

$q^2 = 1/2000$
$q = 1/44$
$2pq = 1/22$ (approximating p to equal 1)

$2pq$ is the frequency of heterozygotes, so starting with the observed frequency of the disease, we can calculate the frequency of carriers. We can perform a similar calculation with autosomal dominant traits, but this time the observed incidence of the disease is $2pq + q^2$.

9.2.2 Changes in allele frequency

Allele frequencies do not always remain constant. If they did, then evolution could not have occurred, because evolution requires such changes. New alleles can only be created by mutation. The classic neo-Darwinain theory of evolution states that alleles change in frequency in response to **selection**, either for advantageous alleles or against deleterious alleles. However, selection is not the only mechanism that operates – migration and random drift of selectively neutral alleles also act to change allele frequencies.

9.2.2.1 Selection

Selection is a consequence of alleles that alter the **Darwinian fitness** of the organism, defined as the number of its offspring reaching reproductive maturity. The rate of change in the frequency of an allele in response to selection can be calculated from its selective coefficient, which is the fractional difference between the Darwinian fitness of an organism carrying the allele and the Darwinian fitness of a reference genotype. Selection is a powerful force when it operates against strongly deleterious alleles from the population, such as those responsible for monogenic disorders. If such a disorder has a selection coefficient of 1, i.e. no affected individuals have children, then selection will ensure that the frequency of homozygous individuals affected by an autosomal recessive disorder will be equal to the frequency with which new alleles are generated by mutation. There is an important corollary to this conclusion: if the frequency of a genetic disease in a population is higher than would be consistent with a reasonable rate of mutation, then some factor must be operating to maintain the allele frequency. We shall consider several such situations in this chapter.

The essence of the theory of Darwinian evolution is that favourable mutations will increase in frequency due to selection. The rate at which a beneficial allele would spread through a population can be calculated from the selection coefficient. One of the best-known examples of positive selection in human populations is the heterozygous advantage, in areas where malaria is endemic, conferred by heterozygosity for sickle cell anaemia and other mutations that affect haemoglobin (discussed in more detail below). The experimentally measured selection coefficient for carriers of such alleles is about 0.1.

Allele frequencies of genes involved in resistance to infectious diseases such as immunoglobulins are highly variable in different parts of the world, which is assumed to reflect different histories of exposure to infectious diseases. As we shall see below, modern humans probably emerged from Africa about 100 kya. If selection had been operating throughout this time, the value of selection coefficients operating on genes of the immune system would be in the order of 0.01.

9.2.2.2 Random drift

Mathematical analysis, by geneticists such as Fisher and Haldane, showed that the effects of selection on Mendelian genes provide a satisfactory mechanism for evolution. This became known as the neo-Darwinian synthesis, and for a while it was thought to be the only mechanism responsible for changes in allele frequency. An important implicit assumption of the neo-Darwinian synthesis is that all, or nearly all, genetic variation is subject to selection – an allele is either advantageous or disadvantageous, it is rarely neutral. During the 1960s, the molecular analysis of gene structure and evolution started to reveal a number of facts that were not readily explicable in terms of selection. Two of the most important observations were:

1. At the level of the amino acid or nucleotide sequence most genes were much more variable than predicted by the neo-Darwinian theory. This fact emerged during the study of protein polymorphisms revealed by electrophoresis. At first it was hotly contested, but the advent of DNA sequencing has shown that this statement is undeniable.
2. The rate of nucleotide or amino acid substitution in a gene was apparently constant in a wide variety of genes examined. This gave rise to the concept of a molecular clock (Figure 9.1).

The first observation was inconsistent with the neo-Darwinian view, because if few alleles were selectively neutral, selection should ensure that the most advantageous allele became fixed in the population. Most genes should therefore be identical. Polymorphisms should be rare and only observed when the frequency of a new and advantegeous allele was being increased by selection. The second observation is surprising because it would be expected that proteins would evolve to a configuration that optimised their function and that subsequent changes would be selected

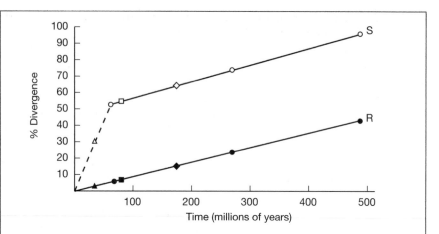

Figure 9.1 The molecular clock of globin evolution. The sequences of α- and β-globin genes were used to measure the divergence between pairs of globin genes. Both silent substitutions (filled symbols) and replacement substitutions (open symbols) accumulate at a constant rate. At first the rate of silent substitutions was 10 times that of replacement substitutions, in agreement with the neutral theory of evolution. Eventually all possible selectively neutral substitutions that could occur, had occurred. Further substitutions were subject to the same selective pressures as replacement substitutions and the rate at which substitutions accumulated slowed down. The data were obtained from following comparisons which can be dated from the archaeological record (circles): intraspecies comparisons of α and β globin genes that were presumed to have diverged 500 mya, comparisons of α- and β-globin genes of mammals with chicken (270 mya), and comparison of α- and β-globin genes of different mammals (85 mya). Further data for which archaeological dates are not available were obtained from comparison of δ versus β (triangles), γ versus ε (squares) and δ or β versus γ or ε (diamonds). For these data the replacement substitutions were plotted on the line R and the silent site divergences plotted directly above. The amount of divergence (corrected for multiple substitutions) was plotted against the time since divergence. Line R passes through the origin. Because the line S does not pass through the origin, and there is no independent verification of the time of the β/δ divergence it is plotted as a dashed line.

against. It is difficult to imagine how over long periods of evolution, new alleles could arise and be selected for with the same selection coefficient.

These observations led Motoo Kimura to propose what is now known as the **neutral theory** of evolution. The foundation of this theory is that many, or even most, alleles are selectively neutral. Allele frequencies change because of random sampling effects between generations. This can be explained by a simple analogy. Consider a bag containing equal numbers of white and red balls. If ten balls were randomly withdrawn from the bag, it would not be surprising if the sample consisted of 4 red and 6 white balls. The frequency of red balls has thus changed from 0.5 to 0.4. However, if we withdrew 1000 balls we would be very surprised to find that the sample contained 400 red and 600 white balls. If the different coloured balls represent two alleles of a gene, two conclusions are evident from this analogy. Firstly, random sampling will result in a change in allele frequency from generation to generation. Secondly, the smaller the population size, the larger the changes are likely to be.

At every generation the frequency of an allele will fluctuate due to this sampling effect, until either the frequency of an allele becomes 100%, (fixation), or it disappears completely from the population (extinction). One of these two outcomes is logically inevitable, but the time it takes depends on population size, occurring more rapidly in smaller populations. Clearly, fixation is a more likely outcome for common alleles, and extinction more likely for rarer alleles. Therefore, the most likely fate for a new allele introduced by mutation is that it will become extinct. However, because a large number of alleles are continually being created by mutation, fixation of some of them will occur by random drift.

It is important to understand that the neutral theory does not state that all alleles are neutral. Firstly, rare advantageous alleles do exist and will increase in frequency as a result of selection. However, they are likely to be rare and make little contribution to overall variability. More importantly, many substitutions will be deleterious. Kimura showed that the rate of evolution can be described by the following equation:

$$K = v_T f_o$$

where k = rate of nucleotide substitutions per year, f_o = fraction of new alleles that are selectively neutral and v_T = mutation rate.

This equation infers that the rate of evolution of a gene, or part of a gene, will be constant and will depend on two variables, the mutation rate and the fraction of selectively neutral mutations. The validity of this equation is readily tested by comparing the rate of nucleotide substitutions in situations where we may be confident about whether or not the result is likely to be selectively neutral. For example, Figure 9.1 compares the rate of two different types of nucleotide substitutions that have occurred in the evolution of the globin genes. **Silent** substitutions do not result in a change in amino acid sequence, because they result in a **synonymous** codon, that is a codon that encodes the same amino acid. **Replacement** substitutions result in a codon that encodes a different amino acid. Clearly, silent substitutions are more likely to be selectively neutral than replacement substitutions. Figure 9.1 shows that this is indeed the case. The initial rate of silent substitutions is 1% divergence per million years, while replacement substitutions occur at about one-tenth of this rate. Silent substitutions in coding regions may still be subject to some selective pressure, for example the secondary structure of the mRNA may be altered. Sequences that are unlikely to be subject to any selection, such as pseudogenes, evolve at a rate of 2% divergence per million years.

Genetic drift and selection are therefore factors that act together. While random drift ensures that substitutions occur at a constant rate, selection controls that rate. Neutral alleles evolve at a rate determined by the mutation frequency, but the protein sequences evolve at a lower rate, determined by the capacity of the protein to accommodate amino acid substitutions. Different proteins evolve at different rates as a result of this constraint. Even different parts of proteins may evolve differentially, for example the active site of an enzyme is likely to be more highly conserved than other parts of the protein. A last point to note is that the selection and drift may operate simultaneously, so that rare advantageous alleles may still become extinct by genetic drift.

9.2.2.3 Migration

The effects of drift and selection described above apply formally to isolated populations. Few populations are entirely isolated – immigration/migration between adjacent populations will result in gene flow that will affect their genetic structure, in particular mitigating the effects of drift. Gene flow between populations as a result of migration is called **admixture**.

9.2.3 Effective population size

The neutral theory of evolution predicts that if alleles in a population are completely neutral and frequencies change only through drift, the number of alleles in the population, known as the population mutation parameter, θ, is related to the effective population size, N_e, according to the equation:

$$\theta = 4N_e\mu$$

where μ is the mutation rate per generation at each one of an infinite number of sites. This equation is valid for autosomal loci. Because there are only three X chromosomes for every four autosomes in the population, the equation for X-linked loci is $\theta = 3N_e\mu$. Similarly for mtDNA the equation is $\theta = 2N_e\mu$.

The effective population size, N_e, is an abstract concept that describes the size of an ideal randomly breeding population in which allele frequencies would drift at the same rate as that observed. In real populations there will be many individuals who are not breeding, for example children and adults past their reproductive age. For this reason, and the fact that real populations do not breed randomly, N_e is about one-half the actual or census population size. N_e also describes the long-term population size. If the population size fluctuates over time, N_e is approximately the harmonic mean of the population size, which means that N_e tends to be closer to the historical minimum rather than maximum population size.

9.2.3.1 Long-term effective population size of human populations

The importance of the equation above is that both θ and μ can be experimentally measured, thus it is possible to estimate the long-term value of N_e. Such estimates have been made from measurements of nuclear protein polymorphisms, minisatellites, microsatellites, mtDNA and Y chromosomes. All the estimates are in remarkable agreement that the long-term effective human population size is about 10 000 individuals. This figure is one of the arguments against the multiregional hypothesis of human evolution, because if this number of humans was distributed world-wide, local populations would be too small to be viable.

9.2.3.2 The mode of population expansion can be discerned in the structure of present populations

The modern human population size is much larger than 10 000, so there must have been an expansion from a very small population. If a population expands, most lineages when traced backwards in time would remain separate after the population expansion and then converge rapidly to a common ancestor who lived just before the expansion. The genealogy of the population is said to have a star-like morphology. The differences between two randomly chosen individuals will be due to the mutations that have occurred since the expansion, because the genealogies were only separate for a short time before the expansion. Mathematical analysis shows that the number of differences between pairs of randomly chosen individuals will follow a smooth Gaussian distribution. If the mutation rate is known, the modal number of mutations can be used to calculate the number of generations since the expansion. Estimates from mtDNA place the expansion about 100 kya. Estimates from the the Y chromosome place the expansion more recently at 50–60 kya.

9.2.3.3 The time to coalescence is proportional to the effective population size

All individuals in a population must ultimately trace their lineage back to a common ancestor, often referred to as the Most Recent Common Ancestor (MRCA). This process is known as **coalescence**. Coalescence takes longer if there are more lineages, so the coalescence time is proportional to the effective population size. The effective population sizes of autosomes, X chromosomes, mtDNA genomes and Y chromosomes are in the ratio of 4:3:2:1. As we shall see below, this is reflected in the coalescence times of an autosomal locus such as β-globin (800 kya), mtDNA (150–300 kya) and the Y chromosome (60 kya).

9.2.4 Population bottlenecks

Human history is marked by large reductions in population size caused by famine, epidemics, climate change and wars. Once the cause of the population reduction is removed, population numbers will recover. Such an event is called a **population bottleneck**. Population bottlenecks reduce genetic variation because while the population is small the effect of drift on allele frequencies will be large, causing some alleles to become extinct, thus reducing variation. Population bottlenecks therefore leave a lasting imprint on a population and will affect the conclusions of a number different types of investigation. For example, the reduction in genetic variation due to a population bottleneck may lead to an underestimate of the effective population size. As illustrated in Figure 9.9 and discussed further below, population bottlenecks also affect the degree of linkage disequilibrium between closely linked loci.

9.2.4.1 Founder effects

A small group that splits off from a central population and moves into a geographically isolated area and then expands is called a **founder** population. For reasons that are similar to population bottlenecks this reduces genetic variability. Although normally thought of in terms of physical migration, a founder group can also arise if it becomes reproductively isolated from the main population because of culture, religion or language. For example, Ashkenazi Jews descend from a founder group that was isolated by religion and culture rather than geography. The effect of drift when the population is small means that the population descended from a founder group may have a quite different genetic structure than the ancestral population. In practical terms this means that they often have a much higher frequency of genetic diseases than the main population because particular disease alleles attain high frequency in the founding group. Examples of this are discussed below.

9.2.4.2 Phylogenetic trees

Phylogenetic trees describe the evolution of a set of genes, populations or species (elements) from a common ancestor. The order of branching of the tree reflects the order in which the different elements split off during evolution. Trees may be rooted or unrooted (Figure 9.2). An unrooted tree describes the relationships between the elements in the tree without specifying the ancestral state. A rooted tree specifies which element is ancestral, usually by reference to an outside element that is more different to the elements in the tree than they are to each other.

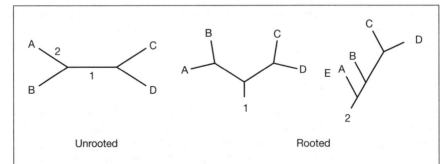

Unrooted   Rooted

Figure 9.2 Rooted and unrooted trees. The unrooted tree on the left describes the similarity relationships between four elements A, B, C, D. However, it does not specify the order in which they branched during their evolution from a common ancestor. To do this the root must be determined. Both of the rooted trees on the right are consistent with the unrooted tree on the left. The number at the base of each rooted tree specifies the position of the root indicated by the numbers in the diagram of the unrooted tree. There are three other possible rooted trees that are not shown in the diagram. The root is determined by reference to an outrider that is more dissimilar to any of the four elements than they are to each other. This allows the deduction that the outrider was the first to branch the lineage from the common ancestor. The element to which to the outrider is most similar places the root.

There are many ways to construct phylogenetic trees and there has been an often acrimonious debate as to which is the most accurate. One point of general agreement is that it is **never** possible to be sure that a deduced phylogenetic tree is actually the true tree. Basically there are two approaches to constructing trees. Numerical methods quantify differences between the elements of the tree and place the elements that are most similar on adjacent branches. **Cladistic** methods seek to identify the tree that depicts the evolution of present-day elements from a common ancestor with the least number of events. A commonly used method of this type is called maximum parsimony. This method requires a large number of possible trees to be examined and is therefore very computer intensive. The number of possible trees that could exist is usually much larger than can be feasibly examined, so only a subset of the possible trees is tested and no unique tree is identified. It is therefore necessary to carry out a statistical test, called bootstrap analysis, to measure the confidence that can be placed in the result. This involves carrying out the analysis on repeated occasions; each time a small number of data are randomly replaced with other data chosen from the dataset. The percentage number of times the computer generates the same tree is a measure of confidence in the result.

9.3 Polymorphism in protein encoding genes

Allele frequencies in polymorphisms of protein encoding genes often vary in different population groups throughout the world. The first observation of this kind was made in 1919 with ABO blood groups (Figure 9.3).

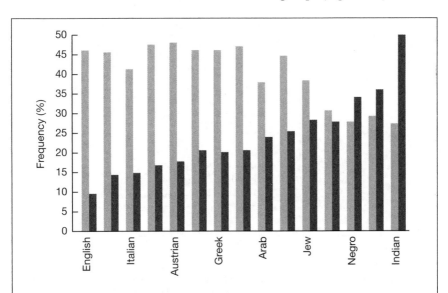

Figure 9.3 World-wide variation in A and B blood groups. The ratio of people with blood group A to those with blood group B varies from 4.5:1 in the English population to 0.5:1.0 in the Indian population.

Subsequently, systematic measurements have been carried out using a wide variety of polymorphisms, which can be detected at the protein level usually by electrophoresis, such as blood groups, serum proteins, human leukocyte antigens, enzymes and other readily detectable proteins. A large and comprehensive study using such markers has been carried out by Luca Cavalli-Sforza and colleagues, using 120 allele frequencies in 42 populations world-wide. In this context 'populations' refer to populations that are probably still similar to those that existed in 1492, before the mass migrations of the last 500 years.

The different allele frequencies in the populations were used to reconstruct population histories. Firstly they were used to construct phylogenetic trees. Figure 9.4. shows a phylogenetic tree of human populations world-

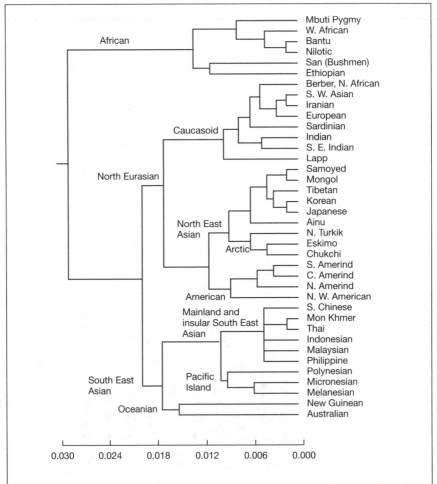

Figure 9.4 Phylogenetic map of human populations based on genetic distances. The order in which the Oceanic group and South East Asia divergence occurred is uncertain. When the genetic distances between the populations are calculated in a different way the Oceanic group splits off before the the South East Asian group. From Cavalli-Sforza, L.C., Meonozzi, P. and Piazza, A. (1994) *The History and Geography of Human Genes.*

wide. The first branch of the tree is between Africans and non-Africans. This is a major conclusion that suggests modern humans originated in Africa and thereafter spread to the rest of the world. The second branch of the tree shows that non-Africans split into Northern Eurasian and South East Asia branches, probably about 60 000 years ago. The Northern Eurasian branch gave rise to Caucasians, Northern Asian and American peoples. The South East Asia branch gave rise to the populations of New Guinea, Australians and Polynesians as well as South East Asians.

The second way of summarising data from different populations is called **principal component** (PC) analysis. It mathematically combines the data from all the experimentally determined gene frequencies and produces a **synthetic map** depicting their geographic variation. PC analysis cannot represent all data simultaneously. It is necessary to represent the data in a series of synthetic maps based on the first, second, and third principal components, each succeeding map utilising the information not used by the preceding maps. The first and second principal component maps between them represent about 50% of the total information and are a convenient way of visualising multivariate data from hundreds of genes. The conclusions from PC analysis show that human variation spreads out from a source in Africa, and thus are consistent with the phylogenetic tree.

Figure 9.5 illustrates how principal component analysis can be used to study the evolution of human populations at a more local level. It shows maps plotting the spread of agriculture, the synthetic map of Europe derived from principal component analysis and the distribution of the ΔF508 allele of the CFTR gene The synthetic map of genetic relatedness shows a close correspondence with the spread of agriculture. This suggests that the population of Neolithic farmers expanded from an origin in the Middle East into lands occupied by Mesolithic hunter–gatherers. The more numerous farmers absorbed the smaller populations of hunter–gatherers. This is known as the wave advance model, which mathematical analysis has shown would generate the gradients in allele frequencies observed. The wave advance model has shown that the previous theory – that agriculture spread by cultural diffusion – is probably incorrect.

We can also see that the distribution of the ΔF508 allele is correlated with the genetic map and the spread of agriculture, suggesting that the allele was present in the Mesolithic inhabitants of Europe before the arrival of agriculture. We shall see below that another line of evidence confirms this conclusion. Interestingly, the highest frequency of the ΔF508 allele is found in the Basque region, which has a non-Indo European language that is not related to any other extant language. It is possible that the present-day Basque population is genetically and linguistically directly descended from the Mesolithic population that existed before the arrival of Neolithic farmers.

Reconstruction of the history of human populations using polymorphisms in protein encoding genes assumes the polymorphisms are neutral. However, it is not possible to be absolutely confident that this assumption is true. This is a major weakness of this type of study. Another weakness of the study is that it is difficult to take account of gene flow through migra-

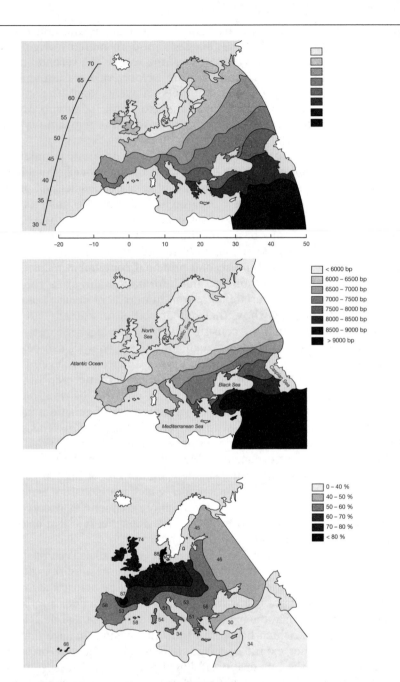

Figure 9.5. The effect of agriculture on the genetic structure of European populations.
Top: The synthetic map of Europe based on the first principal component. The range between
the maximum and minimum values of the principal component has been divided into eight
arbitrary classes denoted by the shading. Middle: Timings of the spread of agriculture based
on the archaeological record. Bottom: Distribution of the ΔF508 allele as a proportion of the
total *CF* alleles. From Cavalli-Sforza, L.C., Meonozzi, P. and Piazza, A. (1994). *The History and
Geography of Human Genes.*

tion between different populations. Nevertheless, these studies provided the first genetic evidence that human populations originated in Africa, and this conclusion has been repeatedly reinforced by other studies where these criticisms do not apply.

Another major conclusion of these studies is that about 90% of genetic variation occurs within rather than between populations. This is most consistent with present human populations originating from the expansion of a small founder population rather than a long period of parallel evolution.

Finally, these studies revealed that compared to a sister species such as chimpanzees, there is very little genetic variation in the species *Homo sapiens*. Essentially it is this lack of variation which suggests that we are a species that originated very recently from a small founding group.

9.4 Mitochondrial DNA polymorphisms

Measurements of mtDNA evolution are usually based on the sequence of the control region (Figure 1.7), which shows greater variation than the rest of the molecule. Variation in mtDNA has proved extremely useful in studying recent human evolution. There are a number of reasons for this:

1. The mitochondrial genome is maternally inherited and there is no recombination between paternal and maternal genomes. This makes the construction of phylogenetic trees straightforward because in the absence of recombination, the number of nucleotide differences between two mitochondrial genomes is a direct reflection of the of the history of these molecules since they originated from a common ancestor.
2. mtDNA evolves 5–10 times more rapidly than nuclear DNA. This makes it suitable for studying recent evolution, because a sufficient number of mutations will have arisen to permit the analysis of population relationships.
3. The phylogenetic analysis was carried out by the maximum parsimony method, which many consider to be a superior method because it is **cladistic**, i.e. the course of evolution is deduced directly from the difference in qualitative characters (see above).

However, mtDNA studies do suffer from two major weaknesses:

1. Because the mitochondrial genome is inherited without recombination, selection that applies to any part of the molecule, whether it is positive or negative, will affect the frequencies of all alleles carried by the genome. This gives rise to the concept of a **selective sweep**, in which a whole mitochondrial genome is lost from a population because of selection against an allele at one locus. This of course will lead to the loss of other genotypes which in themselves are neutral. Of course the reverse can apply, leading to selection for a whole mitochondrial genome because of selection for a single allele.

2. The mitochondrial genome is only a small part of the whole genome and is inherited independently of the nuclear genome through the female line. Stochastic processes play a very important part in evolution, especially when populations are small. Thus any reconstruction of human evolution based on the mitochondrial genome may not be representative of the evolution of the whole genome.

The study using mitochondrial DNA polymorphisms was published in 1987 by Cann, Stoneking and Wilson. It showed that the lineage of all extant mtDNA genomes can be traced back to a single common ancestor who lived in Africa about 150–300 000 years ago. The origin of the most recent common mitochondrial ancestor in Africa is based on three inter-related lines of evidence.

1. Phylogenetic analysis produces trees with two branches (Figure 9.6). One branch contains only African populations, the other contains African and non-African populations. The simplest interpretation of this observation is that the ancestor was African. The phylogenetic trees drawn up from this data were controversial for a while, because it was shown that other types of trees would also be consistent with the data. Further work showed that the original tree, shown in Figure 9.6, is correct.
2. All mitochondrial lineages found outside Africa are a subset of lineages found within Africa. By far the simplest explanation for this observation is that non-African populations are derived from a group that migrated from Africa.
3. Diversity, measured as the number of different mutations in the African populations, is much higher than in non-African populations. Normally the population that has the greater genetic diversity is ancestral to a population with less diversity.

The single common ancestor has been popularised as the 'African Eve'. This appellation has led to the popular misconception that all humans descend from a single woman. Although all human mitochondrial lineages

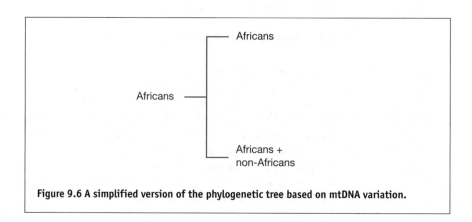

Figure 9.6 A simplified version of the phylogenetic tree based on mtDNA variation.

trace back to a single individual woman, this does not mean that she was the only woman alive at the time when she lived. Women with different mtDNAs were alive at the same time. In each subsequent generation some lineages will be lost, and eventually all lineages but one will be lost. This is illustrated in Figure 9.7. The woman in the ancestral population with this genotype is the so-called mitochondrial Eve; all other mitochondrial genotypes that existed in the other women alive at the same time have become extinct. In Figure 9.7, another lineage could have come to predominate by chance and it might have given a different picture of when and where the mitochondrial Eve lived. Furthermore, note that there is no logical reason why the mitochondrial ancestor has contributed any of her nuclear genes to present-day populations. So the rest of the genome could have had a quite different history. These arguments illustrate why it is unsafe to rely solely on mitochondrial data to reconstruct human population history.

The coalescence time of mitochondrial lineages is critical in distinguishing between the out-of-Africa and the multiregional evolution models of human history, because coalescence must precede the emergence of ancestral population out of Africa and their dispersal around the world. Coalescence time can be simply calculated if two parameters are known: the amount of variation and the rate at which new mutations are generated. The amount of variation has been measured experimentally in the original studies and is not contro-

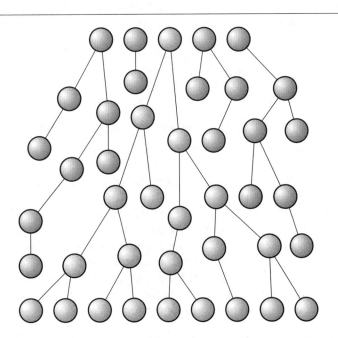

Figure 9.7 The mitochondrial Eve was not the only woman alive. The figure shows how all members of a present-day population ultimately inherit their mitochondria from a single female ancestor shown by the circles filled in blue. However she was not the only woman alive at the time she lived.

versial. The rate of mutation is more difficult to measure. It is calculated from the amount of mitochondrial variation between humans and chimpanzees using the fossil record to estimate the time of divergence. The problem is that the differences between humans and chimpanzees must be corrected for multiple insertions at the same nucleotide site and there is no single reliable method for doing this. One observation that has helped calibrate the clock is that part of the mitochondrial genome has been duplicated and inserted into the nuclear genome on chromosome 11. Because the mutation rate of the nuclear genome is much lower than the mitochondrial genome, the nuclear insertion can serve as a reference sequence to calibrate the mitochondrial clock. There have been a number of different estimates of the age of the most recent common ancestor of mitochondrial genomes, ranging from 100 to 500 kya, but most are between 150 and 300 kya. This figure is consistent with an expansion out of Africa 100 kya, but is inconsistent with the emergence of *H. erectus* from Africa, which took place over 1000 kya.

As in the case of nuclear polymorphisms, the overall level of diversity in the mitochondrial genome can be used to estimate that the size of the founding group was 6000 individuals. This estimate is consistent with the estimate of 10 000 from nuclear polymorphisms. When applied to European populations an extreme bottleneck is revealed. The estimated size of the founding group of European populations is just six individuals! As described in Section 9.2.3.3, the average number of differences in mtDNA sequence between pairs of individuals allows the date of the population expansion to be estimated. The conclusions are that an expansion took place about 100 kya in African populations, about 50 kya in Asian populations and 23 kya in European populations. These estimates have large error margins, so they can only be taken as a guide.

9.4.1 Mitochondrial sequences can be rescued from fossil samples

DNA is chemically stable and in appropriate conditions can survive for long periods. In principle this allows samples from ancient fossils to be amplified by PCR and their sequence determined. The first attempts to do this were hampered by the extreme sensitivity of PCR, which can amplify a single molecule of DNA. As a result the PCR reactions amplified contaminating DNA from the archaeologists who recovered the samples or from the person carrying out the PCR amplification. However, if extreme precautions are taken and appropriate controls used it is possible to overcome this problem. Although DNA is stable, it will eventually fragment, so the primers for PCR must be chosen to amplify small overlapping sections that can be pieced together to provide a longer sequence.

9.4.1.1 Amplified mtDNA from Neanderthal fossils shows that they were not part of the modern human lineage

Using these techniques, DNA sequences were amplified from the original Neanderthal specimen found in Feldhofer in western Germany. In order to maximise the chances of seeing differences between the Neanderthal and modern human sequences, two hypervariable regions were amplified, called HVR I and HVR II. In the combined 600 bp from the two regions, modern humans differ from each other at an average of 11 nucleotides, whereas modern humans differ from the Neanderthal sequence by an average of 35 nucleotides. Thus the difference between the human and Neanderthal sequences is three times the range of variability found within modern human populations. The MRCA of Neanderthals and humans was estimated to have lived 465 kya, three times the age of the MRCA of modern humans. If Neanderthals contributed to the human gene pool, it would be expected that the Neanderthal mtDNA would be most similar to that of modern European populations. However, there was no evidence that this is the case; the Neanderthal sequences were equally different from mtDNA sequences from European populations as they were from mtDNA sequences from Asian and African populations. Subsequently, mtDNA sequences were isolated from two other Neanderthal specimens, from the Mezmaiskaya cave in Russia and Vindija cave in Croatia, and very similar results were obtained. The fact that mtDNA sequences from geographically dispersed sites gave similar results argues against the initial result coming from an individual who happened by chance to come from outside the normal range.

The phylogenetic tree using the three Neanderthal sequences groups the Neanderthals on a branch that diverged from the human lineage before the MRCA of modern humans (Figure 9.8). A further interesting feature of this work is that the variation among the three Neanderthal sequences so far amplified is similar to the variation seen in modern human populations and much less than that seen in gorillas and chimpanzees. As we have seen, the lack of variation in human populations is interpreted as a recent expansion

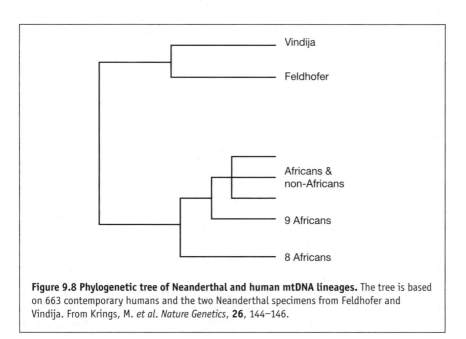

Figure 9.8 Phylogenetic tree of Neanderthal and human mtDNA lineages. The tree is based on 663 contemporary humans and the two Neanderthal specimens from Feldhofer and Vindija. From Krings, M. *et al. Nature Genetics*, **26**, 144–146.

from a small founder group. Although based on a small sample, it could be that a similar conclusion should be made for Neanderthals.

Amplification of Neanderthal DNA provides convincing evidence that modern Euorpeans did not evolve from Neanderthals, or at least that Neanderthals did not contribute to the modern European gene pool. However, it is important to remember that the mitochondrial genome is a small part of the whole. Chance and accident could mean that the Neanderthal mitochondrial genome was lost from the lineage leading to modern Europeans while genes from the nuclear genome survived.

9.4.1.2 Amplified mtDNA from ancient Australians is inconsistent with the Out-of-Africa hypothesis

In contrast to the study of Neanderthal remains, studies of DNA amplified from ancient human remains in Australia are not consistent with the Out-of-Africa hypothesis. DNA was amplified from the fossilised remains of a male skeleton (LM3) excavated in Lake Mungo. The skeleton was dated as being 60 000 years old, but anatomically fell within the range of modern Australian aborigines. The mtDNA sequence was different from any extant human and was most similar to the mitochondrial insert on chromosome 11. The phylogenetic tree showed the LM3 mitochondrial genome and the nuclear insert on a separate lineage that branched off before mitochondrial genomes of modern humans. The conclusion is that the most ancient mitochondrial genome does not belong to an African but to an Australian. Inevitably, such a controversial result has attracted criticism, the main one being that although precautions were taken against contamination, the standard protocols were not followed. Amplification of the nuclear mtDNA insert is known to be a problem in amplification of mtDNA from ancient specimens. Amplification of mtDNA from other ancient specimens is eagerly awaited to see if this result can be substantiated.

9.5 Y chromosome variation

Just as the mitochondrial genome is transmitted through the female line, the Y chromosome is only transmitted from fathers to sons. The human Y chromosome has a complex structure. About 5% of the chromosome recombines with homologous sequences on the X chromosome. This is called the pseudoautosomal region (PAR). The remainder of the chromosome does not recombine with any other chromosome and is called the non-recombining region of the Y chromosome (NRY). The NRY is 60 mb in size, of which only 35 mb is euchromatic. For the most part the euchromatic region of the NRY consists of highly repetitive DNA rich in transposons. However, there are a small number of expressed genes that fall into three classes. There are eight class I genes that have homologs on the X chromosome and are widely expressed throughout the body. There are also eight class II genes; they are expressed only in the testis and do not have homologs on the X chromosome. The five class III genes have homologs on the X chromosome but show tissue-limited expression, which in three cases is confined to the testes. The SRY gene that determines male sexual development is one of the class III genes. It is expressed only in the embryonic bipotential gonad and in the adult testes. Its homolog on the X chromosome is called *SOX3*, is active and so presumably fulfils some other essential function in both males and females.

The Y chromosome shows very little sequence variation. In part this is due to its low effective population size, which is only one quarter that of an autosomal chromosome. Nevertheless, this only partly explains the lack of variation. One possible explanation is that selection operating on an allele of any of the Y genes will also act on the whole Y chromosome. Thus selective sweeps may have acted to reduce diversity. This lack of sequence variation has hindered the use of Y chromosome in population and evolutionary studies. Nevertheless, polymorphisms have now been identified affecting single nucleotides, *alu* elements and microsatellites. These have been used to reconstruct recent human evolution as with other types of variation.

The largest study used a technique to identify variants which is similar to SSLP, described in Chapter 8. PCR-amplified DNA was denatured by heat and sequence variation was identified by the rate of migration during HPLC. Along with some other variants (previously identified by other methods) this provided a collection of 167 markers, which genotyped 1062 individuals representative of world populations. The 167 markers defined 116 separate haplotypes that were used to construct a phylogenetic tree using the maximum parsimony method. The tree allowed the 116 haplotypes to grouped into 10 haplogroups. Figure 9.9 shows a simplified version of the tree, showing these haplogroups.

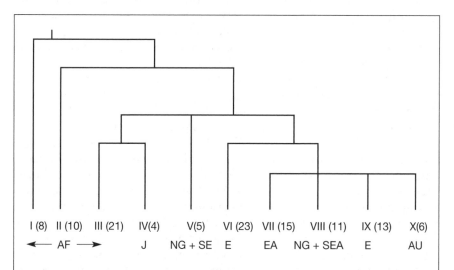

Figure 9.9 Phylogenetic tree based on Y chromosome haplotypes. Each haplogroup is represented by roman numerals, the number in brackets shows the number of individuals who fell within the haplogroup. Haplogroup VIII is the source of Haplogroups VII IX and X. The letters below each haplogroup show the main geographical areas where it is found. AF = Africa, J = Japan, NG = New Guinea, EA = East Asia, SEA = South East Asia, E = Europe, AU = Australia. Redrawn from Underhill, D.A. *et al. Nature Genetics*, **26**, 358-361.

Haplogroup I is defined by three polymorphisms called M42, M94 and M139. In each case members of haplogroup I carry the ancestral alleles while all other males only carry derived alleles. This places the deepest root between haplogroup I and all the other haplogroups. Today haplogroup I is found only in 10–15% of Sudanese, Ethiopian and Khoisan males. This places the root of the Y chromosome phylogeny in Africa and so provides further support for the Out-of-Africa hypothesis. For example, haplogroup I males have an A at the M42 site while all other human males have a T-allele. All primates have an A at the M42 site, so this is presumed to be the ancestral sequence. Thus the M42 A-allele identifies the Y chromosome Adam. At some point one of his descendants underwent a mutation to T at this position. Men who had the A-allele remained in Africa, but others with the T-allele left Africa and spread around the world.

The distribution of Y chromosome mutations is consistent with a rapid expansion from a small founder population. Several studies have placed the time to the MRCA of all Y chromosomes at 50–60 kya, which is considerably later than that based on mtDNA. Using this figure it is possible to work out a mutation frequency for the Y chromosome polymorphisms and use this to calculate the age of the first non-African haplotype. The figure arrived at is 47 kya, which agrees very well with the archaeological evidence for the emergence of modern humans in Europe.

The geographical spread of haplogroups was much more limited than mtDNA types (Figure 9.9). This provides greater resolution in the reconstruction of early population movements. This observation also implies that women move more than men. At first this may seem strange because typically in hunter–gatherer societies men travel more than women. However, men tend to have their families where they are born, while women move to their husband's home to have families. Thus over millennia females spread their genes more widely than men.

Another interesting use of Y chromosome variation is to show that there is a tight correlation between surnames and inheritance of the Y chromosome. According to Jewish tradition, descendents of Aaron, the brother of Moses, were selected to serve as priests known as the Cohanim. Study of Y chromosome microsatellites shows that males of both Ashkenazi and Sephardic Jews with the surname Cohen have strikingly similar Y chromosomes. Moses was a member of the tribe of Levi; but other Levite males have diverse Y chromosomes. The Cohen Y chromosomes, although similar, are not identical – this is due to mutation at the microsatellite loci genotyped. The rate of mutation can be used to date the most recent common male ancestor of Cohens, who by tradition would be Aaron. The result is that there have been about 100 generations since Aaron; assuming 25–30 years per generation this corresponds to 2500–3000 years before present, i.e. during or shortly before the first Temple period, which ended when the temple was destroyed in 586BC. Another study showed that all males in the UK with the surname 'Sykes' have very similar Y chromosomes, suggesting that there was once a man called Sykes from whom all males with the surname Sykes have descended.

The main advantage of mtDNA and Y chromosome polymorphisms is that in the absence of recombination they retain a record of all the genetic changes since divergence from a common ancestor. However, this property is also a major disadvantage, because it effectively makes each of them behave as a single multi-allelic gene and is thus unrepresentative of the genome as a whole. As we have already seen, there is no reason why any of the nuclear alleles of the mitochondrial Eve should have survived in present-day populations. We may expect each segment of the genome to have its own history, with different MRCAs and signatures of demographic history. Only by considering the genome as a whole can we be confident in reconstructing the history of human populations. Single nucleotide polymorphisms (SNPs), described in Chapter 4, provide the means of doing this. On its own, each SNP has limited value, however the haplotypes of closely linked SNPs provide a series of segments dispersed throughout the genome that will have a high level of heterozygosity and will preserve a record of genomic evolution. Taken together they will give us a more comprehensive and more soundly based view of human history.

We saw in Chapter 6 how SNP maps are a critical resource for mapping risk alleles contributing to complex disease. In some cases the SNP itself will be the risk allele, however, more commonly the SNP will be in linkage disequilibrium (LD) with the risk allele. Knowledge of population genetics and evolution is critical in the design and interpretation of searches for risk alleles. One of the main issues is the physical distance of LD from each mapping marker. Population genetics has shown us that the present human population expanded from a small group. Although the date of the expansion has not been exactly determined we can take a rough figure of 100 kya. With 20 years per generation that equates to 5000 generations. The risk alleles that we are concerned with are relatively common but have a small effect. In evolutionary terms most are probably neutral, either because the environmental conditions that contribute to the disease in modern society were not relevant for most of our history as hunter–gatherers; or because the alleles are late-acting, leading to disease after reproductive age, at least at an age that is greater than the life expectancy of most of our ancestors.

The theory of neutral evolution shows that any polymorphism that is common in present-day populations must have been present in the ancestral population, i.e. must be at least 5000 generations old. Therefore we will be searching for LD between the SNP and the disease-causing allele that remains after 5000 generations. Since we know the average recombination rate is about 1 cM/mb, it is relatively straightforward to calculate that we can expect the LD to extend only 3–10 kb from the disease allele. However, this calculation ignores that fact that LD will be affected by demographic effects such as population bottlenecks, that will reduce the number of haplotypes and in effect re-establish LD that has been eroded by recombination (Figure 9.10). A similar argument applies to founder effects. This reasoning suggests that isolated populations that have expanded from small founder groups should be used for LD mapping. However, computer simulations show that in fact this will only be useful when the founding group is very small.

The completion of the draft human genome sequence, the production of SNP maps and the ease of DNA sequencing mean that comprehensive studies are now starting to be published that examine SNP haplotypes and LD throughout the genome. We have already considered one study (Section 6.4.4.5) that showed that LD in two groups of Europeans extended for a much greater distance than at the same loci in a Nigerian group. This led to the conclusion that there was a severe bottleneck in the European lineage 27–53 kya, in which the population number was reduced to as few as 50 individuals. Interestingly, this agrees closely with the conclusions of mtDNA studies that also suggested a similar severe bottleneck in European history. A second conclusion from this study was that the extent of LD was highly locus specific. In some cases no LD was detectable between SNPs that were only 5 kb apart. In other cases perfect association was seen at 160 kb, which was the longest distance examined. This result shows that each segment of the genome has a different history, with different degrees of LD and different coalescence times.

A second recent study carried out a large-scale survey of 313 genes in 82 individuals from US populations that were classified as Hispanic Latino, Caucasian, African–American and Asian. This survey came to the

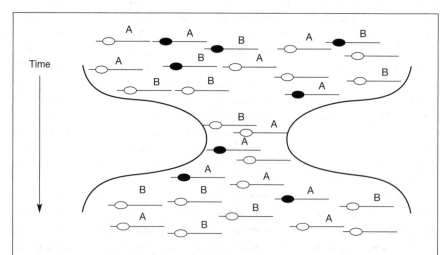

Figure 9.10 A population bottleneck re-establishes linkage disequilibrium between an SNP and a disease allele. The figure shows a representation of population numbers in a bottleneck. In the original population a disease allele (filled symbols) shows no particular association with a nearby SNP (A or B) However, the reduction in diversity in the bottleneck results in a loss of one of the haplotypes, so that now allele A is associated with the disease allowing the disease allele to be mapped.

same conclusion – that LD across the genome varies considerably from locus to locus. One interesting observation that this survey made is that for many genes there is no single dominant haplotype. In other words, the classical genetic concept of a wild-type gene is not really valid for the human genome.

9.7 Disease frequencies in different populations

We saw above that selection would normally act to restrict the frequency of alleles that cause severe monogenic diseases to be close to the mutation rate. There are a number of diseases where in certain populations this is manifestly not the case. Table 9.1 lists some examples. There are two main reasons why disease frequencies in one population may be higher than expected – heterozygous advantage and founder effects/inbreeding.

Table 9.1 Examples of elevated frequencies for monogenic diseases in particular populations. The low frequency of type 1 (insulin dependent) diabetes in Japan is given by way of comparison to the high level in Scandinavia.

Disease/allele	Population	Frequency/10^3 population	Possible cause
Sickle cell disease	Africans	10–20	Heterozygous advantage
Thalassaemia	Africans/Mediterraneans/S.E. Asia	10–20	Heterozygous advantage
Cystic fibrosis	N. Europeans	0.5	Heterozygous advantage?
Tay–Sachs Disease	Ashkenazi Jews	0.17–0.4	Founder effect or heterozygous advantage?
Gaucher's disease	Ashkenazi Jews		Founder effect
BRCA1 185delAG	Ashkenazi Jews	10	Founder effect
BRCA2 617delT	Ashkenazi Jews	10	Founder effect
Porphyria	South African (White)	3	Founder effect
MLH1 (HNPCC susceptibility)	Finnish	14 families	Founder effect
Polydactyly	Amish Community (US)		Founder effect
Type 1 diabetes	Scandinavia	2	Environment and genetic effects
	Japan	0.03	

9.7.1 Heterozygous advantage

9.7.1.1 The haemoglobinopathies

Heterozygous advantage occurs when an allele that is deleterious as a homozygote is advantageous as a heterozygote. This results in a **balanced polymorphism** where selection against the allele in the homozygous state is balanced by selection for the allele in the heterozygous state. The high frequency of haemoglobinopathies in countries where malaria is endemic is a balanced polymorphism. The geographical distribution of malaria correlates with the frequency of sickle cell anaemia and the thalassaemias. Haemoglobin S (HbS) is responsible for sickle cell anaemia (see Chapter 4). HbS differs from the wild-type protein by a substitution of glutamate by valine at residue 6 in the β-globin polypeptide. HbS crystallises at low oxygen tensions and as a result the red blood cells collapse into a sickle shape. HbS heterozygotes are clinically normal but are resistant to malaria for reasons which are not currently known. The allele frequency of HbS has equalised to 0.1 in a number of different geographic locations. This exactly corresponds to the expected value calculated from the observed Darwinian fitness of the three different genotypes.

There are four African regions where the allele frequency is at a maximum: Bantu, Benin, Senegal and Saudi. All HbS alleles are the result of the same nucleotide substitution. However, the linked haplotype is different in each region and corresponds to a haplotype that is locally common in each case. This suggests that the mutation occurred on four separate occasions, although it is possible that it occurred once and has been introduced to the four different genetic backgrounds by migration.

Heterozygous advantage is also likely to be responsible for the elevated frequencies of α and β thalassaemias in areas where malaria is endemic. β thalassaemia is more common in Europe and Africa and less common in Asia. Malaria used to be endemic in Mediterranean countries, although it has now been eradicated. This explains the high frequency of β thalassaemia found around the Mediterranean. Indeed thalassaemia is derived from the Greek word *thalassa* meaning sea.

9.7.1.2 Cystic fibrosis

Among Northern Europeans the incidence of CF is about 1 in 2000. On average 70% of CF mutations are the ΔF508 allele. This allele shows linkage disequilibrium with three nearby microsatellites, which allows the haplotype of the founder chromosome to be deduced. Based on the number of mutations that have occurred in the linked microsatellites, there is a controversial estimate that the mutation occurred 52 000 years ago. The ΔF508 founder haplotype is rare in modern European populations; the next two most frequent CF mutations also occurred on a chromosome with the same haplotype, which is rare in Europe. This suggests that the common CF mutations occurred in a population that had a different genetic background from modern Europeans. The most likely scenario is that they occurred in the first Paleolithic human population in Europe, after the founder group had split off from the main stock of modern humans expanding out of Africa. Eventually, as we saw above, this population was absorbed by the expansion of Neolithic farmers from the Middle East.

Why has the ΔF508 mutation maintained such a high frequency? Founder effects could have been responsible for a high frequency initially, but the minimum age of the mutations means that selection against the homozygote would long ago have reduced the allele frequency below the present figure of 0.02. The likely explanation is heterozygous advantage. It is possible that reduced CFTR function in the heterozygote protects against infantile diarrhoea, because it could reduce water loss across the intestinal mucosa. Recently, there has been support for this hypothesis from experiments in mice. If this is the explanation, it is interesting that more than one CF mutation occurred and was selected for in European Paleolithic populations, but similar mutations were not established elsewhere in the world. Perhaps this reflects differences in the environment or lifestyle of the early Europeans.

9.7.2 Founder effects

Founder effects occur when a small group, separated from the main stock, expands in number. As a result disease alleles present by chance in the founder population may have increased in frequency under the influence of drift and the increased inbreeding that would occur in small isolated groups. This can result in local populations that have a particularly high frequency of a monogenic disorder that is rare elsewhere. Sometimes these populations originated by migration of founder groups in historical times and the identity of the individuals who carried the disease allele is known.

9.7.2.1 The Ashkenazi Jews

Table 9.1 shows that a number of different diseases are more common in Ashkenazi Jews than elsewhere and are maintained at a higher rate than can be explained by mutation. It has been suggested that the high rate of Tay–Sachs disease is a result of carriers being more resistant to tuberculosis, which was endemic in the tenements of Eastern Europe where Ashkenazi Jews lived. An alternative explanation is that the high frequency of all of these diseases is explained by a **founder effect**. The *BRCA1* 185delAG and *BRCA2* 617delT mutations are extremely rare outside this population group and so are almost certainly founder mutations.

The Ashkenazi Jews originated in the vicinity of Strasbourg in eastern France in the early Middle Ages. They expanded eastwards at the time of the Crusades and subsequently to America in the late nineteenth and early twentieth centuries. During this time their culture and religion operated to maintain an effective reproductive isolation from neighbouring populations.

From the time of their migration away from Strasbourg to the present day, their population number would have increased several thousand-fold. They thus conform to a classic founder population. Selection against homozygotes would not have had sufficient time to operate to reduce the frequency of deleterious alleles. Furthermore, the *BRCA1* and *BRCA2* mutations would have limited effect on fitness since they have their effect for the most part after child-bearing age.

9.7.2.2 Complex disease

Table 9.1 also shows that a complex disease such as type 1 diabetes mellitus is unevenly distributed in different ethnic populations. Part of this variation could be environmental in origin. However, we saw in Chapter 5 that the HLA alleles DR3 and DR4 contribute about 40% of the genetic risk. These alleles themselves show variation in allele frequency between populations, so susceptibility to common diseases may vary between populations because of their different genetic structures. It may be that different alleles at the HLA locus are selectively neutral and that ethnic variation reflects the effects of genetic drift. Alternatively, they have been selected for because they conferred resistance to a disease that was historically more common in certain areas.

In the case of diabetes, the situation is even more complex. The DR3/DR4 alleles do not themselves cause the susceptibility to diabetes. They are in linkage disequilibrium with the real culprits. So in people of Northern European extraction the DR3/DR4 alleles act as markers of the chromosomes that carry the susceptibility alleles. In other populations the susceptibility alleles show different linkage configurations and the association between DR3/DR4 and diabetes is not so marked.

9.8 Summary

- Present variation within and between human populations allows the reconstruction of the past history and evolution of human populations.

- A number of genetic observations suggest that present-day human populations are derived from a small population of about 10 000 individuals that lived in Africa about 100 kya.

- Genetic variation in *Homo sapiens* is very low by comparison to other species such as chimpanzees. This lack of variation is seen whether nuclear polymorphisms, mtDNA or Y chromosome sequence is studied. This low level of variation suggests that the long-term effective population size of humans is about 10 000 individuals. A rapid expansion in population numbers occurred 50–100 kya. There was probably a severe bottleneck in the European lineage.

- About 90% of the genetic variation occurs within rather than between populations, consistent with a recent expansion from a single ancestral population. The greatest amount of variation occurs within African populations, suggesting an African origin.

- Phylogenetic trees constructed from nuclear polymorphisms, mtDNA and Y chromosomes all have their deepest root in Africa, suggesting an African origin.

- All non-African mtDNA haplotypes are a subset of a wider range of haplotypes found in Africa.

- PCR amplification of mtDNA from fossilised Neanderthal specimens indicates that they were not ancestral to present-day human populations. However, PCR-amplified mtDNA from a 60 000-year-old Australian specimen shows that it has a sequence outside the range of current mtDNA sequences, suggesting that the most ancient mitochondrial genome does not belong to an African but to an Australian.

- The non-recombining regions of the Y chromosome and mtDNA retain a record of genetic change since they diverged from a common ancestor. This aids the construction of phylogenetic trees. However, selection acting on one allele may lead to selective sweeps that could distort the conclusions. Furthermore, each is only a small part of the whole genome and the stochastic nature of human evolution means that their history may not be representative of overall human evolution.

- The study of SNPs will give us a more comprehensive view of human evolution. Large-scale population surveys of SNPs show that the extent of linkage disequilibrium (LD) varies throughout the genome, emphasising the different evolutionary histories of different genomic segments. A comparison of LD between European and African populations shows that at some loci, LD extends a much greater distance in European populations, suggesting a population bottleneck.

- Many genetic diseases are present in certain populations at rates which greatly exceed the level at which new alleles are introduced by mutation. This can occur for two reasons – heterozygous advantage and founder effects.

- Haemoglobinopathies are present at a high level where malaria is endemic, because carriers are more resistant to the disease. As a result, selection against alleles in the homozygote is balanced by selection for the allele in hetrozygotes.

- Cystic fibrosis is much more frequent in Europe than other areas. The most frequent allele, ΔF508, was present in mesolithic populations absorbed by the advance of Neolithic farmers. It is possible that the high frequency of CF alleles is also due to heterozygous advantage.

- Many populations founded by small groups of migrants have elevated frequencies of otherwise rare diseases. The Ashkenazi Jews have elevated frequencies of several different diseases that probably arose through founder effects. Some argue that Tay–Sachs disease is an example of heterozygous advantage.

- Susceptibility to complex diseases such as diabetes is also affected by population variation in allele frequencies. These may reflect variation of previously neutral alleles due to random genetic drift, or the susceptibility alleles may influence resistance to infectious diseases.

Further reading

General aspects of population genetics

CANN, R.L. (2001) Genetic clues to dispersal in human populations: Retracing the past from the present. *Science*, **291**, 1742–1748.

Reviews how population genetics can be used to reconstruct the detail of past human population movements.

CAVALLI-SFORZA, L.L., MENOZZI, P. and PIAZZA, A. (1994) *The History and Geography of Human Genes*. Princeton University Press, Princeton, NJ.

Detailed description of the theory of population genetics and review of nuclear polymorphisms.

HARPENDING, H.C., BATZER, M.A., GURVEN, M. *et al*. (1998) Genetic traces of ancient demography. *Proceedings of the National Academy of Sciences USA*, **95**, 1961–1967.

Provides a detailed mathematical description of how past demographic changes can be deduced from the genetic structure of present-day populations.

JORDE, L.B., BAMSHAD, M. and ROGERS, A.R. (1998) Using mitochondrial and nuclear DNA markers to reconstruct human evolution. *Bioessays*, **20**, 126–136.

General reviews of what is known about past population histories

WOLPOFF, M.H., HAWKS, J., and CASPARI, R. (2000) Multiregional, not multiple origins. *American Journal of Physical Anthropology*, **112**, 129–136.

Reviews the arguments for the multiregional hypothesis.

The evolution of mitochondrial DNA

ADCOCK, G.J., DENNIS, E.S., EASTEAL, S. *et al*. (2001) Mitochondrial DNA sequences in ancient Australians: Implications for modern human origins. *Proceedings of the National Academy of Sciences USA*, **98**, 537–542.

Describes the amplification of mtDNA from an early Australian that does not easily conform to the Out-of-Africa hypothesis.

CANN, R.L., STONEKING, M. and WILSON, A.C. (1987) Mitochondrial DNA and human evolution. *Nature*, **325**, 31–36.

The original paper that described the Out-of-Africa evolution of early humans based on mtDNA genealogies.

KRINGS, M., STONE, A., SCHMITZ, R.W. *et al.* (1997) Neanderthal DNA sequences and the origins of modern humans. *Cell*, **90**, 19–30.

KRINGS, M., CAPELLI, C., TSCHENTSCHER, F. *et al.* (2000) A view of Neanderthal genetic diversity. *Nature Genetics*, **26**, 144–146.

KRINGS, M., GEISERT, H., SCHMITZ, R.W. *et al.* (1999) DNA sequence of the mitochondrial hypervariable region II from the Neanderthal type specimen. *Proceedings of the National Academy of Sciences USA*, **96**, 5581–5585.

Describes the PCR amplification of mtDNA from the original Neanderthal specimen.

COOPER, A., RAMBAUT, A., MACAULAY, V. *et al.* (2001) Human origins and ancient human DNA. *Science*, **292**, 1655–1656.

HOFREITER, M., SERRE, D., POINAR, H.N. *et al.* (2001) Ancient DNA. *Nature Reviews Genetics*, **2**, 353–359.

Criticises the Adcock *et al.* paper for not following standard precautions when amplifying ancient DNA.

STONEKING, M. (1996) Mitochondrial DNA and human evolution. In *Human Genome Evolution*, Jackson, M., Strachan, T. and Dover, G. (eds), pp. 263–267. Bios Scientific Publishers, Oxford.

Review of the early studies of mitochondrial evolution.

Y chromosome evolution

GIBBONS, A. (1997) Y chromosome shows that Adam was an African. *Science*, **278**, 804–805.

Commentary describing the first reports of Y chromosome genealogies.

LAHN, B.T., PEARSON, N.M. and JEGALIAN, K. (2001) The human Y chromosome, in the light of evolution. *Nature Reviews Genetics*, **2**, 207–216.

Reviews the structure and long-term evolution of the human Y chromosome.

SYKES, B, and IRVEN, C. (2000) Surnames and the Y chromosome. *American Journal of Human Genetics*, **66**, 1417–1419.

Men with the surname "Sykes" share the same Y chromosome.

THOMAS, M.G., SKORECKI, K., BEN AMI, H. *et al.* (1998) Origins of Old Testament priests. *Nature*, **394**, 138–140.

Shows that Jewish males with the surname Cohen descended from a single male ancestor who lived 2500–3000 years ago.

UNDERHILL, P.A., SHEN, P.D., LIN, A.A. *et al.* (2000) Y chromosome sequence variation and the history of human populations. *Nature Genetics*, **26**, 358–361.

Large-scale study of Y chromosome evolution.

Single nucleotide polymorphisms
REICH, D.E., CARGILL, M., BOLK, S. *et al.* (2001) Linkage disequilibrium in the human genome. *Nature*, **411**, 199–204.

STEPHENS, J.C., SCHNEIDER, J.A., TANGUAY, D.A. *et al.* (2001) Haplotype variation and linkage disequilibrium in 313 human genes. *Science*, **293**, 489–493.

Two recent large surveys of SNP diversity.

CHAPTER TEN

DNA fingerprinting

Key topics

- DNA as a unique human identifier
- Minisatellite-based method
  - Multilocus probes
  - Single-locus probes
- Simple sequence length polymorphisms

10.1 Introduction

The human genome contains 3000 Mb of DNA; of this, selection will constrain those sequences (<5% of the total) that encode proteins. While to some extent other factors will act to conserve the sequence of the remainder, we may expect that most human DNA sequences will be highly variable. As a result the exact sequence of each individual human genome is undoubtedly unique (apart from identical twins). Chapter 3 showed how this variation has been exploited in the production of genetic maps.

The individuality of DNA sequences has practical importance in the forensic analysis of biological material left at the scene of a crime or in the resolution of paternity disputes. The DNA sequence within biological specimens, such as blood, semen or hair roots, should unambiguously identify the person from whom they were derived. By analogy to classical fingerprinting, based on the pattern of ridges on the pads of fingers and thumbs, the sequence of the DNA within a biological sample forms a DNA fingerprint that uniquely identifies the person from whom it originates. Thus it could be proved that a semen sample originated from a rape suspect or a blood-stain or hair root came from a murder suspect. In cases of disputed paternity, the children's alleles should be inherited from the parents in a simple Mendelian fashion, so a comparison of the DNA profiles of the children and claimed parents will readily determine the nature of the biological relationship.

To use the potential of DNA sequence variability, a laboratory test will need to fulfil three criteria.

1. It must be simple to carry out by technicians who may not have the level of experience and skill found in a research laboratory.
2. It must provide data that uniquely identify each individual with a clarity that will withstand critical examination in a law court.
3. It must be capable of operating with degraded DNA, since the biological material may be old and not well preserved.

In 1985 Alec Jeffreys, a British geneticist working with a minisatellite sequence located within an intron of the myoglobin gene, realised that minisatellites could be used as the basis of a **DNA fingerprinting** technique that met these criteria. This chapter describes the use of minisatellites in the original DNA fingerprinting procedure and then reviews more recent techniques designed to use the variability inherent in DNA to **profile** individual identity.

10.2 Use of minisatellites for DNA fingerprinting

We considered the general structure of minisatellites in Chapter 3. The myoglobin minisatellite consists of four repeats of a 33-bp sequence. The length of this minisatellite was not polymorphic, but in low **stringency** Southern hybridisation experiments it cross-hybridised to other minisatellites that were polymorphic. A random selection of these were cloned and sequenced. The sequence of the repeat unit in each case contained an 11–16-bp core that was identical or very similar to the sequence GGAG-GTGGGCAGGARG (R indicates a purine). In each case, the remainder of the sequences was different. The myoglobin minisatellite cross-hybridises only weakly with other minisatellites because the 17 bp of non-core sequence interferes with hybridisation. Jeffreys and his colleagues isolated two minisatellites, 33.6 and 33.15, which consisted only of tandem repeats of different versions of the core sequence. These hybridised efficiently to other minisatellites in low-stringency conditions. The pattern of hybridisation was found to be very sensitive to the actual sequence of the repeated element in the probe. Because 33.6 and 33.15 contain different sequences, they each produce a different pattern of hybridisation and hence a different fingerprint from the same individual.

Each probe produced a complex profile of about 36 bands, representing alleles drawn from a pool of about 60 minisatellite loci (Figure 10.1). The variation caused by differences in the number of repeats is maximised by removing as much flanking DNA as possible. This is done by digesting the DNA with the restriction enzyme *Hin*fI. This enzyme cleaves at a 4-bp recognition site that occurs on average every 256 bp. Because the minsatellites are repeats of sequences much shorter than this, the enzyme does not usually cut within the repeat unit (Figure 10.2). All the loci were autosomal,

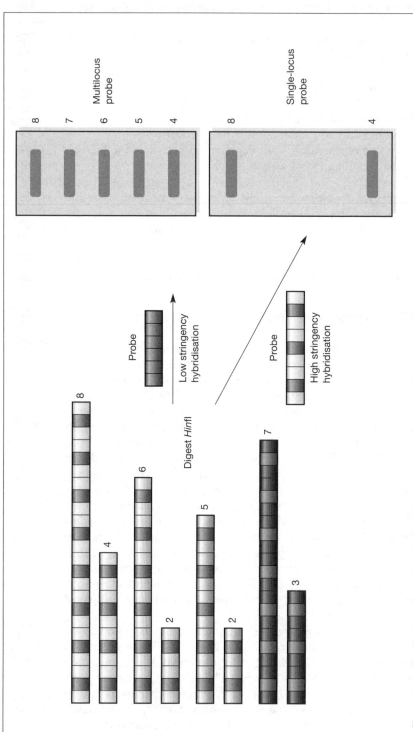

Figure 10.1 DNA fingerprints produced by multilocus and single-locus probes. Four separate loci are shown on the left, each with two alleles that in each case are heterozygous for the repeat length. The number of repeats is indicated by the numbers at the side of each locus. The repeats at each locus contain a core sequence (shown in solid blue) that is found in all the other loci and a sequence that is unique to each locus (indicated by the different shadings). A multilocus probe such as 33.15 consists only of repeats of a version of the core sequence. When it is used as a probe under low-stringency hybridisation conditions, alleles at a number of different loci hybridise and the result is a complex profile of bands that is unique to each individual. Note that alleles smaller than 3 kb have migrated past the end of the gel and are too small to be recognised (shown as alleles with three repeat units or less in this diagram for the purposes of illustration). When a sequence unique to one locus is used as a probe under high-stringency hybridisation conditions (single-locus probe), only the two alleles at that locus hybridise. The rationale for digesting with *Hin*fI is explained in Figure 10.2.

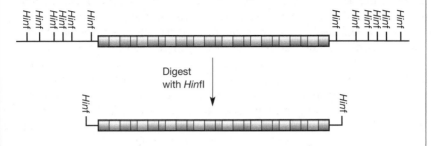

Figure 10.2 Digestion with *Hin*fI trims away flanking DNA sequences. *Hin*fI recognises a 4-bp target sequence that occurs frequently in the sequences flanking a minisatellite. However, it does not occur in the 33-bp repeat so it will not cut anywhere in the whole tandem repeat array. Digestion with *Hin*fI therefore trims away DNA on either side of the minisatellite and so maximises the relative difference in size due to the number of repeats.

so in theory each of the two bands from a heterozygous locus could be visualised. In practice, the difference in the size of the allelic minisatellites is sufficiently great that normally only one of each allelic pair is visible within the size range resolved by electrophoresis.

The average odds that any one band is shared by two unrelated individuals is experimentally observed to be about 0.25. The chance that two unrelated individuals will share all 36 bands is thus $0.25^{36} = 10^{-22}$. Since this is considerably smaller than the reciprocal of the size of the world's population, the profile of each individual may be expected to be unique. This calculation makes the assumption that all individuals are unrelated and that the chance that each band will be shared will always be the same for all individuals. In reality, the world's population is not homogeneous and individuals belonging to the same ethnic population may be more likely to share bands than the average figure for world populations. The extent to which this biases the odds of a chance match has been highly contentious. Because of this, it is necessary to monitor the actual frequency of band sharing in different ethnic groups to validate the high degree of uniqueness claimed for the DNA profiles. This is difficult in multilocus probes because the actual loci hybridising to produce each band in the fingerprint are unknown, so it is not possible to define allele frequencies.

Even taking into account the complications of increased band sharing between members of the same ethnic group, the DNA profile or DNA fingerprint of each individual is unique. The methodology therefore meets the criteria for a simple and robust means of obtaining genetic information.

10.3 Single-locus probes

The limitation of using probes that detect multiple loci is that about 250 ng of DNA is needed for the Southern hybridisation procedure, which is often more than can be obtained from the crime scene. Moreover, since so many different loci are involved, it is not possible to use PCR to amplify small samples of

DNA. Multilocus probes are based on the conserved core sequence, while the remainder of the sequence of each repeat unit would provide a locus-specific probe. It would therefore be possible to visualise the two allelic bands of each locus, providing a much simpler autoradiogram to analyse (Figure 10.1).

A panel of probes is used that are selected to recognise highly polymorphic loci in which common alleles all fall within the range that can be analysed by electrophoresis. The probes are also characterised with respect to the frequency of band sharing between members of different ethnic groups. Since only a small number of bands are analysed a high degree of reliance is placed on each, in contrast to multilocus probes where many bands are produced. The difference between alleles that differ by one repeat unit is small in comparison to the total size range resolved by AGE. Thus it is possible that where the difference in the repeat number is small, two genuinely different alleles may not be resolved. It is therefore necessary to base the analysis on the chance that two bands would migrate to the same region of the gel. Nevertheless, it is estimated that the pattern detected by four locus-specific probes would be matched by chance less than once in a million unrelated individuals. Because single-locus probes do not produce a pattern that is guaranteed to be absolutely unique to each individual, the more neutral term **DNA profile** is now commonly used instead of DNA fingerprint.

The major advantage of locus-specific probes compared with multilocus-specific probes is that they are much more sensitive. Multilocus probes require about 250 ng of DNA, while single-locus probes require only 10 ng of DNA, which allows the analysis of DNA obtained from a blood spot or single hair root. Moreover, a defined locus can be amplified by PCR, increasing its sensitivity still further. A second advantage is that since locus-specific probes recognise both alleles of a defined locus, it is possible to measure the allele frequencies in different ethnic populations, thus providing a precise estimate of the odds that a suspect's DNA profile matches the victim's by chance.

10.4 DNA profiling based on STRs

Since 1994, STRs of the type used to construct genetic maps (see Chapter 3) have become the standard method of producing DNA profiles. The markers used are highly polymorphic and are mostly based on loci with tetranucleotide repeats, each locus commonly having between 5 and 15 alleles. If the loci are carefully chosen, a single multiplex PCR reaction amplifying three to four loci is sufficient to derive a profile. The amplified products are labelled with fluorescent dyes and analysed using an automated DNA sequencer. A number of carefully controlled tests by organisations such as the European DNA Profiling Group (EDNAP) have shown that the use of STR loci is a reliable method to obtain DNA profiles. An example is shown in Figure 10.3. Allele frequencies in different ethnic groups have been measured and a number of loci have been validated for use in forensic casework by the relevant authorities.

Compared with single-locus probes there are a number of advantages to using STR loci for profiling.

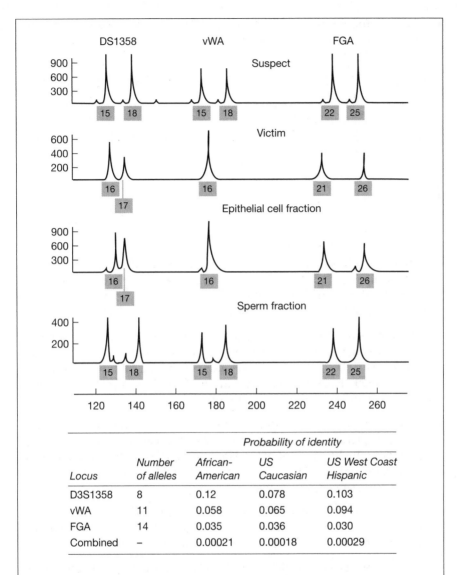

Figure 10.3 STR DNA profiling using STR loci. *Top*: readout from the ABI Prism system of four samples from a rape case. The suspect's profile matches the sperm sample. *Bottom*: ethnic distributions and probability of identity for the STR loci used.

- STR-based tests are quick; in the example shown in Figure 10.3, 32 samples were analysed in 2 hours.

- STR tests are more sensitive, requiring only 0.1 ng of DNA instead of 10 ng for single-locus probes.

- There is no need for the radioisotopes required for the Southern blots used in profiles based on single-locus probes.

- The electrophoresis conditions in the ABI Prism system are much better controlled, leading to little or no variation between lanes.

- Measurements of band migration are precise, allowing unambiguous assignment of genotype. The results can be readily stored in a computerised database.

- It is tolerant of DNA degradation in the sample. One test showed that it could be reliably used to identify victims of a mass disaster whose bodies were badly decomposed.

10.5 Summary

- The sequence of each individual's genome is unique. This provides the potential for a DNA fingerprint.

- About 60 minisatellite loci are made of repeats that have a common core sequence. If this core sequence is used as a probe in a Southern hybridisation experiment, multiple bands are seen that correspond to the alleles of the different minisatellites.

- The repeat lengths of the alleles at a minisatellite locus are polymorphic. The chance of two unrelated individuals sharing a particular band is 0.25. The chances are extremely low of sharing all the 36 bands that can normally be seen. Thus the pattern of bands is effectively unique to each individual and is called a DNA fingerprint.

- Single-locus probes are more sensitive than multilocus probes and give a simpler picture. The pattern resulting from a panel of four such probes provides a high degree of discrimination between individuals but is not necessarily unique. For this reason, it is normal to refer to DNA profiles when using single-locus probes

- There has been considerable debate about whether members of the same ethnic group are more likely to share bands and the extent to which this may erode the discrimination of DNA profiles.

- Since 1994 DNA profiles have been based on STRs. These can be analysed using automatic fluorescent-dye technology of the type used in DNA sequencing or high-density genetic maps.

Further reading

BAR, W., BRINKMANN, B., BUDLOWE, B. *et al.* (1997) DNA recommendations: further report on the DNA commission of ISFH regarding the use of short tandem repeat systems. *International Journal of Legal Medicine,* **110,** 175–176.

CLAYTON, T.M., WHITAKER, J.P., FISHER, D.L. *et al.* (1995) Further validation of a quadruplex STR typing system: a collaborative effort to identify the victims of a mass disaster. *Forensic Science International,* **76,** 17–25.

DEBEBHAM, P.G. (1992) Probing identity: the changing face of DNA fingerprinting. *Trends in Biotechnology,* **10,** 96–102.

FREGEAU, C.J. and FOURNEY, R.M. (1993) DNA typing with fluorescently tagged short tandem repeats: a sensitive and accurate approach to human identification. *Biotechniques,* **15,** 100.

JEFFREYS, A.J. (1987) Highly variable minisatellites and DNA fingerprints. *Transactions of the Biochemical Society,* **15,** 309–316.

LINCOLN, P.J. (1997) Criticisms and concerns regarding DNA profiling. *Forensic Science International,* **88,** 23–31.

CHAPTER ELEVEN

Human genetics and society

Key topics

- Potential dangers of genetics to human society
- Gene testing
  - Prenatal testing
  - Neonatal screening
  - Diagnosis of genetic disease in children after birth
  - Presymptomatic testing for late-onset diseases
  - Presymptomatic testing for predisposition to complex diseases
- Human rights
  - Insurance and employment
  - Privacy and ownership of genetic information
- Patents
- Gene therapy
- Genetic determinism and individual responsibility

11.1 Introduction

This book began with a consideration of the influence that our genetic constitution has over our health. We discovered that single-gene disorders were responsible for much serious childhood ill health, resulting in reduced life expectancy and quality of life for those affected and immense suffering, distress and sorrow both to sufferers and to their families and carers. Furthermore, there are few diseases of later life not influenced in some part by genetic factors. The main part of this book has been concerned with an epic period of scientific advance where, in a few years, we have sequenced the human genome and equipped ourselves with the technology to understand the molecular defects that cause these diseases, to devise efficient means for their diagnosis and to realistically contemplate novel ways to prevent or treat them. Over the next few decades this may well revolutionise medicine and fundamentally alter the way we think of ourselves.

At the same time, these advances in human genetics will have a profound impact on society; therefore, as a society, we must consider carefully how we will respond, ensuring that we use our new-found knowledge to relieve suffering and promote good health whilst avoiding the very real possibility that it could have negative consequences. This chapter explores the nature of these ethical, legal and social issues. It is important to realise that many of these issues are not in any way abstract or far off. They affect us either now or in the very near future.

This chapter is about the ethical issues that arise from human genetics. There has been considerable controversy about the ethical and social implications of human cloning. Although an issue that relates to human genetics, it has developed from separate fields, namely reproductive physiology and developmental biology. It does not use any of the techniques or knowledge described in this book. Although it could be combined with somatic gene therapy at some point in the future, this is not an immediate possibility. Therefore, the possibility of human cloning will not be considered here.

11.1.1 Research and reports into ethical, legal and social issues

There have been a number of formal investigations into these problems by groups comprising scientists, medical practitioners and moral philosophers. One of the goals of the Human Genome Project is to carry out research on the ethical, legal and social issues (ELSI). This has resulted in a number of reports on aspects such as gene testing, employment rights and privacy. These reports are available online from the HUGO website (see Further reading). In the United Kingdom, the House of Commons Select Committee has produced an excellent general report on human genetics and a second specific report on genetic testing and life insurance. Both of these are also available online (see Further reading).

11.1.2 Regulatory and advisory frameworks

In the United Kingdom an extensive framework has been established to provide independent advice to the government on developments in human genetics, to regulate the use to which these developments are put and to provide a mechanism to channel input from different sections of society and interest groups. It is hoped that this will avoid the distrust and problems that were associated with the introduction of genetically modified foods. Details are given in Box 11.1. In the United States, regulation of activities connected with human genetics is less formal, falling largely on local ethics committees. However, the Food and Drug Administration will act to oversee the safety of new treatments as they are introduced.

> ## BOX 11.1: THE REGULATORY AND ADVISORY FRAMEWORK IN THE UK
>
> **The Human Genetics Commission (HGC).** Provides advice to government on over-all issues relating to human genetics. It has a broad remit, incorporating the impact of human genetics on health care and the social, ethical, legal and economic implications of developments in human genetics. It also acts to coordinate the activities of other relevant advisory and statutory bodies, described below. Each of the bodies below includes a member from the HGC. The HGC superseded an earlier body called the Human Genetics Advisory Commission (HGAC) and incorporated a previously independent body that advised separately on gene testing, the Advisory Committee on Gene Testing (ACGT).
>
> **The Gene Therapy Advisory Committee (GTAC).** Provides advice to government and regulates gene therapy. All proposals for gene therapy must be approved by GTAC.
>
> **The Genetics and Insurance Committee (GAIC).** Develops criteria for the use of genetic tests in life insurance and examines proposals from the insurance industry to use specific tests. So far (August 2001), it has only approved one test, which was for the diagnosis of Huntington's disease. Further tests for hereditary breast cancer and early onset Alzheimer's disease are expected in the near future.
>
> Although not directly concerned with human genetics the following bodies are also relevant and coordinate with the HGC.
>
> **The Human Embryology and Fertilisation Authority (HFEA).** Regulates *in vitro* fertilisation and related issues such as human cloning.
>
> **The Committee on Safety of Medicines.** Oversees the safety of medicines, which will be relevant as new treatments based on human genetics are developed.
>
> **The United Kingdom Xenotransplantation Interim Regulatory Authority (UKXIRA).** Regulates the transplantation of animal organs into humans.
>
> Further details of the terms of reference of these bodies may be obtained from the HGC website: http://www.hgc.gov.uk/about_regulatory.htm

11.1.3 Eugenics

Deliberations in this area are coloured by the historically bad record that the discipline of genetics has had in the way it has advised and influenced society. In the early part of the twentieth century, there was widespread concern that the genetic fitness of Western societies was being eroded by the differential reproduction of what were labelled as 'feeble minded' or 'morally defective' classes. The answer to this was seen to be the improvement of the genetic stock by selective breeding, an approach called **eugenics**. We now appreciate that such philosophies are without any scientific justification and are universally regarded with abhorrence. Changes in the human gene pool can only take place over many generations, in response to selective pressures such as plagues, endemic infections and climatic conditions.

It is difficult to appreciate now the extent to which eugenics held sway among both the scientific and political classes. Several academic journals of genetics included the word 'eugenics' in the title. Quotations from many renowned statesmen and scientists show that belief in eugenics was considered normal or even responsible for leading members of society. It was not just in Nazi Germany that eugenics was practised. In the 1920s and 1930s a number of states in the USA passed laws for the forcible sterilisation of the 'feeble minded' and tens of thousands of such operations were actually performed. Recently it has been revealed that Sweden carried out over 60 000 sterilisations for eugenic reasons, ceasing only in the 1970s. The eugenics episode serves as an illustration that the ill-advised application of genetics can be exceedingly harmful and emphasises the importance of careful deliberation about the possible consequences of recent advances.

11.2 Genetic testing

Genetic testing refers to any procedure designed to ascertain the status of any aspect of someone's genotype for purposes of diagnosing or predicting disease wholly or partly caused by hereditary factors. This covers tests based on DNA, RNA or chromosomes but may be extended to include protein-based tests or even more broadly based medical examinations. Genetic tests may be carried out in a variety of contexts:

- prenatal testing;

- neonatal screening;

- diagnosis of genetic disease in children after birth;

- screening prospective parents for the carrier status of genetic disorders common in a particular population;

- presymptomatic testing for a serious late-onset disorder;

- presymptomatic testing for genetic predisposition to a complex disease.

Before we consider these different situations in detail, there are some general points that can be made.

1. It is essential that the tests are reliable and the diagnosis clear. This is not as straightforward as it may at first appear. Some genetic tests use linked markers rather than following the mutation itself. Such tests may be falsified by recombination between the marker and the mutation. In addition, many tests look for common mutations and may fail to detect rarer ones. For example, it is straightforward to detect the *CFTR* ΔF508 mutation (see Chapter 8), but there are hundreds of rarer mutations that might affect the gene. Thus a person who may appear to be a ΔF508 heterozygote may in fact be homozygous for inactivating mutations in the *CFTR* gene. As new tests are developed, there must be a

mechanism to ensure that they are subject to independent review so that their effectiveness is known to doctors and genetic counsellors who will order their use. There is also a need for quality assurance in laboratories that carry out the tests. They will be marketed and carried out by private companies for commercial gain, and we may expect a large number to appear over the coming years. Ensuring quality and reliability and preventing exaggerated claims are major priorities.

2. Experimentation on human subjects must only be carried out with the informed consent of the person being tested. This is an obligation enshrined in international law, known as the Nuremberg Code. It was established after evidence in the Nuremberg trials after World War II in response to revelations of medical experimentation in the concentration camps of Nazi Germany. There is a distinction between research and an established medical procedure. Once the validity of a genetic test has been established by research and it is used to diagnose or counsel individual patients, it is no longer experimental and is not covered by the Nuremberg Code. Nevertheless, the consensus is that the requirement for informed consent should still apply when genetic tests are used in a clinical context. Informed consent requires that the subject fully understands the nature of the test and the consequences that may arise from it. This leads to a general need for counselling, both before the test is undertaken and to explain the results when they are available. In some cases, such as a young child or where there is mental impairment, the person most affected may not be in a position to understand the reasons for testing. There is thus the need for the parents or guardian or some other representative to be able to give consent on their behalf. This consent should only be given for immediate medical reasons. Determining the carrier status of a young girl for a sex-linked disease would be an example where this condition is not met. Children's views can be taken into account, and as they get older and their understanding grows their wishes should be given increasing importance.

   Clear counselling is also essential for informed consent in research procedures. This is particularly important in phase 1 trials of new procedures. These are designed to test the safety of the procedure and to establish a safe dose. They are not designed to test the efficacy, so the subject is unlikely to derive any personal benefit. This may not be clear to the sufferer of the disease, who may undertake the trial in the hope of being the first to receive a new treatment. Because they are designed to test safety, there can be no guarantee beforehand that the trial will be safe. This was tragically emphasised by the death of a volunteer in a phase I trial of a gene therapy procedure (see Section 7.4).

3. Tests should only be carried out when there is some positive action that can be taken as a result. We shall see that these tests can sometimes result in significant detriment to the person tested. Therefore tests can only be justified if they allow the prevention or better treatment of a disease or provide some other genuine benefit.

4. Everybody has the right to know about information that affects their health and has the right to expect that this information will be kept confidential. Thus access to genetic information held on any sort of file must be secure and at the same time made available to the individual concerned. Alongside the right to know, is the right not to know. This not only underlines the need for informed consent to any test but also emphasises the need for vigilance that unwelcome information is not inadvertently obtained or passed on to the individual concerned.

5. Genetic tests have implications for the whole family and may extend beyond the immediate nuclear family. Some instances, discussed below, may lead to situations where one person's right to confidentiality conflicts with another's right to know.

11.2.1 Prenatal testing

Prenatal testing is carried out where there is a known risk of a serious monogenic disorder or in older women where there is an increased risk of Down's syndrome and other chromosome abnormalities. This would normally be done in order to provide the opportunity of termination in the event of an adverse result. Such testing empowers parents, who may not otherwise risk having a seriously affected child, to reproduce. For example, where they have already had one affected child, the woman can become pregnant in the knowledge that the pregnancy can be terminated if the embryo is affected. The positive value of such tests is clear. However there are a number of issues connected with embryo testing.

Prenatal tests can be invasive and carry a certain risk of miscarriage (1% for amniocentesis). The potential benefit gained from the test must be commensurate with this risk. From this it follows that tests with a significant risk should only be carried out where the condition investigated is serious and the possibility that the embryo is affected is significant.

If prenatal tests are made with a view to termination of pregnancy, it is important that the parents fully understand this. There is little point and potentially much harm in testing an embryo if the parents would not wish to terminate the pregnancy whatever the result. This means that the parents should undergo counselling before the tests are carried out. Of course, parents retain the right to proceed with the pregnancy in the event of an unfavourable outcome to the test. Because prenatal testing implies the possibility of a termination, there may be an assumption that members of certain ethnic or religious groups would not wish such testing because of an objection in principle to such procedures. It is for the individual concerned to make such a decision and it is important that the option is always presented.

A number of tests are made routinely in the antenatal clinic. It is possible, even likely, that in such circumstances the prospective mother will not fully understand the nature of the tests and the possible consequences. She may then be unprepared for the unwelcome knowledge from an adverse outcome. This would be particularly serious if she objected in principle to termination and thus would be forced to complete the pregnancy in the knowledge that the child would be seriously disabled. At the very least, she may have wanted the opportunity to decide whether she wanted the test before it was carried out. This is not to say that prenatal tests should not be carried out, just that it is important that time is taken to explain their nature to the mother.

Prenatal testing with a view to termination makes a value judgement about the worth of the unborn child. This has two important consequences. Firstly, if it is agreed that certain conditions are sufficiently serious to warrant termination, then there is a danger that this could be construed as making a judgement about the worth of people who have been born who suffer from the disease. Secondly, since there is a range of disablement that results from genetic disorders, how do we decide what constitutes a sufficiently serious disablement to warrant termination? Most people, but by no means all, agree that a serious condition that will result in a short lifespan with much suffering are sufficient grounds for termination. Many parents would not wish to have a child affected by Down's syndrome. However, parents with such children often enjoy a rich and loving relationship with them. The Down's Syndrome Association in the UK do not consider it as grounds for termination. This seems to be a decision that should properly be left to the parents. How, though, do we react to a diagnosis of a late-acting autosomal dominant disorder where the unborn child can expect to enjoy 40-50 years of normal life? The consensus here is that this is not grounds for termination. If this is the case, it becomes improper to apply the test in the first place, because self-evidently an embryo cannot give informed consent but will have to live many years with the consequences.

The decision to terminate a pregnancy as a result of a genetic test must always be taken by the parents. It is important that the parents do not come under any form of pressure. One such pressure may come from the physician treating them. Another pressure may be the financial consequences of the healthcare needed by the disabled child when it is born. Healthcare is financed in different ways in different countries. In some, such as the UK, it is financed by the National Health Service; in others, such as the USA, it is financed by an insurance-based system. Whatever the system, will it consider the ill health of a child born in such circumstances to be caused by the actions of the parents and this restrict healthcare?

11.2.2 Neonatal screening

Neonatal screening is carried out for the occurrence of a common genetic disorder. Screening for phenylketonuria is already widespread, as is screening for haemoglobin disorders in communities where they are endemic. Screening for CF is starting. The benefit of screening is that early detection leads to early treatment, which would prevent the onset of the disease in the case of phenylketonuria, or can greatly improve the prognosis in the case of CF and the haemoglobin disorders. Like all testing, informed consent is ideal; however such tests are routine and in practice it is unlikely that sufficient time can be devoted to any form of counselling. While this is regrettable, the benefits of such programmes are generally thought to outweigh this problem.

Carrier screening can be very effective in reducing the population incidence of a disorder. This may come about by avoiding marriage between carriers, a decision of two carriers not to have children or by using prenatal testing and termination of affected embryos. Screening for β thalassaemia in Cyprus and Sardinia has led to a marked reduction in the number of cases. Carrier screening is also carried out for Tay–Sachs disease in a number of Ashkenazi Jewish communities. In all these cases, the programme is actively supported by the relevant religious leaders. In some cases, the local religious leader may act as a clearing house by receiving the results of the tests and only acting where both partners are carriers.

The major problem with carrier screening is that it may lead to discrimination against those diagnosed as carriers. This may be in the form of social stigmatisation in finding marriage partners or lead to difficulties in obtaining insurance or employment. A compulsory screening programme for sickle cell anaemia in several American states in the 1970s is widely held to have been disastrous because it led to discrimination against carriers. A common problem in population screening for recessive disorders is a failure to realise that carrier status does not affect the health of the person being tested but will affect the health of any children resulting from union with a partner who is also a carrier. Thus the discrimination against sickle cell carriers was not only unjust but was without any basis in reality.

The need for a clear understanding of the consequences of carrier status means that counselling must be provided to those who are carriers. Since population screening will only be undertaken for common disorders, it is likely that the frequency of carriers will be high, up to 1 in 20 for CF in Europeans. This causes a problem of resources. One solution that has been suggested is to inform people of their carrier status only when their partner is a carrier as well, because only then is there actually a problem. However, in this scenario most carriers will continue to be unaware of their status and, because they have been tested without any adverse result, may assume that they are not carriers. This could cause difficulties if they subsequently had children with another partner who was a carrier.

11.2.3 Diagnosis of genetic disease in children after birth

Genetic tests may be carried out in a child, less commonly in an adult, to confirm a diagnosis of a genetic disease or in some cases to exclude it. Such use does not differ in principle from any other test used to aid medical diagnosis and does not present any ethical problems.

11.2.4 Presymptomatic testing for late-onset diseases

Presymptomatic testing may be carried out for serious late-onset diseases, such as HD, familial Alzheimer's disease, familial colon cancers or familial breast cancer. It is important to distinguish these sorts of tests from novel tests that will become available for predisposition to common diseases, which are considered separately below. For serious late-onset diseases, the presence of the mutation is highly likely to lead to the onset of the disease. However, this is not necessarily inevitable. The penetrance of *BRCA*1 is only 50% by age 50 and 85% throughout life. Thus some women may have the mutation but not suffer the disorder. Equally, even if a woman does not carry the *BRCA*1 mutation there is still a significant lifetime risk that she will develop breast cancer from other causes. The situation is further compounded by current uncertainty about the penetrance of the *BRCA*2 mutation (discussed in Chapter 5). Even where the eventual onset of the disease is virtually certain, as is the case with HD, there is considerable uncertainty about the age of onset and the severity of symptoms.

These tests are unusual because they predict future disease in people who are currently healthy. There can be a number of real and serious consequences from an adverse result. Firstly, such a result can inflict severe distress and psychological damage on a person arising from the knowledge that at some point in the future he/she is likely to suffer from a devastating disorder. Even members of an affected family who test negative can suffer psychologically from feelings of guilt that they have escaped. Secondly, affected people may find it more difficult to obtain insurance and employment. The use of genetic tests in employment and insurance is discussed below.

Because of the possibility of harmful consequences, there have to be clear positive reasons for undertaking such tests. Such reasons do exist. Firstly, there may be some clear action that can be taken to prevent or ameliorate the disorder. In the case of the colon cancers, removal of the colon has been shown to be an effective prophylactic treatment. Here the reason for undertaking the test is obvious. The situation in familial breast cancer is less clear. *BRCA*1 and, to a lesser extent, *BRCA*2 predispose to ovarian cancer at a high frequency and other cancers at a lower frequency. At the very least, prophylactic surgery would require oophorectomy (removal of the ovaries) as well as radical mastectomy. Even then we still do not know whether this will be effective: it is difficult to completely remove all breast tissue and there is still the risk of other cancers. Furthermore, it is the nature of this disease that we are unlikely to know for some years whether prophylactic surgery is effective. In other diseases, such as HD and familial Alzheimer's disease, there is no treatment and it is difficult to see any treatment becoming available in the foreseeable future.

A second advantage from foreknowledge of disease is that it may allow appropriate life choices to be made. This may include decisions about marriage and having children, financial decisions and career choices. People who know they are at risk may make very different decisions if they know for certain whether or not they are affected. One choice that is special to breast cancer is that a woman may choose to have her children at an early age, followed by prophylactic surgery.

In this context of life choices another ethical dilemma arises. These choices are only open to someone who is relatively young and who may therefore decide to be tested. They may have a middle-aged parent who is at risk but who has not yet developed any symptoms. Since there is no positive value to the parent of knowing their status, they may well decide not to be tested. The problem is that if the child is diagnosed as having the disease, this diagnoses the parent as well. How is a balance to be struck between the right of the child to know and the right of the parent not to know?

11.2.5 Presymptomatic testing for predisposition to complex diseases

Presymptomatic testing for predisposition to common complex diseases is not yet widely available but is likely to become available in the next few years. We have seen in Chapter 6 that type 1 diabetes has a strong genetic component and that these components are being successfully identified. However, type 1 diabetes also has a strong environmental component, since the concordance between identical twins is only 30%. It is not difficult to envisage a situation soon where neonatal screening could identify children at genetic risk, whose parents could then be advised on how to avoid environmental risk factors. Alternatively, children could be intensively monitored for early markers of the disease, such as autoantibodies to β cells. This could then lead to early and aggressive treatment to prevent the full-blown disease from developing.

Research is also well developed into genetic factors that predispose to heart disease. One dominantly acting mutation, affecting about 1 in 500 of the population, has been shown to cause familial hypercholesterolaemia. Untreated this is likely to lead to heart disease by the age of 40–50 years. However, this can be prevented by changes in diet and lifestyle or by lipid-lowering drugs. A large number of other alleles have also been identified that alter the risk of heart disease but which can be modified by drugs or by diet and other lifestyle changes.

An important difference between this type of test and tests for late-acting autosomal dominants is that the latter carry either a very high risk or virtual certainty that a disease will occur. In contrast, predisposition indicates an elevated risk but no objective certainty about the outcome. This distinction is very important. Apart from this, many of the issues that were considered above for late-acting dominant disorders apply to predisposition as well. Diagnosis of risk factors comes with potential negatives, such as psychological harm and discrimination in insurance and employment. Normally there should be a clear positive reason for undertaking the test. However, most of these tests are only likely to become widespread if they are coupled to an effective course of action for the prevention of the disease or intensive medical surveillance leading to early and more successful treatment.

The complexity of the balance sheet between the benefits and potential harm arising out of both types of presymptomatic testing underscores the need for effective counselling before the tests are carried out and to explain the implications of the results once they are available. Several research studies have shown that the proportion of at-risk individuals who wish to undertake such tests is reduced following counselling.

The difficulty of providing effective counselling is one reason why there is considerable concern at the prospect of private mail-order testing. The incentive to use such a service is that the result may be kept more confidential than would be possible if it was arranged through a doctor. The desire for confidentiality comes from the penalties that may accrue in respect of insurance and employment.

11.3 Human rights

11.3.1 Insurance

Life insurance pays a benefit to the dependants of the insured upon his or her death. In health insurance the benefit pays the costs of medical treatment in the event of the insured suffering ill health. The size of the premium that the insured pays depends on the risk of the insured event happening. Over the years, insurance companies have built up great expertise in calculating the size of these risks, based on information such as the insured person's age and occupation and lifestyle risks such as smoking. The system cannot work if the insured person knows their personal risk is higher than the insurance company realises. A simple example would be someone who tests positive for HD insuring their life for an abnormally large sum without declaring the result of the test. For the insured, this is betting on a certainty; for the insurance company, this is a certain loss which, if it happens on a large scale, will inevitably lead to higher premiums for all.

To prevent such fraud, insurance companies would like to know the result of any genetic tests that have been carried out. For the individual this is a disincentive to take the test in the first place. The undesirability of this outcome may be seen from consideration of a test for hypercholesterolaemia. A person who tests positive for this condition may save their life by appropriate treatment and lifestyle change, but they will not be insurable. In other words they may take the test and live or not take the test and be insured; but they are unable both to live and be insured at the same time.

There is thus a conflict between the needs of society in general and the interests of insurance companies. Genetic tests have yet to be used on a large scale. Indeed only one test, for HD, has so far been approved in the UK by GAIC (Box 11.1). Nevertheless, many robust tests for single-gene disorders now exist (see Chapter 8) and we may expect tests for susceptibility alleles to come on-stream in the near future. So this issue will grow in importance. In May 2001 the House of Commons Select Committee on Science and Technology issued a report criticising the attitude of the insurance industry to the use of genetic tests and expressed concerns over the constitution and funding of GAIC. It recommended a moratorium on the use of genetic tests. As a consequence, the HGC advised the UK government that there should be a moratorium on the use of genetic tests while a regulatory framework is developed and more data becomes available on the predictive value of the tests. There were two exceptions to this moratorium.

Firstly, those cases where a test could be used to exclude the possibility that someone suffered from a disease for which they were at risk because of family history. Secondly, where cover of over £500,000 was being sought. The UK insurance industry accepted this moratorium voluntarily.

In the United States, 46 states have legislation prohibiting the use of genetic information. Repeated attempts to extend this to a federal level have been blocked in Congress. However, a federal committee called The Secretary's Advisory Committee on Genetic Testing (SACGT) advises the Department of Health and Human Services (DHHS) on the medical, scientific, ethical, legal and social issues raised by human genetics. It recommended in 2000 that there should be legislation to prohibit a requirement for genetic testing by life insurance companies. In 2001 the DHHS issued advice to the insurance and health industries that genetic information should not be used to assess medical care and life insurance premiums.

11.3.2 Employment

An employer may have an interest in genetic tests of prospective or current employees for a number of reasons. Firstly, employers may not wish to invest resources and time in training someone who in the future will be unable to perform their job because of ill health. Ability to perform a job at the present time and not at some point in the future is an implicit assumption of employment laws in most countries. For example, it is not permissible to discriminate against a woman on the grounds that she may become pregnant at some point in the future. It thus seems fairly straightforward to prohibit genetic discrimination on these grounds.

A second reason an employer may wish to apply a genetic test is for safety or other reasons of public interest. An often-quoted example is that of an airline pilot predisposed to sudden heart disease. It is not difficult to imagine other similar examples. When public safety is an issue it may be necessary to override the rights of the individual. However, it is important that if other means are available then they should be used. For example, an employee doing a job that required high-quality eyesight can be checked by simple eyesight tests rather than a genetic test.

Thirdly, an employer may wish to test for sensitivity to some environmental hazard such as a carcinogen, since there is clear evidence that sensitivity to xenobiotics (chemical compounds not found normally in the body) is influenced by genetic factors. We would expect that every practical method would be used to reduce exposure in the first place. Nevertheless, there may be some industries where this is not possible. It would surely be sensible to screen out those who may be particularly sensitive. If this is so, we would then face the problem of how to react to employees who refused such tests. Clearly it would be wrong to force someone to take a test against their will. However, an employer may also wish to be protected against the consequences if the employee who refused the test was to consequently suffer the ill health that could have been avoided.

11.3.3 Confidentiality

Everyone has the right to keep information about their genetic constitution private. There will be many occasions when this right to privacy may come under attack. For example, we have already seen in recent years sensational treatment in the media concerning the HIV status of well-known people. There is far more scope for this sort of problem in genetic testing, e.g. the future health prospects of a politician. Clearly, genetic information will need some form of legal protection

A different problem arises out of the fact that genetic testing is concerned with families as much as individuals. When one member of a family is diagnosed as being a carrier of a genetic disease, it becomes possible that other members of a family are as well. Consider, for example, a woman who has a boy affected by a sex-linked disease and it is confirmed that she is indeed a carrier and that this was not a mutation that arose during gametogenesis in her mother. There will be a 50% possibility that her sister will be a carrier as well. Suppose the sister is about to become pregnant. She clearly needs to be aware of this information so she can ascertain her carrier status. With this knowledge she can make her own decisions about her reproductive choices. She may decide not to have children, or she may go ahead and become pregnant but use prenatal testing to ensure the embryo is healthy. What happens if the first sister refuses permission for the information to be passed on? Here we have a clear conflict between the right of the second sister to know and the right of the first sister to confidentiality. We may hope that counselling will persuade the first sister to divulge the knowledge, but this does not always occur in practice.

11.3.4 Individual responsibility

Research into personality disorders may well lead to the conclusion that certain antisocial actions, such as substance abuse or violent behaviour, are influenced by genetic factors. There has already been a demonstration that a history of violence and criminal activity in an extended Dutch family is due to a single mutation in the monoamine oxidase gene. This leads to two contrasting ethical problems. On the one hand, individuals predisposed to such antisocial behaviour may seek to absolve themselves of responsibility for their actions. Genetic predisposition to criminal behaviour has already been used as a defence in a murder trial in the USA. An opposite problem is that society may seek to stigmatise those found to carry such traits. Since such behaviour may be more common in certain sections of the community and in certain ethnic groups, it is but a short step to label those groups as genetically inferior. The scientific fallacies inherent in this form of genetic determinism are discussed below. It is sufficient to note here that such an approach ignores the overwhelming influence of the environment and poverty on violence and drug addiction.

Another area in which individual responsibility will become important is how we react to warnings that we may be genetically predisposed to a com-

plex disease. For example, it is not difficult to imagine in the future that
neonatal screening may be carried out for predisposition to type 1 diabetes.
Research may also have identified environmental risk factors that trigger the
disease or have led to medical treatments that prevent its onset. Will we hold
parents responsible for ensuring that a child avoids the environmental risks
by keeping to a diet or undergoes suitable medical treatment? Will a disease
be judged self-inflicted if people ignore warnings that they are susceptible?
There is a growing trend to restrict treatment where disease is self-inflicted
as a result of cigarette smoking or substance abuse. We may note here that
the justice of such practices may become even more questionable if it is
shown that some individuals are more susceptible to addiction than others.

11.4 Patents

Much of the current impetus for the advances in human molecular genetics
comes from the private investment of the biotechnological and pharmaceutical industries. This investment is only possible if any invention of commercial
value can be protected by patent, which prevents others from using the
invention for commercial purposes without the permission of the patent
holder. This permission is given in the form of a licence that usually involves
the payment of a royalty. In return for the issue of a patent the inventor must
disclose full details of the invention. In this way society gains the benefit of
the knowledge generated by the research, while the rights of the inventor to
commercially exploit the invention are protected.

Most patents relating to human genetics are filed in one of three patent
offices: the United States Patent Office (USPTO), the European Patent Office
(EPO) or the Japanese Patent Office. International respect for patents comes
through international agreements such as The World Trade Organisation
Agreement on Trade-Related Aspects of Intellectual Property Rights.
Although each patent office has its own policy for granting patents, generally
to be successful a patent application must meet the following criteria:

1. The invention must have 'utility', i.e. it must have a use that is clearly
   specified in the application.
2. The invention must be novel. In the United States, an invention may be
   discussed publicly for a year before filing but in Europe an invention
   may not be revealed publicly before the patent application. As a result
   there is difference between the methods used for deciding priority: in
   the United States priority goes to whoever was the first to invent, in
   Europe it is the first to file. The importance of demonstrating priority in
   the United States has led to the practice of notarised laboratory note-
   books in commercially funded research.
3. The invention must not be obvious. This means that it is not an
   improvement that would be routinely made by an experienced and well
   trained person in the field. The phrase 'skilled in the art' is often used to
   describe such a person.

4. The patent must be sufficiently detailed for the invention to be reproduced by a person skilled in the art.

5. There are often further restrictions connected with morality and public good. For example, the European Patent Convention prohibits the issue of patents to inventions that offend morality or public order, for example a letter bomb or an instrument of torture. In the United States, Congress prohibited patents on nuclear weapons.

A patent application may be provisional. In such cases the applicants have up to a year to make a full application, in which they must substantiate claims made in the provisional application. The patent officer examines the application to see if it meets the above criteria and, if satisfied, the patent is granted. The USPTO pronounces in three years, the EPO within 18 months. The decision of the patent officer may be challenged in court. Once granted, a patent lasts for 20 years from the date of application, which in practice means 17 years if the time taken for the USPTO to pronounce on the patent is taken into account. Drug companies point out that the time taken for development and safety trials is often as much as 10 years, so that the time they have to commercially exploit their invention is limited. However, they often find ingenious ways to extend their period of protection by applying for additional patents relating to a successful drug.

In 1980 in the United States a precedent was set for patenting living organisms in the case of Diamond vs Chakrabry, which was decided by the Supreme Court. The case concerned a bacterium genetically modified to dissolve oil. The court ruled that it was patentable as it did not occur in nature. Since the Diamond vs Chakrabry case, over 3 million genome-related patents have been filed in the United States, but only a few patents have actually been issued by the USPTO. Some of the applications relate to genes that have been cloned, fully sequenced, their biological role established and a clear utility described, such as a diagnostic test or a screen for a drug that may interact with the encoded protein to modify its function. On the whole these have not been controversial. However, most of the applications have related to ESTs that are only gene fragments – they encompass as little as 20% of the gene from which they were derived – and for which the function could only be guessed. As genome sequencing developed there was widespread concern that many parts of the human genome would be patented before it came into the public domain. This was one of the main driving forces for the accelerated completion of the publicly funded genome project and the central ethos that sequence should be placed on the Internet within 24 hours of it being obtained. By placing the sequence in the public domain it was hoped to limit the scope for patents based solely on sequence without further work to establish function and the development of a useful application. Further applications related to SNPs for which patents were sought on the basis that they could be used in diagnostic tests (See Chapter 6). The SNP consortium patented all SNPs identified but only on a defensive basis, to prevent others from patenting. The declared policy of the consortium is not to enforce these patents.

There is general agreement that it is right that commercial companies should be able to patent in the field of human genetics. However, the attempt to patent gene sequences without further work to establish their function and utility has provoked a fierce controversy. A number of issues have been raised:

1. The rush to patent genomic sequences has been likened to a land grab, where commercial organisations rush to stake claims on parts of the human genome without having fully characterised them or establishing utility.

2. A patent on a gene sequence that forms the basis of a diagnostic test gives the patent holder a monopoly position on all future tests based on that gene. If it turns out that the gene sequence has a totally different application than that envisaged in the initial patent, the patent holder still controls the use of the sequence for the new application, even if this application was quite unimaginable at the time of patenting.

3. Cloning a gene and determining its nucleotide sequence is a discovery not an invention. Patent lawyers argue that cloning a gene or cDNA is like purifying a substance from nature that has value, which is recognised in patent law. Opponents of this view argue that the value lies in the gene sequence, not the physical DNA whose isolation is incidental to the process.

4. The essence of human genetic research is that it is co-operative. The process of cloning a disease gene may start with individual family members collecting information, which may be assembled into family trees by local research laboratories and physicians. Mapping, and finally cloning the gene, requires the use of dense maps and probably nucleotide sequence data obtained from publicly accessible databases. Both the maps and the databases were assembled with great effort by many different laboratories. It is often remarked that genes seem to be independently cloned by several groups within a few weeks. There is a reason for this: once the information has been collected that allows the gene to be located, there are many competent laboratories that can make the last step. Granting a patent to any one of them provides a reward to one party for the collective labours of many others.

5. The practice of seeking patents for gene fragments means there may be a conflict when the gene is fully characterised by others. If a patent is granted for an EST sequence, does that extend to sequences in the gene that were not present in the original EST? Moreover, different patent applications may unwittingly overlap. For example, a patent for an SNP may involve a sequence that is part of an EST. This may mean that there are a number of different patents controlling the use of a single gene. This causes what is known as patent stacking, which multiplies the royalties that must be paid by a third party wishing to use the gene. This might ultimately discourage product development because of the royalty costs.

6. Patent applications remain secret until the patent is granted. This may mean that companies may invest in a product only to find later that the patent is held by another party.

Despite the philosophical objections to patenting gene sequences, it is clear that patent law allows it in principle and that this is unlikely to change in the foreseeable future. However, in December 1999 the USPTO altered its policy to tighten the requirement to identify utility. Patents relating to gene sequences must have 'substantial real-world utility'. It was envisaged that this would disqualify many EST applications.

11.5 Gene therapy

Gene therapy promises for the first time to cure a genetic illness rather than ameliorate the symptoms. It may be carried out using two fundamentally different strategies. In somatic gene therapy, the genetic constitution of a person is modified or supplemented in such a way that the change is limited to that person's own body and is not passed on to the next generation. In germline therapy, a person's germ cells, or cells that will become germ cells, are changed so that the change is passed on to the next generation.

Somatic gene therapy is theoretically no different from any other form of medical treatment that seeks to rectify a physiological dysfunction. The view that it is similar to an organ transplant is not entirely accurate because gene therapy requires no donor other than the cells from which the original DNA was cloned. Provided that the gene therapy is directed towards treating a disease, rather than enhancing natural characteristics, somatic gene therapy requires no further control than would normally be applied to any experimentation involving human subjects, the ethical principles of which are well established. If the research is successful, then marketing of gene therapy products will be subject to the same controls as any new drug, i.e. clinical trials will be needed to prove that it is both safe and effective.

Germline gene therapy in principle offers many advantages over somatic gene therapy. Once a reliable method is found of manipulating the germline then that method would be generic for all genes, whereas somatic therapy relies on different *ad hoc* methods of introducing genes to different tissues according to which is affected by a particular mutation. For example, somatic gene therapy for CF requires genes to be delivered to cells lining the airways of the lung. Even if this is successful, other affected cell types such as those of the vas deferens or the pancreatic duct are not treated. Somatic gene therapy for DMD requires genes to be delivered to muscle cells and for HD would require genes to be delivered to brain cells. Each disease presents its own special problems and will require its own research programme to solve them.

Germline therapy would make it much simpler for a couple at risk of having a child affected by a genetic disease to have normal children. At present, each pregnancy must be tested and affected embryos aborted. Sometimes the laws of chance operate unkindly and it is not unknown for a series of pregnancies to have an unfavourable outcome. It has been argued that it is better to fix the problem before conception than to terminate the life of an embryo. In this context it should be noted that preimplantation testing (see Chapter 8) may provide a more realistic way of solving this problem than germline gene therapy.

There are considerable practical difficulties to be overcome before germline gene therapy can be contemplated. Somatic gene therapy aims to supplement the defective gene by introducing the functioning gene in such a way that it does not integrate into the genome. Germline therapy will require the replacement of the defective gene with the functioning copy. If the new gene is not precisely targeted to the location of the defective gene, two deleterious consequences may follow. Firstly, the original mutation may reappear in subsequent generations due to reassortment of the genetic material at meiosis. Secondly, insertion of new DNA at another site could be mutagenic: it may either inactivate another gene with unforeseeable consequences or change the pattern of expression of another gene perhaps with oncogenic consequences. Techniques are not currently available for this type of targeting. A second technical difficulty is the current lack of any method for delivering genes to the germline. These difficulties are likely to prevent germline gene therapy for some years. Nevertheless, we may reasonably expect these problems to be solved at some time in the future.

There are multiple ethical objections to germline gene therapy. Firstly, it is perceived by many to be 'playing God' by interfering with the basic genetic constitution of the human species. Secondly, it opens up the road to manipulation for eugenic reasons, i.e. manipulation designed to enhance characteristics such as intelligence, musical or sporting ability, etc. Thirdly, it is irreversible; if for some reason a change turns out to be in error it cannot be removed from the human gene pool. Fourthly, its development will require a large amount of experimentation on human embryos and, at some point, will lead to experimental humans who will be at considerable risk of suffering from a genetic dysfunction. By definition, such humans will be unable to give their prior informed consent.

The combined practical and ethical objections have meant that germline therapy is prohibited in all countries that have the research capability to attempt it. Nevertheless there are many scientists, genetic interest groups and even religious leaders who are of the opinion that the benefits outweigh the drawbacks. Should the technical problems be solved the controversy is likely to resurface.

11.6 Genetic determinism

One of the themes that runs throughout this book is that genetic factors influence the risk of common diseases and psychiatric disorders. It is also highly probable that our personalities and abilities are also influenced by genetic factors. We have seen that the means have been developed to identify these genetic factors and that we may expect that over the next few years these techniques will be applied to the analysis of the causes of many common diseases and psychiatric disorders. Inevitably they will also be applied to an attempt to improve our understanding of the origin of fundamental human characteristics.

In doing this, there is a danger that we will come to regard our health, personalities and abilities to be mechanically determined by our genetic constitution. Such a view is scientifically wrong. It is important to be fastidious about the relationship between genotype and phenotype because it is easily caricatured by both sides of the political spectrum. Susceptibility to common diseases and the development of phenotypic traits are likely in each particular case to be controlled by several genes that interact with each other and with the environment. These interactions will be complex and are unlikely ever to yield simple predictions about the outcome. Moreover, because the contribution of the environment is nearly always significant, the outcome will usually be modifiable by environmental change.

It is worth considering what we know about the genetic predisposition to IDDM because it is instructive in these matters. Recall that the concordance in identical twins is 30%. Immediately, it is clear that the environment plays the larger part in the onset of the disease. It is not easy to be sure what this environmental contribution could be, since we would imagine that identical twins would be exposed to the same obvious environmental factors, such as diet, etc. It may be very subtle; for example, an important factor in the development of multiple sclerosis, another autoimmune disease with a clear genetic component, is thought to be a virus infection. The locus labelled *IDDM1* in the HLA region is thought to contribute about 40% of the genetic risk. In European populations, the *IDDM1* genotype DR3/DR4 predisposes towards diabetes, but this can vary in other ethnic groups. The concordance of DR3/DR4 is very high in affected sib-pairs. However, whereas 2% of European populations have the DR3/DR4 haplotype, only 0.1% succumb to diabetes. Moreover, there are some who have IDDM who do not have the DR3/DR4 genotype. Thus the great majority of people who have the risk genotype do not succumb to the disease and some who have the disease do not have the predisposing genotype. It would clearly be wrong to label *IDDM1* as the 'diabetes gene'. Nevertheless, many independent lines of research confirm that *IDDM1* contributes towards the risk of diabetes and we can even start to formulate plausible models of its action.

11.7 Summary

- It is important that advances in human genetics lead to the relief of suffering and the promotion of good health. There are clearly foreseeable ways in which they may have a negative impact on society through genetic discrimination and loss of privacy.

- Genetic tests allow diagnosis of genetic illness. They can be used in a variety of situations to prevent, or allow the early treatment of, diseases that have a genetic component.

- Genetic testing can have negative consequences. It can cause psychological harm or it may lead to actual discrimination in insurance or employment.

- Genetic tests should only be used when a clear benefit can be derived. They must only be used after the person tested has given their informed consent. Informed consent requires counselling.

- Everyone tested has the right to expect confidentiality. This can lead to some conflicts of interest when there are good reasons for other family members to know the result of a test.

- As well as the right to know comes the right not to know. Care must be taken to ensure that unwelcome information is not given inadvertently.

- Insurance companies wish to know about the results of a genetic test to avoid fraud. This results in a disincentive to genetic tests that may be important for someone's health.

- Employers may wish to discriminate on the basis of genetic tests to avoid hiring or promoting those with poor health prospects. Genetic tests may sometimes be justified on the grounds of public safety or to protect those who are especially sensitive to environmental hazards.

- Confidentiality may be compromised by media interest.

- Predisposition to antisocial behaviour raises issues of individual responsibility for criminal actions or substance abuse. Equally, it may lead to particular groups being labelled as genetically inferior. In such arguments, we must never lose sight of the large contribution made by the environment.

- Gene therapy offers the possibility of a cure for genetic diseases. Somatic gene therapy is not thought to raise any ethical problems. Germline gene therapy may have considerable advantages over somatic gene therapy. However, practical difficulties prevent it being used at present and there are powerful ethical arguments against it ever being used; at present germline gene therapy is prohibited.

- Patents in human genetics are required to allow private investment to fund research. A debate over whether genes and DNA sequences can be patented has been settled in favour of allowing such patenting.

- Genes influence our health, personalities and aptitudes, but genetic determinism is misguided. These characteristics come about through the interaction of many genes with each other and with the environment. These interactions are complex and unlikely ever to yield simple predictions about their outcome.

Further reading

General

House of Commons Science and Technology Committee (1995) *Third Report – Human Genetics: the Science and its Consequences. Volume 1. Report and Minutes of Proceedings.* HMSO, London.

An excellent and well balanced introduction to the issues. Includes first-hand evidence from all interested parties including patient groups and physicians.

The HUGO website
http://www.ornl.gov/hgmis/research/research.html

Follow the link to ethical, legal, and social issues. Presents good introductions to issues such as patenting (see below), and contains many links to other articles and reports.

The Human Genetics Commission
http://www.hgc.gov.uk/

Describes the regulatory network in the UK and links to reports from government advisory committees.

Gene testing

Promoting Safe and Effective Genetic Testing in the United States. Principles and Recommendations. Task Force on Genetic Testing of the NIH–DOE Working Group on Ethical, Legal, and Social Implications of Human Genome Research
http://www.med.jhu.edu/tfgtelsi/promoting/

Adequacy of Oversight of Genetic Tests. Preliminary Conclusions and Recommendations of the Secretary's Advisory Committee on Genetic Testing.
http://www4.od.nih.gov/oba/sacgt/reports/adequacy_of_oversight_of_geneti.htm

This report was requested by the US Surgeon General and published in April 2000.

Genetics and insurance
The House of Commons Science and Technology Committee. *Fifth Report – Genetics and Insurance.*
http://www.parliament.the-stationery-office.co.uk/pa/cm200001/cmselect/cmsctech/174/17402.htm

HGC statement on the use of genetic information in insurance
http://www.hgc.gov.uk/business_publications_statement_01may.htm

KEEFER, C.M. (1999) Bridging the gap between life insurer and consumer in the genetic testing era: the RF proposal. *Indiana Law Journal*, **74**, 1375–1395.
http://www.ornl.gov/hgmis/resource/keefer.html

Analyses of the legal position of genetics in insurance in the United States
REILLY, P.R. (1997) Fear of genetic discrimination drives legislative interest. Ownership, predisposition major issues. *Human Genome News*, **8** (3&4).
http://www.ornl.gov/hgmis/publicat/hgn/v8n3/01fear.html

Patents
BOBROW, M. and THOMAS, S. (2001) Patents in a genetic age. *Nature*, **409**, 763–764.

ENSERINK, M. (2000) Patent office may raise the bar on gene claims. *Science*, **287**, 1196–1199.

HUGO project information – Genetics and Patenting
http://www.ornl.gov/hgmis/elsi/patents.html

Glossary

λ 1. Relative risk, see Section 5.5.

2. Bacteriophage λ. A commonly used vector for gene cloning.

ab initio Analysis solely from theorectical considerations without the use of experimental data.

acrocentric A chromosome where the centromere is nearer one end.

additive interaction Where the effect on phenotype of multiple alleles is the sum of their individual effects.

admixture Gene flow between populations as a result of migration.

aetiology The causes of a disease.

affected pedigree member Two individuals affected by a disease in a pedigree who are genetically related.

affected sib-pair Two sibs affected by a disease.

allele One of several alternative forms of a gene or DNA sequence.

allele frequency The frequency in a population of each allele at a polymorphic locus.

Alu The most common repeated sequence in the SINE category.

amniocentesis Sampling of cells for genetic testing from the amniotic fluid surrounding the developing foetus. The procedure involves inserting a large tube through the mother's abdomen. These cells are cultured *in vitro* for about three weeks to increase numbers. The procedure cannot be attempted before the 16th week of pregnancy.

anaemia Blood disorder characterised by deficiency of red blood cells.

annotate Identify the features of a genome, including the location of genes and information concerning their likely function.

anticipation The effect by which the severity or penetrance of a disease apparently increases with each succeeding generation.

antigen The protein recognised by an antibody.

anti-oncogene See tumour suppressor.

antisense mRNA An RNA molecule or oligonucleotide that is complementary to an mRNA molecule. It can thus form a duplex with the mRNA molecule, interfere with its translation and provoke its destruction by ribonucleases specifc for double-stranded RNA molecules.

APC **(adenomatous polyposis coli)** The gene that is dysfunctional in familial adenomatous polyposis.

ApoE Lipoprotein found in blood plasma. Different alleles at the *ApoE* gene have been demonstrated to be risk factors for both cardiovascular disease and Alzheimer's disease.

apoptosis Programmed cell death provoked by irreparable genetic damage.

APP (amyloid precursor protein) Precursor of the amyloid protein found in senile plaques in the brains of Alzheimer's disease patients.

archaic humans Members of the species *Homo sapiens* that predated anatomically modern humans.

ARMS (amplification refractory amplification system) A commercial multiplex PCR system for genetic testing.

Ashkenazi jews A particular population group that originated in eastern France in the early Middle Ages, who subsequently migrated to Eastern Europe and later to the USA and Israel. By virtue of culture and religion they have been largely reproductively isolated from surrounding populations.

ASO (allele specific oligonucleotide) An oligonucleotide that will only anneal to one allele, either mutant or wild-type.

association Simultaneous occurrence of a disease phenotype and an allele at a frequency that is statistically significant.

autoantibody An antibody produced by an individual's own immune system that recognises antigens in their own body.

autoantigen The protein or part of a protein recognised by the body's immune system in an autoimmune disease.

autoimmune Attack by an individual's own immune system on some part of the individual's body.

autosomal dominant A trait showing a characteristic pattern of segregation indicating that the gene in question is located on an autosome and that the mutant allele is recessive to the wild-type allele. Thus the disease is manifested when one copy of the mutant allele is inherited.

autosomal recessive A trait showing a characteristic pattern of segregation indicating that the gene in question is located on an autosome and that the mutant allele is recessive to the wild-type allele. Thus the disease is only manifested when two copies of the allele are inherited.

autosome A chromosome that is not an X or Y chromosome. There are 22 autosomes in the human chromosome complement.

BAC (bacterial artificial chromosome) A cloning vector that uses the origin of replication of the F plasmid. Can take inserts of 100–300 kb and is thought to be less prone to cloning artefacts, such as chimeric inserts or rearrangements, upon propagation within the host.

balanced polymorphism A polymorphism that is maintained in the population because the heterozygous state has a greater Darwinian fitness than either homozygous state. Normally, one homozygote is at a severe selective disadvantage, but the allele is maintained in the population at a significant frequency because of the selective advantage of heterozygotes.

balanced translocation When translocated chromosomes are inherited together so there is no change in genomic information content, apart from at the breakpoint.

banding Any process that produces a discrete pattern that can be used to identify individual chromosomes and chromosomal regions. See G-bands and Q-bands.

biallelic A locus with two alleles.

bivalent A chromosome with two chromatids.

buoyant density ultracentrifugation A process of centrifugation that separates molecules according to density rather than mass. Analysis of DNA molecules involves mixing the DNA with a solution of CsCl whose concentration is chosen to closely match the average density of DNA. The solution is centrifuged at very high speed (40 000–50 000 g) for 48 hours. The centrifugal force generates a gradient in CsCl concentration. The DNA molecules move to an equilibrium position where their density is matched by the density of the CsCl at a particular point in the gradient. The density of DNA is determined by its base composition. Any fraction of the genome whose DNA content differs from the average will form a separate or satellite peak.

CAAT box The sequence GGCCAATCT that can be found in the promoter region of genes and which binds the transcription factors CTF and NF1 to stimulate transcription.

cap The covalent modification of the 5' end of an mRNA molecule during transcription.

capsid The protein coat of a virus particle.

case control study Study of the association of a disease with particular alleles in which each case of the disease has an unaffected control matched for age, sex and ethnicity.

cDNA library Gene library composed of cDNA inserts synthesised from mRNA using reverse transcriptase.

Celera Genomics Private company founded by Craig Ventnor to carry out genomic research. Undertook the sequencing of the human genome.

cell cycle engine Molecular mechanism that controls the passage of cells through the cell cycle in eukaryotic cells. It consists of one or a family of protein kinase subunits whose activity is regulated by association with different regulatory subunits called cyclins.

centimorgan Unit of genetic map distance corresponding to a recombination fraction of 0.01. Named after Thomas Hunt Morgan.

centromere A constriction visible in metaphase chromosomes where the chromosome is attached to the mitotic or meiotic spindles.

CEPH families Set of reference families for genetic mapping maintained by the Centre d'Étude Polymorphism Humain in Paris.

checkpoint A mechanism that prevents a cycle event from occurring if a previous event is not completed.

chorionic villous sampling The chorion is derived from the zygote but is not part of the developing embryo. It is a membrane with projections or villi that surround the embryo. Chorionic villous sampling is carried out by introducing a catheter through the vagina until it touches the chorion. It can be safely carried out in the eighth or ninth week of pregnancy. It thus has an advantage over amniocentesis because it allows an earlier termination of pregnancy should this be indicated by the results of the genetic test.

chromatid A chromosomal strand consisting of a complex of a single DNA molecule and its associated protein and RNA components.

chromatin The complex of DNA, RNA and protein that makes up a chromosome.

chromosome jumping A method used in the course of a chromosome walk that allows tens or hundreds of kilobases to be covered in a single operation.

chromosome walking A method of moving from a linked marker to a gene.

cis See phase.

cladistic Method of phylogenetic classification that traces organisms back to a common ancestor using qualitative characters to determine the order of descent.

clone-by-clone Strategy adopted by IHGSC for sequencing the human genome in which the genome was cloned into BAC vectors with an average insert size of 100–200 kb. The clones were ordered by *Hin*DIII fingerprinting. Each BAC clone was then shotgun sequenced. The final sequence was then assembled by merging the contigs from adjacent BAC clones.

clone library A collection of clones that ideally contains all possible sequences from a donor genome.

clone map A physical map based on determining overlap between clones.

coalescence time The average number of generations to the common ancestor of any two random members of a population.

coding sequence A length of DNA or RNA whose sequence determines the sequence of amino acids in a protein.

complement fixation The binding of a group of blood globulin proteins to cause lysis of foreign cells after they have been coated with antibody.

complex disease Disease whose aetiology consists of a mixture of environmental and genetic factors.

compound heterozygote An individual that lacks a gene function because each copy of the gene is inactivated by different mutations.

congenital A disorder that is present at birth.

congenital bilateral absence of the vas deferens (CBAVD) Often one of the symptoms of cystic fibrosis; results in male sterility.

consanguineous The individuals concerned are related and therefore have a proportion of their genes in common.

consensus sequence An idealised sequence that represents the nucleotides most often found at each position when a number of different sequences are compared.

contig A DNA region represented by a group of clones whose relationship one to another is defined by overlap between the sequences of the inserts.

cosmid Cloning vector that takes the form of a plasmid that contains the *cos* site from phage λ. This allows recombinant DNA molecules of ~40 kb to be packaged into λ phage particles *in vitro*. Upon infection into a host cell the recombinant molecule replicates as a plasmid.

coverage The average number of times a nucleotide is sequenced.

CpG island A region of approximately 1 kb in which the CpG dinucleotides are not methylated and occur at higher frequency than that found in the rest of the genome, being equal to that expected from the percentage (G+C) of human DNA. CpG islands often span the promoter region of transcriptionally active genes and sometimes the start of the coding sequence.

cystic fibrosis transmembrane conductance regulator (CFTR) The protein affected by mutation in cystic fibrosis patients. Also used to describe the encoding gene.

cytogenetic A characteristic of the karyotype that is revealed by examination of metaphase chromosomes.

cytokines Hormone-like factors that regulate the activity of the immune system.

Darwinian fitness The number of an organism's offspring reaching reproductive maturity.

deletion mutations The loss of one or more nucleotides.

diabetes mellitus Syndrome caused by failure of cells to take up glucose.

dinucleotide Two successive nucleotides on the same DNA strand, written with the 5' nucleotide first.

distal Away from the centromere compared to the chromosomal point of reference.

dizygotic twins Twins derived from different fertilised eggs. Otherwise known as non-identical twins.

DNA chip A high-density miniaturised array of oligonucleotides attached to a silica or glass substratum. They are being developed to scan genes for the presence of mutations. It may be possible to use them for automatic sequencing.

DNA methylation The addition of a methyl group to cytosine to form 5-methylcytosine in CpG dinucleotides.

DNA profile Features of a DNA sample that can be used to identify the individual from which it originated. Differs from genetic fingerprinting in that it does not purport to uniquely identify the individual from all other humans.

domain Part of a protein where a continuous length of the polypeptide chain folds to form a discrete globular structure that may have a defined function.

dominant negative mutation A mutation that prevents another wild-type protein in the same cell from functioning. Commonly acts by producing an altered polypeptide that prevents the assembly of a multimeric protein.

downstream A region of the DNA molecule that lies 3' to the point of reference.

draft sequence Preliminary form of sequence from the human genome project where there are gaps and the sequence is not edited to ensure the highest possible accuracy.

dynamic mutation A trinucleotide repeat expansion mutation that changes in size during meiosis or, in some cases, mitosis.

dystrophia myotonica See myotonic dystrophy.

dystrophia myotonica protein kinase (DMPK) Implicated in myotonic dystrophy, as the trinucleotide repeat expansion responsible for the disease occurs in the 3' untranslated region of this gene.

dystrophin The protein affected by mutation in Duchenne muscular dystrophy. Also used to describe the encoding gene.

endocytosis A process by which eukaryotic cells take in material from the outside. The cell membrane invaginates to form intracellular vesicles, which then fuse to form the endosome.

endosome See endocytosis.

enhancer A *cis*-acting DNA sequence that increases the expression of a gene upon binding a transcription factor. The action of an enhancer is not critically dependent on its position or orientation and in many cases exerts its influence at a considerable distance.

ePCR Using a computer to carry out a conceptual PCR reaction using a known nucleotide sequence.

episome A DNA molecule that can stably replicate. The significant point is that it does not have to be integrated into a chromosomal location to be propagated in daughter cells.

epistasis Originally defined as the interaction between two genes so that the expression of one is controlled by the other. Now more loosely applied to situations where the effect on phenotype of alleles at two genes is different from that which would be expected by combining the individual effect of each allele.

erythropoiesis Production of new red blood cells.

euchromatin The main fraction of chromosomal DNA that is not heterochromatin. It is uncoiled during interphase, and contains transcriptionally active regions.

eugenics 'Improvement' of the human genetic stock by selective breeding.

exon The part of a gene that is transcribed and remains in an mRNA molecule after splicing. This mainly consists of protein-coding regions, but it also includes 5' and 3' untranslated regions of the mRNA molecule.

exon trapping Special technique used to search for exons in genomic clones.

expressed sequence map A map that plots the position of DNA sequences expressed in mRNA. Based on EST markers.

expressed sequence tag A special form of STS that is generated from cDNA clones and thus identifies sequences expressed in mRNA.

expressivity The difference in the severity of a disorder in individuals who have inherited the same disease alleles.

ex vivo **gene therapy** Where cells are removed from the body, manipulated *in vitro* and then reintroduced to the patient.

familial Where a disease is transmitted in families as opposed to its sporadic occurrence.

familial adenomatous polyposis Cancer of the colon and rectal areas. Inherited as an autosomal dominant trait due to mutations in the *APC* gene.

fingerprint clone contig Contig formed by BAC clones using the *Hind*III fingerprint of each clone to detect overlap.

fingerprinting Any method that identifies unique features of a clone that can be used to determine overlap between other clones in a library.

finished sequence Final form of sequence from the human genome project containing less than 1 error in 10 000 bp.

fluorescence *in situ* hybridisation *In situ* hybridisation in which the probe is labelled with a fluorophore. Upon examination with a fluorescence microscope, the chromosomal region to which the probe is binding can be visualised.

fluorophore A chemical moiety that fluoresces when stimulated by light of a particular wavelength, usually in the ultraviolet range.

founder effect The effect that results in the genetic structure of the population that is derived from the founder group being different from the ancestral population.

founder group A small group of individuals that splits off from the main population and settles in a previously uninhabited territory.

founder mutation A disease-causing mutation on chromosomes in different individuals that has descended from an ancestral chromosome on which the original mutation occurred.

fragile site Non-staining gap in a chromosome visible in metaphase spreads of cells cultured under conditions such as folate deprivation or chemical inhibition of DNA synthesis.

frameshift mutation Deletion or insertion of bases that alter the reading frame.

framework See scaffold.

functional genomics The study of the function of genes in a genome.

G_1 (Gap 1) The period between nuclear division and DNA synthesis (S phase) in the cell cycle.

G_2 (Gap 2) The period between DNA synthesis (S phase) and nuclear division.

gametogenesis The process by which gametes are produced.

ganglioside A group of complex glycolipids found chiefly in nerve membranes, containing sphingosine, fatty acids and an oligosaccharide chain containing at least one acid sugar such as N-acetylglucosamine, N-acetyl neuramic acid or N-acetylgalactosamine.

gap junction A small area where the plasma membranes of two adjacent cells are separated by a narrow gap.

G-bands Bands produced in metaphase chromosomes using Giemsa.

GC box The sequence GGGCGG found in the promoter region genes, which binds the transcription factor SP1 to stimulate transcription.

gene A DNA sequence that contributes to the phenotype of an organism in a way that depends on its sequence. Normally this refers to a protein-coding sequence, together with any sequences that are required for expression. Some genes encode RNA that is directly functional and is not translated.

gene families Similar but not identical genes that have arisen by a process of gene duplication and divergence. Gene families can be dispersed through the genome or exist as gene clusters at one location.

gene library See clone library.

general transcription factor A protein required for the formation of the transcription complex. Normally designated in the form TFIIX, where II indicates the RNA polymerase (in this case RNA polymerase II) and X denotes the particular protein.

gene therapy Treatment of a disease by genetic modification of the patient's cells.

Genethon Laboratory in Paris funded by the French Muscular Dystrophy Association. Constructed the standard genetic map of the human genome.

genetic drift The change in allele frequencies from one generation to another due to sampling effects.

genetic fingerprinting A process that uniquely identifies an individual according to some feature of their genome.

genetic heterogeneity Where apparently clinically similar disorders are caused by mutations in different genes (non-allelic), or where mutations in the same gene result in clinically diverse conditions (allelic).

genetic map A genomic map based on the order of genetic mapping markers and the genetic distance between them measured by the recombination frequency.

genome The sum total of the genetic material.

genome scan A systematic survey to discover if a phenotypic trait or genetic disease is linked to a genetic mapping marker.

genomics Study of a genome by considering all the genes together rather than the properties of individual genes.

genomic imprinting A mechanism whereby cells preferentially express an autosomal gene inherited from one parent.

germline The cells that will give rise to gametes. Mutations in germline cells will be inherited by the next generation.

germline gene therapy Where the genetic modification may affect the germline and be passed on to succeeding generations.

Giemsa A dye that binds DNA.

glycolipid A lipid containing a carbohydrate chain.

Golgi apparatus A membranous cellular compartment that forms part of the secretory pathway.

growth factor A specific factor that must be present in the environment before a cell can proliferate. Usually a protein or peptide that interacts with a growth factor receptor at the cell surface.

haemoglobinopathies A group of genetic diseases that affect the production or function of haemoglobin.

haplotype A set of closely linked alleles that tend to be inherited together, i.e. not separated by recombination at meiosis.

Hardy–Weinberg distribution The mathematical description of the distribution of locus genotypes in a population.

hemizygous The presence of only one copy of a gene. Usually applied to genes on the X chromosome in males.

hereditary breast cancer (HBC) A subgroup of breast cancer cases where the cancer is caused by the inheritance of an autosomal dominant allele of genes such as *BRCA1* and *BRCA2*.

hereditary non-polyposis colorectal cancer (HNPCC) Cancer of the colon and rectal areas. Inherited as an autosomal dominant trait due to mutations in genes mediating mis-match repair.

heritability The proportion of total phenotypic variation in a population that is due to genetic variation.

heterochromatin Chromosomal regions that are late replicating, transcriptionally inactive and stain more densely with Feulgen, a dye that binds DNA.

heterogeneous nuclear RNA (hnRNA) Primary transcripts before processing containing introns and exons. Initially named because of its variable size and location.

heteroplasmic Where both mutant and wild-type mitochondrial genomes are found in the same individual.

heterozygous An individual who has two different alleles at a specified locus.

heterozygous advantage When an allele that is deleterious in the homozygous state is advantageous in the heterozygous state.

highly repetitive DNA Simple sequences repeated up to 10^6 per genome.

histones Highly conserved basic proteins that bind DNA and organise the formation of nucleosomes.

homologs Genes that are similar by virtue of descent from a common ancestor.

homoplasmic Where the mitochondrial genomes within an individual are all identical.

homozygous An individual with identical alleles at a locus.

housekeeping gene A gene that encodes a protein performing a basic function common to most cells.

human leucocyte antigens (HLA) Antigens found on the surface of cells that signal whether the cell is self or non-self.

huntingtin The protein affected by mutation in Huntington's disease. Also used to describe the encoding gene.

hybridisation The formation of a hybrid DNA molecule (or RNA/DNA duplex) by the base pairing of complementary strands that originated from different sources.

hydrops fetalis A condition lethal at or before birth caused by a complete absence of α-globin.

IDDM See insulin-dependent diabetes mellitus.

identical by descent (IBD) A situation where two individuals in a kindred have inherited the same alleles at a locus from a common ancestor.

identical by state (IBS) A situation where two individuals in a kindred have the same alleles at a locus but it cannot be demonstrated that they are identical by descent.

immortal The property of a cancer cell line that enables it to be propagated indefinitely in culture.

informative meiosis A meiosis where it is possible to distinguish between parental and recombinant chromosomes in the progeny.

initial gene index Name given by the IHGSC to the collection of sequences identified or suspected of forming genes.

in silico Analysis carried out by computer.

in situ **hybridisation** Hybridising a nucleic acid probe to chromosomes spread on a slide. Usually a metaphase spread is used but recently also carried out with interphase spreads to increase resolution.

insulin-dependent diabetes mellitus Also known as type 1 diabetes. Form of diabetes mellitus that responds to insulin. Usually shows juvenile onset and is caused by the autoimmune destruction of the Islets of Langerhans in the pancreas.

intermediate repeated DNA DNA sequences that are repeated between 100 and 10^5 times per genome.

interphase The period of the cell cycle between nuclear divisions.

intron The portion of a gene that is transcribed but removed during splicing.

in vivo **gene therapy** Where the transgene is introduced directly into the target cells of the patient's body.

karyotype The number, size and shape of chromosomes in a somatic cell.

kb Kilobase. A thousand base pairs.

ketone bodies Group of chemical compounds, such as β-hydroxybutyrate, acetoacetone and acetone, that are metabolites of acetyl coA and which accumulate in diabetes.

ketonemia Accumulation of ketone bodies in the bloodstream in diabetes.

ketonuria Accumulation of ketone bodies in the urine in diabetes.

ketosis Accumulation of ketone bodies in diabetes.

kindred A group of people that are related genetically or by marriage.

kinetechore A protein structure that assembles at the centromere to which the mitotic or meiotic spindle attaches.

knockout mice Mice where specific genes have been deleted.

LINE1 (L1) A repeated sequence in the LINE category.

lariat A splicing intermediate where the intron resembles a stem–loop structure or lasso.

liability The combined total of genetic and environmental effects that predispose an individual to a multifactorial disease.

linkage disequilibrium The tendency of particular alleles at one locus to occur with particular alleles at a second closely linked locus.

linkage The propensity of two genetic markers to be coinherited through meiosis. It indicates that the markers are located close together on the same chromosome.

lipoplex Complex formed between cationic lipids and DNA used for gene therapy.

liposome Spheres consisting of lipid molecules surrounding an aqueous interior. Used to encapsulate DNA for gene therapy.

locus A unique chromosomal region that corresponds to a gene or some other DNA sequence.

LOD score The logarithm of the ratio of odds that two loci are linked with a specified recombination fraction θ, to the odds that they are unlinked. An LOD score of 3 or more is required for linkage to be significant.

long interspersed elements (LINEs) A category of intermediate repeated sequence.

long-range restriction map See rare cutter.

long terminal repeat (LTR) Repeated sequences at each end of a retrovirus.

loss of heterozygosity (LOH) When a region of a chromosome becomes homozygous for alleles that were heterozygous.

Lyonisation See X-chromosome inactivation.

lysosome A membrane-bound organelle that contains digestive enzymes that break down material taken in by phagocytosis or recycle cellular components.

major disorders Disorders that are both relatively common and severe in their consequences.

maximum likelihood score (MLS) The most likely of a series of alternatives judged by which gives the highest LOD score. For example, the MLS for different values of recombination fraction, θ, indicates the most likely map distance between two loci.

Mb Megabase. One million base pairs.

meconium ileus Obstruction of the bowel in the newborn. Often one of the symptoms of cystic fibrosis.

meiosis A process consisting of two successive nuclear divisions which results in reduction of chromosome number from the diploid chromosome complement to the haploid complement. During the first meiotic prophase recombination takes place between homologous chromosomes.

Mendelian inheritance The pattern of segregation of a disease in a family or kindred that conforms to Mendel's laws of inheritance. Normally taken to indicate that the disease is monogenic.

mesolithic A period of the Stone Age between the Paleolithic and Neolithic periods.

metacentric A chromosome that has a centrally located centromere, such that both arms are of equal size.

microsatellites Tandem repeats of a short sequence 2–4 nucleotides in length found at many different locations in the genome.

minisatellites A class of highly repetitive sequences that consist of sequences between 10 and 100 bp long repeated in tandem arrays that vary in size from 0.5 to 40 kb. They tend to occur near telomeres although they have been found elsewhere. Also known as variable number tandem repeats (VNTRs).

missense mutation Nucleotide substitution that results in an altered amino acid sequence in the encoded protein.

mitogen A substance that stimulates cell division.

mitosis The process of nuclear division in which the daughter cells contain the same number of chromosomes as the mother cell.

mitotic crossing-over Recombination that occurs in mitosis of somatic cells. Results in homozygosity in daughter cells of the region of the chromosome distal to the cross-over point.

mobile genetic element A sequence of DNA that can move from one chromosomal location to another by transposition. Characteristically they are flanked by short direct repeats.

model organism Any organism used because features of its biology, life cycle or existing knowledge make experimentation easier than a more complex organism. Use of model organisms is based on the supposition that fundamental biological processes are conserved in evolution.

MODY (maturity onset diabetes of the young) A form of insulin-dependent diabetes that affects teenagers and young adults.

monogenic See single-gene defect.

monosomic A somatic cell with one copy of a chromosome.

monozygotic twins Derived from the same fertilised egg – otherwise known as identical twins.

mortal The propensity of human cells to die after about 50 generations in tissue culture.

mosaic Where cells in the same organism have a different genetic constitution.

mosaicism When not all cells in the body are genetically identical. This may come about through a mutation in early development and may result in either the germline or somatic cells being affected.

multifactorial disease A disease caused by the interaction of the environment and polygenes or oligogenes.

multiplex PCR Where several PCR reactions are carried out together using more than one pair of primers.

multiplicative interaction Where the effect on phenotype of multiple alleles is the product of their individual effects. Multiplicative interaction is a special case of epistasis.

multipoint mapping Testing for linkage between genetic markers and a disease by using more than one marker at a time.

multiregional evolution hypothesis Theory of human evolution that states that *Homo sapiens* evolved independently from *Homo erectus* on multiple occasions in different parts of the world.

myoblasts Mononucleate cells that are normally quiescent but can divide and fuse with muscle fibers to repair damage.

myotonic dystrophy (MD) Neuromuscular disorder caused by a trinucleotide repeat expansion.

neolithic Sometimes called the New Stone Age. A period of human history which started about 10 000 years ago, marked by the appearance of agriculture, settled communities and polished stone tools.

neural crests A ridge of ectoderm that forms above the neural tube during early embryogenesis. Subsequently the cells migrate and develop into the dorsal root ganglia of the sensory nervous system, the adrenal medulla and some skeletal elements of the face.

neutral allele One of the alleles of a neutral polymorphism.

neutral molecular polymorphism A polymorphism for a DNA sequence that has no effect on the genotype.

neutral polymorphism A polymorphism in which the different alleles do not have a discernible effect on the fitness of the organism.

neutral theory of evolution Theory that postulates that the main mechanism that generates nucleotide substitution is the fixation of neutral alleles through the effect of genetic drift.

non-disjunction The failure of homologous chromosomes to segregate at meiosis, resulting in one daughter cell with two copies and one daughter cell with no copies of the chromosome in question.

non-insulin-dependent diabetes mellitus Form of diabetes that does not respond to insulin (type 2 diabetes). Typically shows middle-aged onset.

non-obese diabetic mouse (NOD) A mouse that spontaneously develops a disease that closely resembles IDDM in humans.

non-parametric A method of mapping disease genes that does not require the formulation of models with specified parameters.

nonsense codon A codon that does not encode an amino acid; causes translation to terminate.

nonsense mutation Nucleotide substitution that results in a nonsense codon that results in a truncated protein. Also result from frameshift mutations, as the incorrect reading frame will contain nonsense codons.

northern blotting A procedure similar to Southern blotting but RNA not DNA is fractionated and hybridised to the probe. Allows the size and expression of mRNA molecules to be studied.

nucleosome Basic unit of DNA packing in which DNA winds around a tetramer of the histones H1, H2A, H2B, H3 and H4 to form a beads-on-a-string appearance in the electron microscope.

Oct box The sequence ATTTGCAT that can be found in the promoter region genes, which binds the transcription factors Oct-1 or Oct-2 to stimulate transcription.

oligogene A gene which makes a large contribution to the aetiology of a disorder.

oligogenic disorder A disorder caused by the action of a few genes.

oligonucleotide A polymer of nucleotides of small size (5–100 nucleotides).

oncogene A gene the mutation of which is implicated in the aetiology of cancer.

oogonia The diploid precursors of female gametes.

open reading frame (ORF) A significantly long sequence of codons in one of the six possible reading frames in a DNA sequence that does not contain a nonsense codon.

ortholog Genes in different organisms that are similar by virtue of having descended from a common ancestor. Usually assumed to carry out similar functions.

OTTO System developed by Celera Genomics to identify sequences in the genome that are genes.

ordered clone library A clone library with overlapping inserts where overlap between all the clones has been determined.

Out-of-Africa hypothesis Theory of human evolution that states that modern humans are descended from a small group that emerged from Africa 100 kya.

oxidative phosphorylation The production of ATP at the inner membrane of the mitochondria. Electrons are passed along the respiratory chain in the mitochondria eventually interacting with oxygen and protons to form water. The energy released in the process is used to pump protons across the inner mitochondrial membrane. As the protons re-enter the mitochondrial matrix the energy of the proton gradient drives the synthesis of ATP.

p arm The shorter arm of a chromosome in a metaphase spread.

P1-derived artificial chromosome A cloning vector based on generalised transduction using phage P1. Can accept inserts of approximately 100 kb and is liable to fewer cloning artefacts than YAC vectors.

paleolithic Sometimes called the Old Stone Age. Phase of human history prior to the advent of agriculture and settled communities. Marked by a hunter–gatherer lifestyle and chipped stone tools.

paralog Gene copies formed by duplication within the genome of the same organism.

parametric A model of inheritance that can be tested by LOD score analysis that requires the specification of parameters.

parasitic DNA See selfish DNA.

pathophysiology The biochemical or cellular changes that cause the symptoms of a disease.

pedigree A representation of the ancestral relationship between individuals related genetically or by marriage.

penetrance The frequency with which a particular genotype manifests itself in the phenotype.

phase Specifies whether particular alleles at adjacent loci are on the same (*cis*) or different (*trans*) chromosomes.

phenocopy The occurrence of a disorder casued by an environmental factor with the same symptoms as an inherited disorder.

phylogenetic tree An attempt to describe the history of populations since they diverged from a common ancestral population. The order in which the branches split describes the order in which populations diverged, and the length of the branches describes the time since the divergence.

physical map A genomic map based on the actual location of DNA sequences.

polygene One of several genes each of which makes a small contribution to the aetiology of a disorder.

polygenic disorder A disorder caused by the action of several genes.

polymerase chain reaction (PCR) A technique for amplifying DNA *in vitro* using a thermostable DNA polymerase and primers that anneal at sites which flank the region to be amplified.

polymorphic A locus that is heterozygous at a significant frequency in a population.

polymorphism information content (PIC) Measure of the informativeness of a genetic marker.

population bottleneck When a previously large population contracts to a small number and then expands to its previous numbers. As a result the gene pool may be less diverse, and gene frequencies may be altered because of the effects of sampling and genetic drift.

population stratification The presence in a population of different ethnic subgroups that may have different susceptibilities to complex disease.

positional candidate cloning Cloning a disease gene by identifying a biologically relevant gene located near the map position of the disease gene.

positional cloning The cloning of a gene using no other information apart from its chromosomal location.

power The probability that a study will identify the involvement of a risk allele, with a given locus-specific λ or genotype relative risk, to a defined statistical threshold.

premutation A mutation which has no phenotypic effects but which increases the odds of a trinucleotide repeat expansion in a subsequent generation.

primary RNA transcript The RNA molecule before splicing that contains both introns and exons.

primers Oligonucleotides that anneal to template DNA to prime synthesis mediated by DNA polymerase.

principal component analysis Mathematical procedure to represent multivariate data by producing a synthetic map.

private mutation A mutation that is unique to a particular family or individual.

proband An individual suffering from a genetic disease who is the first member of a pedigree to come to medical attention.

processed pseudogene A pseudogene that lacks introns and is therefore thought to have arisen by retrostransposition.

prodrug A form of a drug which requires some kind of modification before it is active.

promoter Part of the gene upstream of the transcribed region that contains the elements necessary for a basal level of transcription.

protein kinase An enzyme that phosphorylates another protein and so regulates its activity.

proteome The complete set of proteins that can be encoded by a genome.

proximal Towards the centromere compared to the chromosomal point of reference.

pseudogene A DNA sequence that closely resembles a functional gene but has been rendered non-functional by an inactivating mutation.

pulsed field gel electrophoresis (PFGE) A special adaptation of agarose gel electrophoresis, where the direction of the electric field is periodically changed. Used to construct long-range restriction maps. See also rare cutter.

q arm The longer arm of a chromosome in a metaphase spread.

Q-bands Bands produced in metaphase chromosomes using the fluorophore quinacrine.

quantitative trait locus (QTL) A locus contributing towards a quantitative phenotype.

quinacrine A dye that binds DNA in chromosomes and fluoresces in UV light.

radiation hybrid mapping Method of mapping that is based on the frequency with which markers are co-retained on the same genomic fragment after fragmentation by X-rays.

rare cutter A restriction enzyme that has target sites in human DNA that are more widely spaced than normal. Used in conjunction with PFGE to construct long-range restriction maps, that is, maps that extend over hundreds of kilobases.

ras A 21 kilodalton protein with GTPase activity that acts as a molecular switch in signal transduction pathways involving external mitogenic stimulation. The *ras* gene is a protooncogene because mutations that lock it into its active state cause cellular transformation. Mutations to this gene have been implicated in a wide variety of cancers.

reassociation The reformation of a duplex DNA molecule after the original molecule has been denatured by heat or other physical treatment.

recombinant protein A protein produced by the expression of a gene in a heterologous host cell.

recombination fraction The proportion of chromosomes that are recombinant, i.e. of a different constitution from parent chromosomes.

recurrence risk See relative risk.

relative risk (λ) The ratio of the frequency of a multifactorial disease in the relatives of an affected person compared to its rate in the general population.

replacement substitution A nucleotide substitution that does result in an amino acid substitution in the encoded protein.

response element A region upstream of a gene that activates its transcription in response to a general environmental influence such as heat, serum or cAMP.

restriction fragment length polymorphism (RFLP) A genetic marker based on the presence or absence of a target for a restriction enzyme due to a polymorphism at a single base pair.

retinoblast Retinal epithelial cells that give rise to neuroblasts and neuroglial precursor cells.

retinoblastoma A tumour of the eye originating from retinal cells. May be sporadic or inherited due to mutation of the *RB* gene.

retrotransposition A process of transposition in which RNA serves as a template for reverse transcriptase. The resulting cDNA is then inserted into the genome at a new location.

retrovirus A virus that has an RNA genome. The mature virion contains reverse transcriptase encoded by the RNA genome. This copies the RNA into cDNA which is then inserted into the host genome.

reverse transcriptase (RNA-dependent DNA polymerase). An enzyme that copies RNA into DNA; the product is known as cDNA.

RH mapping See radiation hybrid mapping.

risk allele Allele that increases the risk of a complex disease.

RT-PCR A form of PCR that amplifies RNA by first converting it to cDNA using reverse transcriptase.

S phase The period in the cell cycle during which nuclear DNA is replicated.

sarcolemma The flexible sheath that surrounds muscle fibres.

satellite DNA See highly repetitive DNA.

satellite peak See buoyant density ultracentrifugation.

scaffold (in sequencing) Use of long-range maps or other features to link together adjacent contigs and locate them along the length of a chromosome.

segregation The transmission of alleles through a pedigree in a way that follows Mendel's laws of inheritance.

selection The process which leads to an increase in the frequency of a favourable allele and a decrease in the frequency of a deleterious allele.

selfish DNA DNA that does not contribute to the phenotype of the organism, but evolves to increase its copy number in the genome by transposition.

sequence tagged site (STS) A site that is identified from its sequence using a PCR reaction. Can be used as a marker in physical, genetic and genomic maps.

serotype The subdivision of bacteria or viruses based on their anitigenic properties.

short tandem repeat (STR) An array of short sequences each normally 2–4 nucleotides in length. See also minisatellite.

shotgun sequencing Method of sequencing DNA by first generating random fragments. These are sequenced separately before assembly of the whole sequence by overlapping the sequences in the individual fragments.

sibs Individuals who have the same parents, e.g. brother and sister.

silencer A region upstream of a gene that turns off transcription in response to a particular signal.

silent substitution A nucleotide substitution that does not result in an amino acid substitution in the encoded protein because of the redundancy in the genetic code.

SINES Short interspersed elements, a category of intermediate repeated sequence.

single-gene defect A disorder or disease that is caused by a mutation in a single gene.

single nucleotide polymorphism (SNP) Single nucleotide that shows significant variation (above 1%) in a population. They occur approximately every 1.3 kb in the human genome.

single sequence DNA A sequence of DNA that is not repeated in the genome.

somatic Cells in the body that are not part of the germline.

somatic gene therapy Where the attempt to repair the genetic defect is restricted to the somatic cells of a person affected by the disease.

Southern blotting A form of hybridisation where DNA from the target is digested with restriction enzymes and fractionated by agarose gel electrophoresis. The DNA is transferred by capillary action to a nitrocellulose or nylon membrane. The membrane is then hybridised to a radioactively labelled probe and the molecular weight of the hybridising fragments revealed by autoradiography.

spacer region Tracts of DNA that apparently have no function and which separate one gene from the next.

splice acceptor site The junction between the dinucleotide AG at the end of an intron and the start of the next exon.

splice donor site The junction between the end of an exon and the dinucleotide GT at the start of the next intron.

splicing The process by which introns from the primary transcript are removed in the nucleus. The product of splicing of an RNA polymerase II transcript is the mRNA molecule, which only contains introns.

stringency During Southern hybridisation duplex formation can still occur even if there is a mismatch between the sequence of the probe and the target sequence. The amount of mismatch that can be tolerated is dependent on the physiochemical conditions during the hybridisation process. For probes longer than about 200 bp low salt ($0.1 \times$ SSC) and high temperature ($65°C$) are high stringency conditions which only allow duplex formation if the match between the probe and target is perfect or near perfect. Higher salt and lower temperature allow a greater degree of mismatch to be tolerated and are said to be low stringency conditions. If the probe is an oligonucleotide the exact conditions for high stringency hybridisation are calculated from its length and base content.

structural genomics The study of the structure of the proteins encoded by a genome.

STS content map A map based on the order of STSs in an ordered clone library.

submetacentric A chromosome with the centromere nearer one end compared to the other. Often appears J-shaped in metaphase spreads.

susceptibility allele See risk allele.

synteny Conservation of gene location on chromosomes of different organisms.

synthetic map See principal component analysis.

tandem repeat array An identical or near identical sequence repeated so that each copy lies immediately adjacent to the next and arranged so that the end of one unit abuts the start of the next.

TATA-binding protein (TBP) The protein that recognises the TATA box to initiate formation of the transcription complex.

telocentric A chromosome where the centromere is at one end.

telomerase The enzyme that adds the telomere. It contains an RNA molecule that serves as the template for DNA synthesis.

telomere A special structure at the ends of chromosome that in humans consists of a tandem repeat of the sequence TTAGGG and ends in a 3' extension.

thalassaemias Anaemias caused by an imbalance of globin synthesis.

tiling path or array The set of clones that represents the sequence of a region or entire chromosome with the minimum overlap.

topology The geometrical arrangement of sequences in clones.

trans See phase.

transcription factor A protein that binds upstream of a gene to facilitate or stimulate its transcription.

transcriptome The sum total of all the DNA sequences in a genome that are transcribed.

transgene A DNA molecule introduced into a cell to alter its genetic constitution.

transgenic An organism that has a genome modified by genetic engineering.

translocation An exchange of segments between non-homologous chromosomes.

transmission disequilibrium test (TDT) Ascertains whether a parent who is heterozygous for an associated and a non-associated allele transmits the associated allele more often to affected offspring.

transplacement An organism where genes have been replaced with genes carrying specific mutations.

transposition The process by which a mobile genetic element copies itself and inserts the copy in a new location.

transposon A sequence of DNA that has the capacity to move to a new position in the genome. Usually the process involves duplication, so that a copy of the sequence remains in the original location. Thus transposons are normally repetitive elements.

trinucleotide repeat expansion (TRE) A mutation caused by the increase in the number of copies of a repeated trinucleotide.

trisomic A somatic cell with three copies of a chromosome.

tropism The propensity of certain viruses to infect particular cell types.

tumour suppressor Genes which negatively regulate cell division. Mutations to the genes are recessive at a cellular level but show a dominant pattern of inheritance.

type 1 diabetes See insulin-dependent diabetes.

type 2 diabetes See non-insulin-dependent diabetes.

unbalanced translocation The non-reciprocal duplication of a chromosome segment.

univalent A chromosome with one chromatid.

untranslated region The region of the mRNA molecule that is not translated. The 5' UTR is the part of the mRNA molecule upstream of the AUG start codon. The 3' UTR is the part of the mRNA molecule downstream of the stop codon.

upstream A region of the DNA molecule that lies 5' to the point of reference.

variance A parameter that measures the total phenotypic variation in a population.

vehicle The physical means by which a transgene is introduced into a target cell in gene therapy.

virion A complete virus particle.

X chromosome One of the pair of sex chromosomes. XX individuals are female, XY are male.

X-chromosome inactivation The clonal inactivation of one of the X chromosomes in females during development.

X-linked A trait showing a characteristic pattern of segregation indicating that the gene in question is located on the X chromosome. See Section 1.3.

yeast artificial chromosome (YAC) A cloning vector using a yeast host cell that can accept very large inserts of DNA (~1 Mb).

zoo blotting Southern blotting where human DNA is used to hybridise DNA from a range of other mammals.

Index

AAV 235, 240, 245
ABC membrane transporter 147
ADA deficiency 226, 241
additive alleles 191, 221, 334
adenoassociated virus, *see* AAV
adenovirus 233–5
admixture 278, 334
adoption studies 183
affected pedigree member 198, 334
affected sib pair 198–202, 205–6, 211, 219, 221, 334
African Eve 286–7
ageing 21–22
agriculture 283–4
alchoholism 14
α-1 antitrypsin deficiency 4, 6
α-globin 5–6, 33, 37, 39–40, 50, 159–63, 176, 178,
allele frequencies 190–1, 195, 205, 211, 221, 273–85, 334
allele specific oligonucleotide *see* ASO
Alu element 38, 48, 50, 76–7, 105, 107–10, 334
Alzheimer's disease 8, 14, 181, 216–21, 223
amniocentesis 254, 317, 334
amplification refractory mutation system, *see* ARMS
amyloid plaques 216
amyloid precursor protein 216
anencephaly 14
Angelman syndrome 13
anticipation 13, 23, 154, 156, 335
anti-oncogene, *see* tumour suppressor

antisense mRNA/oligonucleotide 228–43, 335
APC (adenomatous polyposis coli) 8, 172, 340
ApoE (apolipoprotein E) 211–3, 218–20,
ApoE 195, 211, 213, 218–20, 223, 335
apoptosis 165, 334
APP 195, 217
Arabidopsis thalania 103
ARMS test 255, 257–8, 335
Ashkenazi Jews 4, 6, 9, 171, 297–8, 319, 300,
ASO (allele specific oligonucleotide) 255–6, 269, 335
association studies, *see* population association studies
AT (ataxia telangiectasia) 173–4, 179
asthma 14, 180
autism 184–5
autoimmune disease 207
autosomal dominant disorders 2, 7–9, 23, 335
autosomal recessive disorders 2–6, 23, 335
autosome 42, 48, 336

BAC vectors (bacterial artificial chromosome) 72, 75, 94–7, 336
balanced polymorphism 296, 336
balanced translocation 15–16, 336
banding (of chromosomes) 42–4, 336
  G-bands 43, 48, 142, 341
  Q-bands 43, 142, 350

Basque region 283
Baylor College of Medicine, Human
    Genome Sequencing Center 92
Becker muscular dystrophy, *see* BMD
β-globin
    β-thalassaemia 5–6, 159–64, 176, 178,
      296–319
    constituent of haemoglobin 5, 34, 159
    δ globin 160–3
    DNA chip 266
    further reading 178
    γ-globin 160–3
    gene cluster 34, 159–64, 176, 178
    molecular clock 276
    mutations and deletions 160–3
    rabbit 145
    sickle cell anaemia/disease 4–6,
      159–60, 296 319
bias of ascertainment 171, 217
bivalent 43, 168, 336
blood groups 281
BMD (Becker muscular dystrophy) 12,
    152, 154, 239
bootstrap analysis 281
bottleneck, *see* population bottleneck
bouyant density ultracentrifugation
    39, 336
*BRCA*1 8, 170–2, 176, 179, 184, 195, 197,
    264–6, 270, 295, 297, 320
*BRCA*2 8, 170–2, 176, 179, 184, 195, 197,
    264–6, 295, 297, 320
breast cancer, *see also BRCA*1, *BRCA*2,
    8–9, 169–72

CA repeat loci 41, 65–6
CAAT box 30, 336
Caenorhabditis elegans 103
cancer 164–75
candidate gene 145–6, 194, 221
cap 31, 336
carrier screening 318–9
Cavalli-Sforza, Luca 282
CBAVD (congenital absence of the *vas*
    deferens) 5, 147
CD4 233
CDK inhibitors 165–6, 169
CDK2 (cell dependent protein kinase)
    165–6
CDK4 165–6
cDNA library 79, 141, 145, 149, 336
Celera Genomics 87–8, 92, 94, 97–100,
    104–5, 112, 115–16, 118, 135, 137
cell cycle 165–6
cell cycle engine 165, 337
CENP-B binding protein 40
centimorgan 58, 337

centiray 78
centromere 39–40, 43, 337
CEPH families 59–60, 66, 337
CF (cystic fibrosis, CFTR) 4,5, 146–52,
    177, 237–9, 250, 256–9, 283–4,
    295–7
CFTR, *see also* CF, 147, 150–2
Charlie transposons 105, 108
checkpoints 165–6, 337
chemical mismatch cleavage 263
chorionic villus sampling 254, 337
chromatid 43, 168, 337
chromatin 45–8, 51, 337
chromosomal defects, imbalances and
    mutations 1, 14–17, 24
chromosome
  acrocentric 43
  banding 42–4
  evolution 111–13
  metacentric 43
  nomenclature 42–4
  paints 45, plate 3
  structure 42–7
  submetacentric 43
  tracking 251, 268
  walking 143, 148–50
chronic granulomatous disease 153
CIP1 165–6
cladistic analysis 281, 337
cleft lip/palette 14
clone map 67–77, 81–2, 84, 94–7, 337
CNGA3/CNGB 121
Collins, Francis 150
colorectal cancer 172–3
Committee on Safety of Medicines 314
complex disease 13–14, 23, 26,
    180–224, 298, 338
compound heterozygote 257, 337
concordance 183, 337
congenital absence of the *vas deferens*,
    see CF
congenital disorders/malformations
    1–2, 14, 338
congenital heart disease 14
congenital muscular dystrophy 154
contig 69–70, 76, 92, 94–9
continuous variation 186–8
Cooperative Human Linkage Centre
    67–8
COS cells 145
cosmid 70, 144, 337
cousin marriages 3–4
CPEO 19, 21
CpG island 32, 80–1, 101–2, 105, 107,
    146, 149, 156, 338
cri-du-chat syndrome 16

cyclin dependent protein kinase, *see* CDK
cyclins A/D/E 165–6
cystic fibrosis, *see* CF
cytogenetic rearrangement 142
cytokine 205, 338

Darwinian fitness 274, 338
DCC 172
denaturing gradient gel electrophoresis,
    see DGGE
Dentatorubral-pallidoluysian atrophy 155
DGGE (denaturing gradient gel
    electrophoresis) 262–3, 269
diabetes mellitus, *see also* IDDM,
    MODY and NIDDM, 203–4,
    207–8, 339
disease frequency 2, 4, 8, 10, 16, 295–8
displacement (t) 191–2, 196, 221
dizygotic twins 183–4
DMAHP 159
DMD (Duchenne muscular dystrophy)
    10–11, 152–4, 175, 178, 239, 260
DMPK 159, 339
DNA
  highly repetitive 29, 39, 339
  intermediate repetitive 29, 36–9, 344
  loops 30
  methylation 32, 80, 156, 339
  mtDNA 17–22, 26, 272, 285–90, 299,
    301–2
  parasitic 28, 36
  rDNA 142
  repair 166
  repetitive 28–9, 36–9, 105, 107–10
  satellite 39–41, 50–1
  sequencing for mutation detection
    266–7, *see also* Human Genome
    Project
  selfish 28, 36, 49, 352
  single sequence 28, 353
DNA chips, *see* oligonucleotide
    microarrays
DNA fingerprinting/profiling 304–11, 339
DOE (Department of Energy, USA) 22
DOE Joint Genome Sequencing Center 92
domains (protein) 34, 119–120
dominance 191–2, 195–6
dominant negative mutations 243, 339
Down's syndrome 15–16, 318
Drosophila melanogaster, *see also* fly, 103
drug
  discovery 129–30
  individual variation in response to
    treatments 130–5
  target hits 129
  target validation 130

Duchenne muscular dystrophy, *see* DMD
dynamic mutations, *see* TRE
dyslexia 14, 184
dystrophia myotonica, *see* MD
    (myotonic dystrophy)
dystrophin, *see also* DMD, 33, 153–4

elastin plate 1
ELSI (ethical legal and social issues),
    see also ethical issues, 313
EMBL 102
endocytosis 238, 339
endosomes 238, 340
enhancer 29–31, 339
Ensembl website 101–3
environmental effects 12, 181, 183,
    186–93
enzyme mismatch cleavage 263–4, 269
episome 228, 340
epistasis 191, 202, 221, 340
Escherichia coli 103
EST 55, 79, 84, 86, 97, 114–15, 327, 340
ethical issues 312–33
  carrier screening 317–18
  confidentiality 324
  employment 323
  eugenics 314–15
  gene testing 315–17, 332
  gene therapy 328–9
  individual responsibility 324–5
  informed consent 316
  insurance 322–3, 333
  late-onset disorders 319–21
  neonatal screening 318–19
  patents 325–8, 333
  prenatal testing 317–18
  presymptomatic testing 319–22
euchromatin 47–8, 98, 340
exon trapping 145, 157, 340
exons 31, 105, 124–7, 340
  trapping 144–5
expressed sequence tag, *see* EST
expression map 79
expression profiling 125–7
expressivity 12, 23, 340
extraversion 185

factor VIII 33, 141, 144, 240–1
factor IX 240
familial adenomatous polyposis, *see* FAP
familial Alzheimer's disease, *see*
    Alzheimer's disease
familial hypercholesterolaemia 8, 195,
    321
family studies 182
Fanconi's anaemia/syndrome 226

FAP (familial adenomatous polyposis)
8, 172, 195, 340
favism 159
Feldhoffer 288–9
Feulgen stain 47, 340
fingerprint clone contig 94, 97
fingerprinting
clone fingerprinting 74–7, 94–7
DNA/genetic fingerprinting 304–11
Finnish population 214–15
FISH (fluorescence *in situ*
hybridisation) 44–5, 68, 94, 96,
341, plates 1–3
fluorophore 45, 341
fly (*Drosophila melenogaster*) 103, 116,
118, 119–21
FMR1 156
founder effects/population 280, 297–8,
300, 341
founder mutation, 209–10, 215, 296–7
fragile site 155–6
FRA16A 155–6
FRAX11B 155–6
FRAXA 155–6
FRAXE 155–6
FRAXF 155–6
fragile-X syndrome 10–11, 155–6
framework 69–70, 92
Fugu rubripes 103

GAIC (Genetics and Insurance
Committee) 314, 322
gallstones 14
gametogenesis 77, 341
ganciclovir 242
ganglioside 6, 341
gap junction 237, 242, 341
Gaucher's disease 226
GC box 30, 341
Gelsinger, James 234
gene 29–32
defined 29, 341
duplication 34
expression 29–32
families 33–34, 342
housekeeping 31, 343
identification 113–15, 124–7, 144–7
knockout 104, 345
non-protein coding 35
number 32–3, 115–16
structure 33, 105
size 32–3, 105
gene testing/screening 249–70, 315–22
gene therapy 140, 225–48, 328–9, 342
Gene Therapy Advisory Committee
(GTAC) 314

Genethon laboratory 22, 65, 95
genetic drift (random drift) 275–8, 279,
343
genetic heterogeneity 11
genetic map 58–69, 81–2, 84–5, 110, 342
correlation of physical and genetic
distance 110
Genetics and Insurance Committee, *see*
GAIC
genome, *see also* Human Genome
Project, 342
GC content 106
maps 52–86
sequence 87–139
statistics 105
structure 27–51, 104–20
genome browsers 101–3
genome scan 198–202, 205–6, 218–19,
221, 224, 342
genomic imprinting 13, 23, 342, plate 1
genomics 120, 342
functional 120, 341
structural 120, 353
genotype relative risk, *see* GRR
GENSCAN 102, 114
giemsa 42–3, 342
glucose-6-phosphate dehydrogenase
deficiency 133, 159
glycolipid 6, 342
Golgi appatatus 152
growth factors/growth factor receptors
164, 342
GRR (genotype relative risk) 193, 205,
211
GTAC (Gene Therapy Advisory
Committee) 314
GTPase–activating protein (GAP) 174
guanosine exchange factor (GEF) 174
Guthrie 6-day blood test 250, 254

$5HT_{3A}$ 13
haemochromatosis 4, 6
haemoglobinopathies 5–6, 159–64, 176,
296
haemoglobin 5, 159
Hb A 162
Hb anti-Lepore 164
Hb Barts 163
Hb F 162
Hb Gower 162
Hb H 162–3
Hb Indian 161
Hb Lepore 161, 164
Hb Portland 162
Hb S 296
haemophilia A and B 10, 240–1, 245, 248

haplotype 59, 208–10, 343
happiness 185
Hardy–Weinberg equilibrium 2, 215,
    273–4, 343
HBC (hereditary breast cancer), *see*
    breast cancer
HD (Huntington's disease) 7–8, 157–8,
    209, 267–8
heart disease 195
hereditary breast cancer, *see* HBC
hereditary non-polyposis colorectal
    cancer, *see* HNPCC
Hereditary persistence of foetal
    haemoglobin, *see* HPFH
heritability 186–7
  locus specific 192–3
herpes 244
heterochromatin 47, 98, 343
heteroplasmy 19–20, 343
heterozygous advantage 4–6, 160,
    296–7, 343
HFEA (Human Embryology and
    Fertilisation Authority) 314
HGC (Human Genetics Commission)
    314, 322, 332–3
histones 45–6, 343
HIV 242
HIV *TAT* intron 145
HLA locus 141, 144–6, 150–1, 154, 171,
    201–5, 207, 240, 243–4, 288, 298,
    343
HNPCC (hereditary non-polyposis
    colorectal cancer) 8, 172–3,
    195, 343
Homo erectus 272–3
Homo habilis 272–3
Homo sapiens 272–3
Homo sapiens neaderthalensis 272–3,
    288–90
homolog 118, 120–1, 130
homoplasmy 19–20
HPFH 163
HUGO, *see* Human Genome Project
human artificial chromosomes 40, 238
Human Embryology and Fertilisation
    Authority, *see* HFEA
human evolution 281–95
Human Genetics Commission (HGC)
    314, 322, 332–3
Human Genome Project 22–3, 87–139
  accuracy, 87, 98,
  analysis of sequence 104–20
  annotation 101–2
  clone-by-clone sequencing strategy
    88, 94–8
  coverage 88, 95, 98, 100, 102, 111
  completeness 87, 98
  draft sequence 87–8, 97–8

finished sequence 87–8, 97–8
  goals 22–3
  HUGO (Human Genome
    Organisation) 313, 332
  sequencing factories 91–2
  sequencing technology 89–91
  shotgun sequencing strategy 88–9,
    92–3, 98–100
Huntington's disease, *see* HD
Hurler's syndrome 226
hybridisation (to detect clone overlap)
    74–6
hydrops fetalis 160, 162, 344
hypertension 180, 189,
hyperthyroidism 14
hypophosmataemic rickets 9

IBD (identical by descent) 192,
    199–202, 344
identical by descent, *see* IBD
IBS (identical by state) 199, 344
IDDM (insulin dependent diabetes
    mellitus or type 1 diabetes) 184,
    188, 201–5, 207–9, 298, 325
IDDM1 202, 330
IDDM2 202, 209
IDDM4, 5 202
IGF2 209
IGI (initial gene index) 115, 344
IHGSC 88, 92, 94–8, 100, 104–5, 109,
    115, 118, 124, 130, 135
immortal cells 42, 344
informative meioses 58, 344
INK4a/b/c/c 165–6, 169
INS (insulin locus), *see* IDDM2
insulin 202–3
insulin dependent diabetes mellitus, *see*
    IDDM
insulin locus (*INS*), *see* IDDM2
integrated genomic map 81–3, 84–5
intelligence 185
introns 31, 105, 145, 344
IRP 148–9
ischaemic heart disease 14
islets of Langerhans 202–3
IT15 (huntingtin) 157–8, 343, *see also* HD

karyotype 43–5, 344
Kearn's Sayre syndrome 19, 21
ketone bodies 202, 344
ketonemia 202, 344
ketonuria 202, 344
ketosis 202, 344
Kimura, Motoo 276–7
kinetochore 43, 344
KIP1/2 165

Kleinfelter's syndrome 16
Knudson, Alfred 167
KRAS 172

λ, *see* phage λ and relative risk
L1, *see* LINES 37, 345
Lake Mungo 290
language 283
lariat 32, 345
LD, *see* linkage disequilibrium
LDL (low-density lipoprotein) receptor
    195
Leigh's syndrome 19–20
lentivirus 233
LHON 19–20
liability 189–90, 345
library, *see* cDNA library
Li-Fraumeni syndrome 169
limb girdle muscular dystrophy 154
LINES 37, 105, 107–8
linkage, guidelines for 201
linkage disequilibrium (LD) 208–16,
    221, 224, 293–5, 299, 303, 345
lipoplex 236, 238
liposome, *see also* gene therapy 236, 238
LM3 290
LOD score analysis 60–3, 195–8, 345
LOH, *see* loss of heterozygosity
long-range restriction map 68, 80–1, 86,
    148–9
loss of heterozygosity, (LOH) 167–70
Lyonisation 9, 142, 345

M13 vector 57
Machado–Joseph disease 155
major affective disorder 184
malaria 6, 160, 296
male breast cancer 9, 170
MalR 105
manic depression 180, 184
maps 52–86, *see also* expression maps,
    genetic maps, integrated genomic
    maps, physical maps, radiation
    hybrid maps
maturity onset diabetes of the young,
    see MODY
maximum likelihood score, *see* MLS
maximum parsimony 281
MD 8–9, 158–9, 347
MDR 232, 243
melanoma 169, 244
MELAS 19, 21
MER transposons 105, 108
mesolithic period 283, 346
microarray, *see* oligonucleotide
    microarray

microsatellites, *see also* STR, 40–1, 48,
    64–7, 84, 172, 199–200, 266–8, 346
migraine 14, 180
migration 278
minisatellites 40–1, 48, 63–4, 208, 305–7
MIR/MIR3 38
mismatch cleavage detection 263–4, 269
mismatch repair 172–3, 176
mitochondrial disorders 1, 17–22, 24–6
mitochondrial DNA polymorphisms
    285–90
mitochondrial genome, *see also* mtDNA
    under DNA, 18–22
mitogens 164, 346
mitotic crossing over 167–8, 346
MLH1 173, 195, 295
MLS (maximum likelihood/LOD score)
    62–3, 199–202, 345
mobile genetic elements 36–9, 346
model organisms 103–4
MODY (maturity onset diabetes of the
    young) 195, 203
molecular clock 275–6
monogenic disorders 1–13, 23, 25,
    140–79, 346
monozygotic twins 183–4, 217, 346
mortal cells 44
mosaicism 12, 23, 346
mouse 103, 108, 111–13, 115, 135
MRCA (most recent common ancestor)
    279, 289, 292–3
MSH2 173, 195
mtDNA, *see* DNA and mitochondrial
    genome
multifactorial disease/inheritance
    13–14, 180–224, 347
multiple sclerosis 14, 180
multiplex PCR, *see* PCR
multiplicative interactions 191, 347
multipoint mapping 347
multiregional hypothesis 272–3, 287, 347
mus musculus, *see* mouse
muscular dystrophy, *see* MD
mutation
    nomenclature 251–3
    screening and testing, *see* gene testing
    types 253
mutS 173
myoblasts 239, 347
myotonic dystrophy, *see* MD

NARP 19–20
National Center for Biological
    Information (NCBI) 25
Neanderthal man, *see* Homo sapiens
    neaderthalensis

neonatal screening 250, 318
neural crests 173, 347
neural tube defects 14
neurodegenerative disorders 157–8
neurofibrillary tangles 216
neurofibroma 173
neurofibromatosis type 1 8, 173, 176, 179
neurofibromin 173
neuroticism 185
neutral alleles/polymorphism 253, 271,
    276–8, 347
neutral molecular polymorphism 65, 347
neutral theory of evolution 276–8, 347
NF1 173
NIDDM (non-insulin dependent
    diabetes mellitus or type II
    diabetes) 14, 195, 203, 347
NIH (National Institute of Health,
    USA) 22
non-disjunction 14
non-insulin dependent diabetes mellitus,
    see NIDDM
non-obese diabetic mouse (NOD) 347
non-parametric analysis 194, 198–206,
    221 348
non-shared environment 183
nucleosomes 45–6, 48, 348
Nurembourg code 316

obesity 189
obligate carrier 10
OCT box 30, 348
oligogenes 13, 193, 348
oligonucleotide arrays 121–36, 138,
    265–6
oncogenes 164–5, 348
Online Mendelian Inheritance in Man
    (OMIN) 2, 25
ortholog 117–19
OTTO 115, 348
out-of-Africa hypothesis 272–3, 285–95,
    348
ovarian cancer 9, 170, 320
oxidative phosphorylation 17, 348

p53 166, 172, 174, 243
P450 131
PAC vector (P1-derived artificial
    chromosome) 72, 74, 349
pancreatic exocrine deficiency 5, 147
paralog 101, 118, 121, 130
parametric linkage analysis analysis
    194–8, 219, 221, 349
PCR 55–8, 349
    ePCR 95
    multiplex PCR 255, 347
    RT-PCR 145

Pearson's syndrome 19, 21
penetrance 11–12
peptic ulcer 14, 184
PFGE, see pulsed field gel electrophoresis
phage λ 73
pharmacogenetics/pharmacogenomics
    129–35, 138–9
phase 59, 349
phenocopy 12, 23, 349
phenylketonuria 4, 6, 12, 250, 318
phylogenetic trees 280–1, 282, 286
physical maps, see also clone maps,
    67–78
    correlation of physical and genetic
        distance 110
PIC (polymorphism information
    content) 64
PMS1 173, 195
PMS2 173, 195
POINTER 196
polycystic kidney disease 12
polydactyly 295
polygenes/polygenic inheritance 13,
    188–93, 221, 349
polymerase chain reaction, see PCR
polymorphism information content,
    see PIC
polymorphisms, see also SNPs 271–2
    mitochondrial DNA 285–90
    neutral 253, 271, 276–8, 347
    protein-coding genes 281–5, 299
    Y chromosome 290–2
population
    association studies 206–11, 221, 242
    bottleneck, 20, 212, 215, 279, 293–5
    coalescence time 279, 287
    effective size 278
    Finnish 214–16
    Icelandic 214–16
    isolated 214–16
    mode of exapansion 279
    stratification 206–8
    Yoruban 212
porphyria 295
positional candidate cloning 146, 175,
    350
positional cloning 52, 144–6, 175, 350
Prader Willi syndrome 13
premutation 156, 350
presenilin I/II 195, 217–18
primary transcript 31, 350
principle component analysis 283–4, 350
private mutations 251, 350
processed pseudogene 39, 110, 350
profound deafness 12
promoters 29–31, 350

protein truncation test 264–5, 269
proteome 116–20, 350
pseudogenes 34, 160, 277, 350
psychiatric disorders 14, 184
pUC vectors 70
pufferfish (*Fugu rubripes* and *Tetraodon nigroviridis*) 103, 115
pulsed field gel electrophoresis 68, 81, 86, 350
pyloric stenosis 182, 189

q-arm 43, 350
quantitative character 189
quinacrine 43, 351

radiation hybrid map, *see* RH map
random genetic drift, *see* genetic drift
ras 165, 173–4, 243, 351
rare cutting enzymes 80, 144, 351
rat 103
RB1 167, 169
reassociation 28, 351
recombinant protein 181, 351
recurrence risk, *see* relative risk (λ)
RefSeq database 103, 105, 114–15
relative risk (λ) 188, 351
    locus specific 192
repetitive DNA, *see* DNA
replacement substitution 277, 351
respiratory chain subunits 17–18
response element 29–31, 351
    CRE 31
    HRE 31
    SRE 31
restriction fragment length polymorphism, *see* RFLP
retinitis pigmentosa 12, 19
retinoblastoma, *see also* RB1, 8, 166–9, 176, 351
retinoblasts 167, 351
retrotransposition, *see* retrotransposon
retrotransposon 37–9, 109–10, 351
retrovirus, 38, 230–3, 351
RFLP (restriction fragment length polymorphism) 63, 144, 149, 250, 351
RHmap (radiation hybrid map) 67, 78, 81–4, 86, 351
rheumatoid arthritis 14, 180, 207
risk allele 189
RNA
    5s RNA 29, 35
    7SLRNA 35, 38
    18S RNA 35
    28S RNA 35
    45S RNA 35
    HnRNA 115, 343
    non-coding functional RNA 29, 35, 48
    processing 31
    rRNA 29, 35, 48
    splicing 35, 5
    tRNA 29, 38
RNA polymerase I 29, 35
RNA polymerase II 29, 35
RNA polymerase III 29, 35
rRNA (rDNA) 35, 142
rubella 204

Saccharomyces cerevisiae, *see also* yeast, 103
SACGT (Secretary's Advisory Committee on Genetic Testing) 323
Sanger Centre 92, 101, 103
sarcolemma 153–4, 352
scaffold, *see also* framework, 92, 352
schizophrenia 14, 180, 183
SCIDS (Severe Combined Immunodeficiency Syndrome) 241, 245, 248
Secretary's Advisory Committee on Genetic Testing, *see* SACGT
selection 274–5, 352
selective sweep 285, 291, 299
sequence tagged site, *see* STS
sex-linked disorders, *see* X–linked disorders
sickle cell disease/anaemia 4–6, 12, 275, 296, 319
signal recognition particle 38
signal transduction pathways 164–5
silencer 31, 352
silent substitution 277, 352
SINES, *see also alu* elements, 37–8, 105, 352
single-gene defects 1–13, 140–79, 352
single nucleotide polymorphisms, *see* SNPs
single stranded conformational polymorphism, *see* SSCP
small lung cell carcinoma 169
SNPs 100–3, 121–2, 127, 131–6, 211–14, 220–1, 293–5, 303
snRNP 31
Southern blot 146, 353
spacer region 25–6, 353
spina bifida 14
spinal and bulbar muscular atrophy 155
spinocerebellar ataxia 155, 261
splice donor/acceptor site 31–2, 105, 353
spliceosome 31–5
splicing 31
split hand syndrome 12

SSCP (single stranded conformational polymorphism) 261-2
statistical analysis, *see also* LOD score analysis and MLS, 199, 201
STR (short tandem repeat) *see also* microsatellite 304-11, 353
STS, 53, 55-8, 65, 74-9, 81-2, 84, 86, 97, 101, 352
surnames 292
susceptibility allele 189
SV40 145
SwissProt database 102
Sykes 292
synonymous substitution 277
synteny 111, 353
synthetic maps, *see* principle component analysis

tandem repeat array 35, 354
TATA box 29
Tay Sachs disease 4, 6, 295, 297, 300
TDT (transmission disequilibrium test) 205-6, 208-9, 354
telomerase 41-2, 354
telomere 41-2, 48, 51, 64, 354
TH 209
thalassaemias 4, 159-64, 296, 354
    α-thalassaemia 5-6, 159-63, 176, 296
    β-thalassaemia 5-6, 159-63, 176, 296, 319
threshold model 188-92, 221
tiling array 125, 127-8, 345
TNF (tumour necrosis factor) 204-5, 244
Tourette's syndrome 14
TP53, *see also* p53, 169
transcription factor 29-31, 354
transcription 29-32, 37
    CTF 30
    E2F 165-6
    general 29, 342
    NF1 30
    Oct1 and Oct2 30
    SPI 30-1
    TBP 29, 354
    TFIID 29
transcriptome 120
transgene 226, 354
translocation 14-17, 354
    balanced 15, 336
    unbalanced 15, 336
transmission disequilibrium test, *see* TDT
transposition, *see* transposons
transposons 36-9, 107-10, 354

TRE (trinucleotide repeat expansion) 8-9, 141, 154-9, 175-6, 178, 251, 256, 260-1, 354
trisomy 14-17, 354
TSC (the SNP consortium) 100-1
Twin studies 183-5
    tuberculosis 184
tumour suppressor 166-76, 176, 354
Turner's syndrome 16-7
type 1 diabetes, *see* IDDM
type 2 diabetes, *see* NIDDM

unbalanced translocation 15-16, 355
Unigene database 102
United Kingdom Xenotransplantation Interim Regulatory Authority (UKXIRA) 314
University of California at Santa Cruz 101
upstream regions 29-31, 355

vectors, 70-4, *see also* BAC vectors, cosmid, PAC vectors, phage 1 and YAC
Vindija 289
vitamin C 22
VNTR, *see also* minisatelllites, 41, 64-5, 208-9

Washington University Genome Sequencing Center 92
wave advance model 283
Wellcome Trust 23, 88
Whitehead Institute, Center for Genome Research 92
William's syndrome plate 1
worm (*Caenorhabditis elegans*) 103, 116, 118, 119-21, 136

x-chromosome 17, 42-4, 101, 105, 116, 142, 355, plate 2
X-chromosome inactivation 9, 355
X-linked disorders 2, 9-11, 355
xenobiotics 323

Y-chromosome 42, 85, 101, 105, 116, 272, 290-2, 298-9, 302-3, plate 2
YAC (yeast artificial chromosome) vectors 53, 70-3, 76, 81-2, 84, 355
    EST markers in YAC clones 79
    STS markers in YAC clones 53, 57, 76
yeast (*Saccharomyces cerevisiae*) 103, 111-12, 116, 118, 119-21, 136

zebrafish 103-4
zoo blot 146, 149, 355